New Mineral Species and Their Crystal Structures

New Mineral Species and Their Crystal Structures

Special Issue Editors

Irina O. Galuskina
Igor V. Pekov

MDPI • Basel • Beijing • Wuhan • Barcelona • Belgrade

MDPI

Special Issue Editors

Irina O. Galuskina
University of Silesia in Katowice
Poland

Igor V. Pekov
Lomonosov Moscow State University
Russia

Editorial Office
MDPI
St. Alban-Anlage 66
4052 Basel, Switzerland

This is a reprint of articles from the Special Issue published online in the open access journal *Minerals* (ISSN 2075-163X) from 2018 to 2019 (available at: https://www.mdpi.com/journal/minerals/special_issues/new_mineral)

For citation purposes, cite each article independently as indicated on the article page online and as indicated below:

LastName, A.A.; LastName, B.B.; LastName, C.C. Article Title. *Journal Name* **Year**, *Article Number*, Page Range.

ISBN 978-3-03897-688-2 (Pbk)
ISBN 978-3-03897-689-9 (PDF)

Cover image courtesy of Ivano Rocchetti.

Contents

About the Special Issue Editors

Irina O. Galuskina is an Associate Professor at the Faculty of Earth Sciences of the University of Silesia, Poland. She received a PhD (1998) and habilitation (2011) in Earth Sciences in the discipline of geology at the University of Silesia. She teaches crystallography, mineralogy, genetic mineralogy, and methods of investigation. Her research interests focus on the crystal chemistry of minerals and mineralogy of rodingites, high-temperature skarns, and pyrometamorphic rocks. She is the author and co-author of more than 50 new mineral species, and she has published 72 scientific papers in international journals.

Igor V. Pekov is a Professor of Mineralogy at the Lomonosov Moscow State University, Russia. He graduated in Geology and Geochemistry in 1989 at this University, where hd received a PhD in Mineralogy and Crystallography in 1997 and a DSc in 2005. He was elected a Corresponding Member of the Russian Academy of Sciences in 2016 and was elected a member of the Academia Europaea in 2018. His main research interests are in the fields of mineralogy and crystal chemistry of rare elements; mineralogy and geochemistry of alkaline rocks, volcanic exhalations, and oxidation zone of ore deposits; crystal chemistry of oxysalt minerals; crystal chemistry and properties of mineral-like microporous materials; and history of mineralogy. To date, he has discovered 145 IMA-approved new mineral species as a senior author and more 98 new mineral species as a co-author. He has authored more than 200 articles in international journals, as well as eight books.

Preface to "New Mineral Species and Their Crystal Structures"

The discovery of any new mineral is a significant event in fundamental science. It enriches mineralogy and geology as well as crystallography, solid-state physics, and chemistry. Some new minerals are interesting for material science and engineering due to their useful physical properties. The number of discoveries of new mineral species is the one of the most important quantitative parameters of success in mineralogical science. The USA and Russia are at present the record-holders in the number of new mineral found: About eight hundred mineral species have been discovered on the territories of each of these two countries. Italy and Germany (350 each) are next, followed by Canada (230), Sweden (183), Australia (162), Japan (143), Chile (133), China (132), the Czech Republic (127), Great Britain (126), France (119), and Namibia (104), which have also made significant contributions to world mineralogy.

This Special Issue of *Minerals* is devoted to new mineral species and their crystal structures. Fourteen articles were published therein. We are very thankful to all authors, who published the data of their recent investigations in this issue.

All articles were carefully reviewed by experts in mineralogy and crystallography. We are grateful to all reviewers who found the time to help to improve the submitted manuscripts. Further, of course, we would like to express our deepeset appreciation to the staff of Minerals, as well support staff, for the big and necessary work on all stages of the manuscript preparation for production.

Irina O. Galuskina, Igor V. Pekov
Special Issue Editors

minerals

MDPI

Editorial

Editorial for Special Issue "New Mineral Species and Their Crystal Structures"

Irina O. Galuskina [1,*] **and Igor V. Pekov** [2]

1   Faculty of Earth Sciences, University of Silesia, Będzińska 60, 41-200 Sosnowiec, Poland
2   Faculty of Geology, Moscow State University, Vorobievy Gory, Moscow 119991, Russia; igorpekov@mail.ru
*   Correspondence: irina.galuskina@us.edu.pl

Received: 17 January 2019; Accepted: 1 February 2019; Published: 13 February 2019

Mineralogy is the oldest and one of the most important sciences of the geological cycle. Minerals, the basis of overwhelming mass of solid matter in the universe, are direct subjects of investigation in mineralogy. Minerals, or mineral species, are generally solid crystalline substances. Their definition indicates that, they are: (1) naturally occurring; (2) belonging to the distinct structural type; (3) stable, varying merely in the relatively small limits of chemical composition. If a given mineral differs from other known species in its structure (2) and/or composition (3) then it can be considered as a new mineral species.

According to the data of the Commission on New Minerals, Nomenclature and Classification (CNMNC) of the International Mineralogical Association (IMA) (http://nrmima.nrm.se/), there are currently about 5600 known mineral species. The number of minerals increases steadily year by year. Before the 2000s the number of minerals approved by the IMA Commission varied from 50 to 70 species per year. In 2010s this number reached 100–160 species per year, and at present we observe an increasing tendency again, i.e., in 2018, almost two hundred new mineral species were approved by the IMA CNMNC. Is it a lot? It depends on what one compares it to. For instance, the number of biological species known at present is close to two million, and the number of synthetic chemical compounds, including organic substances, comes nearer to ten million. In comparison with these numbers, the species diversity of the Mineral Kingdom is relatively small, so a discovery of every new mineral species is a significant event in science. Investigation of new minerals, many of which possess a unique crystal structure and unusual properties, has a great significance for the understanding solids' structure and processes in the interior of the Earth, on the Earth's surface and in the Universe.

Articles submitted to the Special Issue of *Minerals* effectively demonstrate the great chemical, structural and genetic diversity of new mineral species of the present time, as well as the geography of discoveries and the variety of analytical methods used in studies of new minerals. Recently an increasing number of new mineral species discovered in volcanic fumaroles has been observed [1]. In the present issue, there are two articles devoted to the two new volcanic minerals: Verneite, $Na_2Ca_3Al_2F_{14}$, described simultaneously from Eldfell and Hekla volcanoes in Iceland and Vesuvius in Italy [2], and thermaerogenite, $CuAl_2O_4$, from the Tolbachik volcano at Kamchatka, Russia [3]. In these articles, readers could not only find the high level of the analytical investigation, but also the authors' imagination with regards to the choice of names for the new minerals:

- "Verneite is named after Jules Verne (1828–1905). In his novel Voyage au centre de la Terre (1864), Verne describes a group of characters descending through a crater of a quiescent volcano in Iceland (Snæfell) and, after an adventurous journey through exciting Earth's underground, finally being ejected in South Italy with the eruption of a volcano (Stromboli). Therefore, we consider the name verneite appropriate for a mineral found and described by the same team of researchers on the best-known Icelandic and Italian volcanoes [2]".

- "The name thermaerogenite (spinel group member) is constructed based on the combination of Greek words $\theta\varepsilon\varrho\mu\acute{o}\varsigma$, hot, $\alpha\acute{\varepsilon}\varrho\iota\upsilon\nu$, gas, and $\gamma\varepsilon\nu\acute{\eta}\varsigma$ that means "born by". Thus, in whole it means *born by hot gas*, that reflects the fumarolic origin of the mineral" [3].

The next two new minerals, sharyginite, $Ca_3TiFe_2O_8$ (perovskite supegroup member) [4] and nöggerathite-(Ce) $(Ce,Ca)_2Zr_2(Nb,Ti)(Ti,Nb)_2Fe^{2+}O_{14}$ (zirconolite-related mineral) [5], were discovered in the Eifel region, Rhineland-Palatinate, Germany. Volcanic rocks of Eifel are a unique source of new minerals: more than fifty mineral species were discovered there [6]. In Germany local collectors of minerals provide substantial assistance in collecting of specimens and so they are rightly the co-authors of the new mineral descriptions (Christof Schäfer, Bernd Ternes, and Willi Schüller). New data on the composition and structure of rusinovite $Ca_{10}(Si_2O_7)_3Cl_2$, found in altered xenoliths of Eifel and Southern Ossetia, are presented by Środek et al. [7]. Until now rusinovite was known only from xenoliths within ignimbrites of the Upper Chegem Caldera at Northern Caucasus, Russia [8].

More than twenty new mineral species were discovered in pyrometamorphic rocks of the Hatrurim Complex in the Dead Sea rift in the last eight years [9]. In this issue the new mineral ariegilatite, $BaCa_{12}(SiO_4)_4(PO_4)_2F_2O$ (with intercalated antiperovskite structure) is described [10]. It was collected from spurrite rocks in the Negev Desert, Israel and has also been found in several localities in the Palestinian Autonomous Territory and Jordan [10]. Previously, barioferrite, $BaFe_{12}O_{19}$, a new mineral of the magnetoplumbite group, was described in rocks of the Hatrurim Complex. However, due to the small size of crystals, its structure could not be studied [11]. Krzątała with co-authors reported the structure of barioferrite in a different article [12].

Pieczka and co-authors described the Ca-Mn-ordered new mineral of the apatite supergroup, parafiniukite, $Ca_2Mn_3(PO_4)_3Cl$, from the Szklary pegmatite in Lower Silesia, Poland. Szklary is the type locality for previously discovered lepageite, niboholtite, titanoholtite and szklaryite [13].

Ore minerals in this Special Issue are represented by gold amalgam aurihydrargyrumite, Au_6Hg_5, found on gold particles in the Iyoki deposit at Shikoku Island, Japan [14]; oyonite, $Ag_3Mn_2Pb_4Sb_7As_4S_{24}$, a new sulphosalt of the lillianite homologous series from the Uchucchacua deposit in Oyon district, Peru [15]; and cerromojonite, $CuPbBiSe_3$, a new selenide of the bournonite group from the El Dragón mine in Potosí, Bolivia [16].

Some new supergene minerals were reported from Italy. Demartin and co-authors present fiemmeite, $Cu_2(C_2O_4)(OH)_2\cdot 2H_2O$, from Val di Fiemme in Trentino. It occurs in coalified woods which were permeated by mineralizing solutions containing Cu, U, As, Pb and Zn. The oxalate anions have originated from altered plant remnants included in sandstone [17]. Biagioni with co-authors studied Si-analogue of chalcophyllite from the Cretaio Prospect in Grosseto, which was named tiberiobardiite in the honour of Tiberio Bardi, a mineral collector who found a specimen which became the holotype of tiberiobardiite with the simplified formula $Cu_9Al(SiO_3OH)_2(OH)_{12}(H_2O)_6(SO_4)_{1.5}\cdot 10H_2O$ [18].

Repeated studies of early investigated mineral species using modern analytical methods to clarify their formal position in the actual mineralogical classification is an important aspect of modern mineralogy. Pankova with co-authors reported the results of a structural investigation of kurchatovite and clinokurchatovite, two modifications of $CaMgB_2O_5$, from their type localities: Solongo in Buryatia, Russia, and Sayak-IV in Kazakhstan, respectively [19]. As a conclusion of the comparative study of kurchatovite and clinokurchatovite, the authors stated: "kurchatovite and clinokurchatovite are not polytypes, but polymorphs, and therefore re-consideration of their status as of separate mineral species is not warranted. However, the structures of the two minerals are closely related: the crystal structure of kurchatovite may be considered as a derivative of clinokurchatovite through the modular approach".

We hope that the present Special Issue presents an interesting read not only for mineralogists and geochemists but also for scientists who work in the fields of crystallography, chemistry, solid-state physics and materials science, on synthesis and on crystal chemical studies of novel technological materials related to minerals. We also hope that research articles on new mineral species attract the attention of museum curators and mineral collectors.

Author Contributions: I.G. and I.P. wrote the paper.

Conflicts of Interest: The authors declare no conflict of interest.

References

1. Pekov, I.V.; Koshlyakova, N.N.; Zubkova, N.V.; Lykova, I.S.; Britvin, S.N.; Yapaskurt, V.O.; Agakhanov, A.A.; Shchipalkina, N.V.; Turchkova, A.G.; Sidorov, E.G. Fumarolic arsenates—A special type of arsenic mineralization. *Eur. J. Mineral.* **2018**, *30*, 305–322. [CrossRef]

2. Balić-Žunić, T.; Garavelli, A.; Pinto, D.; Mitolo, D. Verneite, $Na_2Ca_3Al_2F_{14}$, a new aluminum fluoride mineral from Icelandic and Vesuvius fumaroles. *Minerals* **2018**, *8*, 553. [CrossRef]

3. Pekov, I.V.; Sandalov, F.D.; Koshlyakova, N.N.; Vigasina, M.F.; Polekhovsky, Y.S.; Britvin, S.N.; Sidorov, E.G.; Turchkova, A.G. Copper in natural oxide spinels: The new mineral thermaerogenite $CuAl_2O_4$, cuprospinel and Cu-enriched varieties of other spinel-group members from fumaroles of the Tolbachik Volcano, Kamchatka, Russia. *Minerals* **2018**, *8*, 498. [CrossRef]

4. Juroszek, R.; Krüger, H.; Galuskina, I.; Krüger, B.; Ježak, L.; Ternes, B.; Wojdyla, J.; Krzykawski, T.; Pautov, L.; Galuskin, E. Sharyginite, $Ca_3TiFe_2O_8$, a new mineral from the Bellerberg Volcano, Germany. *Minerals* **2018**, *8*, 308. [CrossRef]

5. Chukanov, N.V.; Zubkova, N.V.; Britvin, S.N.; Pekov, I.V.; Vigasina, M.F.; Schäfer, C.; Ternes, B.; Schüller, W.; Polekhovsky, Y.S.; Ermolaeva, V.N.; et al. Nöggerathite-(Ce), $(Ce,Ca)_2Zr_2(Nb,Ti)(Ti,Nb)_2Fe^{2+}O_{14}$, a new zirconolite-related mineral from the Eifel volcanic region, Germany. *Minerals* **2018**, *8*, 449. [CrossRef]

6. Engelhaupt, B.; Schüller, W. *Mineral Reich Eifel*; Christian Weise Verlag: München, Germany, 2015; p. 340.

7. Środek, D.; Juroszek, R.; Krüger, H.; Krüger, B.; Galuskina, I.; Gazeev, V. New occurrence of rusinovite, $Ca_{10}(Si_2O_7)_3Cl_2$: Composition, structure and Raman data of rusinovite from Shadil-Khokh Volcano, South Ossetia and Bellerberg Volcano, Germany. *Minerals* **2018**, *8*, 399. [CrossRef]

8. Galuskin, E.V.; Galuskina, I.O.; Lazic, B.; Armbruster, T.; Zadov, A.E.; Krzykawski, T.; Banasik, K.; Gazeev, V.M.; Pertsev, N.N. Rusinovite, $Ca_{10}(Si_2O_7)_3Cl_2$: A new skarn mineral from the Upper Chegem caldera, Kabardino-Balkaria, Northern Caucasus, Russia. *Eur. J. Mineral.* **2011**, *23*, 837–844. [CrossRef]

9. Galuskin, E.V.; Galuskina, I.O.; Gfeller, F.; Krüger, B.; Kusz, J.; Vapnik, Y.; Dulski, M.; Dzierżanowski, P. Silicocarnotite, $Ca_5[(SiO_4)(PO_4)](PO_4)$, a new 'old' mineral from the Negev Desert, Israel, and the ternesite-silicocarnotite solid solution: indicators of high-temperature alteration of pyrometamorphic rocks of the Hatrurim Complex, Southern Levant. *Eur. J. Mineral.* **2016**, *28*, 105–123. [CrossRef]

10. Galuskin, E.V.; Krüger, B.; Galuskina, I.O.; Krüger, H.; Vapnik, Y.; Wojdyla, J.A.; Murashko, M. New mineral with modular structure derived from Hatrurite from the pyrometamorphic rocks of the Hatrurim Complex: Ariegilatite, $BaCa_{12}(SiO_4)_4(PO_4)_2F_2O$, from Negev Desert, Israel. *Minerals* **2018**, *8*, 19. [CrossRef]

11. Murashko, M.N.; Chukanov, N.V.; Mukhanova, A.A.; Vapnik, E.; Britvin, S.N.; Polekhovsky, Y.S.; Ivakin, Y.D. Barioferrite $BaFe_{12}O_{19}$: A new mineral species of the magnetoplumbite group from the Haturim Formation in Israel. *Geol. Ore Depos.* **2011**, *53*, 558–563. [CrossRef]

12. Krzątała, A.; Panikorovskii, T.L.; Galuskina, I.O.; Galuskin, E.V. Dynamic disorder of Fe^{3+} ions in the crystal structure of natural barioferrite. *Minerals* **2018**, *8*, 340. [CrossRef]

13. Pieczka, A.; Biagioni, C.; Gołębiowska, B.; Jeleń, P.; Pasero, M.; Sitarz, M. Parafiniukite, $Ca_2Mn_3(PO_4)_3Cl$, a new member of the apatite supergroup from the Szklary Pegmatite, Lower Silesia, Poland: Description and crystal structure. *Minerals* **2018**, *8*, 485. [CrossRef]

14. Nishio-Hamane, D.; Tanaka, T.; Minakawa, T. Aurihydrargyrumite, a natural Au_6Hg_5 phase from Japan. *Minerals* **2018**, *8*, 415. [CrossRef]

15. Bindi, L.; Biagioni, C.; Keutsch, F.N. Oyonite, $Ag_3Mn_2Pb_4Sb_7As_4S_{24}$, a new member of the lillianite homologous series from the Uchucchacua Base-Metal deposit, Oyon District, Peru. *Minerals* **2018**, *8*, 192. [CrossRef]

16. Förster, H.J.; Bindi, L.; Grundmann, G.; Stanley, C.J. Cerromojonite, $CuPbBiSe_3$, from El Dragón (Bolivia): A new member of the bournonite group. *Minerals* **2018**, *8*, 420. [CrossRef]

17. Demartin, F.; Campostrini, I.; Ferretti, P.; Rocchetti, I. Fiemmeite $Cu_2(C_2O_4)(OH)_2 \cdot 2H_2O$, a new mineral from Val di Fiemme, Trentino, Italy. *Minerals* **2018**, *8*, 248. [CrossRef]

18. Biagioni, C.; Pasero, M.; Zaccarini, F. Tiberiobardiite, $Cu_9Al(SiO_3OH)_2(OH)_{12}(H_2O)_6(SO_4)_{1.5} \cdot 10H_2O$, a new mineral related to chalcophyllite from the Cretaio Cu prospect, Massa Marittima, Grosseto (Tuscany, Italy): Occurrence and crystal structure. *Minerals* **2018**, *8*, 152. [CrossRef]
19. Pankova, Y.A.; Krivovichev, S.V.; Pekov, I.V.; Grew, E.S.; Yapaskurt, V.O. Kurchatovite and clinokurchatovite, ideally $CaMgB_2O_5$: An example of modular polymorphism. *Minerals* **2018**, *8*, 332. [CrossRef]

Article

Verneite, Na$_2$Ca$_3$Al$_2$F$_{14}$, a New Aluminum Fluoride Mineral from Icelandic and Vesuvius Fumaroles

Tonči Balić-Žunić [1],*, Anna Garavelli [2], Daniela Pinto [2] and Donatella Mitolo [3]

[1] Department of Geosciences and Natural Resource Management, University of Copenhagen, Øster Voldgade 10, DK-1350 København K, Denmark

[2] Department of Earth and Geo-environmental Sciences, University of Bari "A. Moro", via E. Orabona 4, I-70125 Bari, Italy; Anna.garavelli@uniba.it (A.G.); Daniela.pinto@uniba.it (D.P.)

[3] Autorità di Bacino Distrettuale dell'Appennino Meridionale Sede Puglia, Str. Prov. Per Casamassima km 3, I-70010 Valenzano (BA), Italy; dmitolo@libero.it

* Correspondence: toncib@ign.ku.dk; Tel.: +45-35322337

Received: 26 October 2018; Accepted: 21 November 2018; Published: 28 November 2018

Abstract: Verneite, Na$_2$Ca$_3$Al$_2$F$_{14}$, is a new mineral first discovered in fumarolic samples from both Hekla, Iceland and Vesuvius, Italy. Additional occurrences are so far from Eldfell and Fimmvörduhals, both on Iceland. Verneite is cubic, $I2_13$, a = 10.264(1) Å, V = 1081.4(3) Å^3, Z = 4, and corresponds to the known synthetic compound. The empirical formula is Na$_{2.01}$Ca$_{2.82}$Al$_{2.17}$F$_{14.02}$ (scanning electron microscopy with energy dispersive spectrometer from an unpolished sample). It appears in crystals up to 20 μm in diameter, with {110}, {100}, and {111} as the main forms. In the crystal structure of its synthetic analogue, Na is coordinated by 7 F atoms in the form of a capped octahedron, Ca with 8 F atoms in the form of a bisdisphenoid, and Al with 6 F atoms in the form of an octahedron. The crystal structure of Na$_2$Ca$_3$Al$_2$F$_{14}$ contains sinuous chains of Ca coordination polyhedra interlacing with similarly sinuous chains of Na coordination polyhedra and forming together with them layers parallel to {100}. The intersecting layers parallel to three equivalent crystallographic planes form a three-dimensional mesh with Al coordinations imbedded in its holes. The characteristics of Ca coordinations in fluorides, as well as their relations to other ternary Na–Ca–Al fluorides are discussed. Verneite is named after Jules Verne.

Keywords: verneite; new mineral; crystal structure; Hekla; Vesuvius; Eldfell; aluminofluoride

1. Introduction

The new mineral verneite, Na$_2$Ca$_3$Al$_2$F$_{14}$, was discovered among sublimates collected from fumaroles on the Eldfell and Hekla volcanoes, where a considerable number of new fumarolic minerals has been observed [1]; six of them have so far been fully described [2–7]. The same mineral phase was identified at approximately the same time in a sublimate sample from Vesuvius, belonging to the Pelloux collection, which is stored in the mineralogical section of the Museum of Earth Sciences at Bari University. The original label of this Museum sample, with its 1925 date, indicates "Avogadrite from Vesuvius".

Sveinn P. Jakobsson from the Icelandic Institute of Natural History collected the verneite type specimen on 16 September 1992 on Hekla. He also collected the cotype specimen from the Eldfell volcano in 1988. The holotype and the cotype are kept in the mineral collection of the Icelandic Institute of Natural History, Garðabær, Iceland, under sample numbers NI 15509 and NI 12256, respectively. The mineral was also registered in the samples NI 15518 and NI 17046 from Hekla, and NI 24457 and NI 24565 from Fimmvörduhals, kept in the same museum, and in samples E4-1A, E4-2A and E4-2B, collected on Eldfell in 2009 and presently kept at the Department of Geosciences and Natural Resource

Management of the University of Copenhagen. The Vesuvius sample is kept in the mineralogical collection of the Department of Earth and Geo-environmental Sciences, University of Bari.

Verneite is named after Jules Verne (1828–1905), the famous French author of novels, poetry, and plays, best known for his adventure novels and his profound influence on the literary genre of science fiction and, through it, the promotion of science, especially among young people. In his novel *Voyage au centre de la Terre* (1864), Verne describes a group of characters descending through a crater of a quiescent volcano in Iceland (Snæfell) and, after an adventurous journey through exciting Earth's underground, finally being ejected in South Italy with the eruption of a volcano (Stromboli). Therefore, we consider the name verneite appropriate for a mineral found and described by the same team of researchers on the best-known Icelandic and Italian volcanoes. Both the mineral and the mineral name have been approved by the Commission on New Minerals, Nomenclature and Classification of the IMA (no. 2016-112).

In the present work, we give a detailed description of the occurrences of verneite, the morphological and chemical analysis by scanning electron microscopy with energy dispersive spectroscopy (SEM–EDS), and the crystallographic analysis by Powder X-ray Diffraction (PXRD) and discuss its crystal structure details.

2. Materials and Methods

The geological settings and a description of the Hekla and Eldfell fumaroles where the new mineral was found are given in the papers mentioned in the Introduction [1–7]. Recently, the mineral has also been identified in samples originating from fumaroles on Fimmvörduhlas, Iceland, active during and after the eruption in 2010 [8]. Verneite occurs in medium to low temperature (170 °C at the time of sampling) fumaroles, as white-yellowish to brown crusts and massive aggregates up to several mm in size, sometimes also in transparent, colorless to pale yellowish crystals. In the sample from Eldfell, crystals up to 20 μm in diameter with a rhombic dodecahedral habit have been observed (Figure 1a), whereas in the Vesuvius sample, smaller (up to 10 μm) crystals having a combination of {100}, {110}, and {111} forms have been noted (Figure 1b). Verneite from Vesuvius was found during a reexamination of a sublimate sample belonging to the "Alberto Pelloux mineralogical collection", housed at the "Palace of the Earth Sciences" of Bari University. The original label gives the following indication in Pelloux's own handwriting: "Avogadrite from Vesuvius collected on 15 July 1925—avogadrite or malladrite?" We conclude, therefore, that the mineral originates from fumaroles formed after the violent eruption of 1906, which was the last prior to the date reported in the label. Considering that avogadrite was discovered in 1926 by Professor Ferruccio Zambonini of Naples University, we conclude that the acquisition of the sample (from the well-known mineral salesman "Roberto Palumbo", according to the indication on Pelloux's label) happened after 1926. As indicated on the label, Pelloux himself pointed to a need for further investigation of the sample. This was initially conducted by C.L. Garavelli and coworkers in the 1960s, who reported in it the presence of ralstonite, matteuccite, avogadrite, malladrite, and of a probably new mineral, $MgSiF_6 \cdot 6H_2O$, suitable for further studies (C.L. Garavelli, unpublished documents). We could not confirm this last phase during the present investigation.

Verneite in Hekla samples forms mixtures with ralstonite and hematite, sometimes also with jakobssonite and "mineral HB" [1] with a still unknown composition, but known PXRD data. The other minerals, which appear together with verneite in the type specimen and other samples from Hekla, are leonardsenite, heklaite, malladrite, opal, and fluorite. In the samples from Eldfell, where the cotype stems from, verneite is associated with jakobssonite, "mineral HB", anhydrite, leonardsenite, ralstonite, jarosite, and meniaylovite. In the present investigation of the sample from Vesuvius, we found verneite associated with ralstonite and, to a lesser degree, to hieratite and knasibfite.

For the determination of the chemical composition, samples of verneite were analyzed by SEM-EDS. The Eldfell sample was analyzed by a S 360 Cambridge SEM, coupled with an Oxford-Link Ge ISIS EDS equipped with a Super Atmosphere thin window, whereas a 50XVP LEO SEM and

Oxford AZtec system with an Oxford SDD XMax (80 mm^2) detector were used for the Vesuvius sample. The samples were sputtered with a 30 nm thick carbon film before analysis. As we had to measure inclined surfaces, a "noncritical" working distance was utilized [9,10]. X-ray intensities were converted to wt % values by the ZAF4/FLS quantitative analysis software support of Oxford-Link Analytical. For standards, we used synthetic LiF (F), albite (Na), wollastonite (Ca), corundum (Al), and orthoclase (K).

The crystallographic data were obtained by PXRD on diffractometers with Bragg–Brentano geometry, first the Panalytical (formerly Philips) PW3710 diffractometer with a long fine focus Cu sealed tube, secondary-beam graphite monochromator, and a variable-slit for the beam divergence. Subsequently, a Bruker-AXS D8 diffractometer with a ceramic Cu tube, primary-beam Ge111 monochromator, fixed divergence slit, and Lynx-Eye silicon strip detector was used. Bruker–AXS program Topas was used for the Rietveld refinement.

The crystallographic data for verneite and other compared crystal structures were calculated by program IVTON [11].

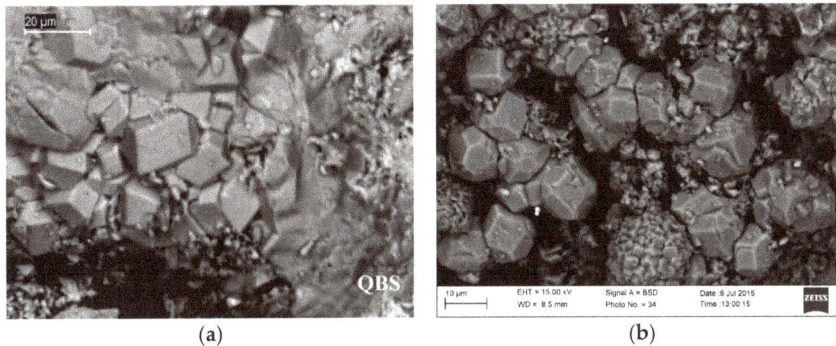

Figure 1. (**a**) SEM image of the crystals of verneite in association with fine-grained jakobssonite in the cotype sample from Eldfell. (**b**) SEM image of verneite crystals from the Vesuvius sample.

3. Results

3.1. Chemical Formula and Physical Properties

The empirical formulae (based on 7 cations *pfu*) are: $Na_{2.01}Ca_{2.82}Al_{2.17}F_{14.02}$ for the Eldfell sample and $(Na_{1.47}K_{0.09})_{\Sigma1.56}Ca_{3.25}Al_{2.19}F_{14.33}$ for the Vesuvius sample. The ideal formula is $Na_2Ca_3Al_2F_{14}$, which requires: F = 54.71, Na = 9.46, Ca = 24.73, and Al = 11.10 wt %.

The calculated density of verneite, from the empirical formula and the unit–cell data, is 2.974 g/cm^3. The calculated refractive index using the Gladstone–Dale constants of Mandarino [12] is 1.357.

The cleavage, hardness, streak, and lustre of verneite could not be accurately determined due to the minute size of the crystals and the admixture with other minerals. No fluorescence was observed on the investigated samples, either under short-wavelength or long-wavelength ultraviolet radiation.

It could be expected that verneite is piezoelectric due to its space group symmetry.

3.2. Crystal Structure Data

Verneite is analogous to synthetic $Na_2Ca_3Al_2F_{14}$ investigated by Courbion and Ferrey [13]. It is cubic, $I2_13$, $a = 10.264(1)$ Å, $V = 1081.4(3)$ Å^3, $Z = 4$. The atomic parameters and a list of bond lengths and angles are given in Reference [13].

Verneite was identified by PXRD in samples from all localities mentioned above. Rietveld refinements of verneite using the atomic parameters of Courbion and Ferrey [13] match the observed data very well in intensities. Due to this, and the fact that it was impossible to obtain a pure diagram of the mineral, which was always mixed with at least three other components in all the investigated samples, a full crystal structure refinement from powder diffraction data, with the inclusion of atomic parameters, was not attempted. Table 1 presents the PXRD data for the sample from the type locality (Hekla), containing verneite, ralstonite, hematite, and jakobssonite, as well as a minor undetermined amount of the still not fully investigated "mineral HB" [1].

Table 1. X-ray powder diffraction data of verneite from Hekla (recorded with automatic variable divergence slit) compared with the experimental diagram of synthetic $Na_2Ca_3Al_2F_{14}$ (PDF 36-1496). HB = mineral HB [1]; R = ralstonite; J = jakobssonite; H = hematite.

hkl/Mineral	d (Å) [1]	I/I_0 % [1]	d (Å) [2]	I/I_0 % [2]
0 1 1	7.24	17.4	7.24	20
R	5.72	39.0	-	-
0 0 2	5.11	17.6	5.11	14
2 1 1	4.18	76.2	4.18	91
?	3.84	15.3	-	-
H	3.67	25.6	-	-
0 2 2	3.62	54.7	3.62	55
HB	3.54	22.4	-	-
?	3.30	25.6	-	-
0 3 1	3.23	68.1	3.24	60
HB, J	3.17	28.7	-	-
R	2.99	56.9	-	-
2 2 2	2.95	100.0	2.96	85
R	2.87	41.6	-	-
3 2 1	2.73	38.2	2.74	24
H	2.70	74.6	-	-
H	2.512	69.1	-	-
4 1 1	2.414	40.5	2.413	33
?	2.349	20.0	-	-
4 0 2	2.288	40.5	2.289	21
H	2.201	35.7	-	-
3 3 2	2.184	78.3	2.183	72
HB	2.127	20.5	-	-
4 2 2	2.088	20.2	2.090	5
R, J	2.042	22.1	-	-
3 4 1, 4 3 1	2.009	98.2	2.008	100
R	1.915	37.1	-	-
2 5 1	1.871	75.1	1.877	72
H	1.840	37.9	-	-
0 4 4	1.811	84.1	1.810	72
4 3 3 + R	1.755	40.5	1.756	9
0 0 6	-	-	1.708	7
H	1.697	51.3	-	-
6 1 1, 5 3 2, 3 5 2	1.663	66.2	1.661	55
0 6 2	-	-	1.620	1
H	1.607	23.0	-	-
4 5 1	1.582	28.4	1.581	8
6 2 2	1.545	45.9	1.544	38
3 6 1	1.512	30.6	1.510	10

[1] verneite Hekla. [2] PDF 36-1496.

Table 2 gives the results of the Rietveld refinement. The refinement shows that the sample is made up by 53(1) wt % of verneite, 30(1) wt % of ralstonite, 15.9(6) wt % of hematite, and 1.6(4) wt % of jakobssonite, as well as a minor, not determined quantity of "mineral HB".

Table 2. Rietveld refinement results. "Mineral HB" [1], also present in the sample, was not included in refinement because structural details are unknown. Average crystallite size modelled by Lorentzian function. Global parameters: R_{exp} = 6.58%, R_{wp} = 11.6%, GoF = 1.76, profile function: Fundamental parameters, background: Chebyshev polynomials.

	Verneite	Ralstonite	Hematite	Jakobssonite
average crystallite size (nm)	121(13)	35(2)	56(5)	fixed to 200
a (Å)	10.264(1)	9.963(2)	5.035(1)	8.63(3)
b (Å)	-	-	-	6.36(2)
c (Å)	-	-	13.824(4)	7.25(2)
β (°)	-	-	-	114.4(5)
atomic parameters	fixed [13]	fixed [14]	fixed [15]	fixed [4]
R-Bragg	4.7%	6.5%	5.8%	7.2%

4. Discussion

4.1. Description of the Crystal Structure

The original description of the crystal structure [13] presents it in an unconventional form as a combination of the cation-centered [AlF$_6$] groups (octahedra) and anion-centered [FNaCa$_{3/2}$]$_2$ framework. This helps in relating the structure to some complex oxide structures presenting it as their "negative" (with the roles of cations and anions in frameworks exchanged), but ignores the coordinations of Na and Ca and makes the comparison with other fluorides difficult. Here, we give another view on the crystal structure, based solely on cation coordinations.

As expected from the Al:F ratio, verneite is an aluminofluoride with isolated [AlF$_6$] octahedral groups and additional F atoms not bonded to Al. The arrangement of the six F atoms around the Al site is almost perfectly octahedral with a perfect sphericity [16] and a volume distortion (υ) [17] of only 0.0018. The Al atom sits on the three-fold axis 0.036 Å from the centroid of coordination, with eccentricity [16] of 0.0232. The average Al–F bond distance is 1.804 Å [13].

Ca atoms have an eight-fold coordination in the form of a bisdisphenoid (Figure 2). This type of coordination can achieve a configuration with the minimum ratio of the volume of a circumscribed sphere and the volume of the polyhedron for the coordination number 8; in other words, it is a maximum-volume polyhedron for this coordination number [18]. The Ca coordination polyhedron in verneite does not completely fulfill the conditions of a maximum-volume polyhedron because the four F atoms that form the shortened-disphenoid part of the coordination approach a square-planar arrangement (equatorial atoms on Figure 2). Consequently, its υ parameter (or volume distortion compared to the maximum-volume bisdisphenoid) is larger than zero (Table 3). This type of coordination is unique among the Ca–F coordinations in mineral fluorides and related synthetic compounds represented in Table 3.

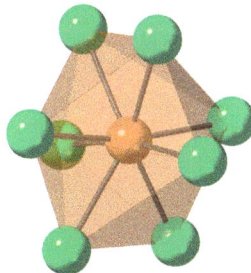

Figure 2. Bisdisphenoidal coordination of Ca in verneite. Projection based on crystal structure data from Reference [13].

Table 3. The coordination parameters of [CaF$_n$] coordination polyhedra in fluoride minerals and related synthetic compounds; υ = volume distortion; O. = Octahedron; P.b. = Pentagonal bipyramid; S.o. = Split octahedron; Bis. = Bisdisphenoid; S.a. = Square antiprism; C. = Cube; T.t.p. = Tricapped trigonal prism; T.c. = Tricapped cube.

Compound	Site	CN	<Ca–F> (Å)	υ	Asphericity	Eccentricity	Polyhedron	Ref.
CaLiAlF$_6$	Ca1	6	2.281	0.0038	0	0	O.	[19]
Na$_4$Ca$_4$Al$_7$F$_{33}$	Ca1	6	2.246	0.0232	0	0	O.	[20]
β-NaCaAlF$_6$	Ca1	6	2.308	0.0439	0.0049	0.023	O.	[20]
KCaAl$_2$F$_9$	Ca1	6	2.302	0.0703	0.0348	0.0165	O.	[21]
	Ca2	6	2.292	0.0559	0.0426	0.017		
CaAlF$_5$	Ca1	7	2.318	0.0265	0.0335	0.0521	P.b.	[22]
Ca$_2$AlF$_7$	Ca1	7	2.316	0.0344	0.0217	0.0248	P.b.	[23]
	Ca2	7	2.346	0.1391	0.0112	0.0404	S.o.	
α-NaCaAlF$_6$	Ca1	7	2.351	0.0943	0.0331	0.054	transitional	[24]
	Ca2	7	2.343	0.0869	0.0405	0.0467	P.b./S.o.	
Na$_2$Ca$_3$Al$_2$F$_{14}$	Ca1	8	2.374	0.0234	0.0254	0.0286	Bis.	[13]
BaCaAlF$_7$	Ca1	8	2.372	0.0467	0.0309	0.0361	S.a.	[25]
	Ca2	8	2.372	0.0569	0.0271	0.0362		
Ba$_2$CaMgAl$_2$F$_{14}$	Ca1	8	2.371	0.0565	0.0453	0.0349	S.a.	[26]
Ca$_2$PbAlF$_9$	Ca1	8	2.360	0.088	0.0181	0.0237	transitional	[27]
	Ca2	8	2.361	0.0899	0.0234	0.0241	C./S.a.	
CaNa$_3$Mg$_3$AlF$_{14}$	Ca1	8	2.421	0.1136	0.0583	0	distorted C.	[28]
CaF$_2$	Ca1	8	2.365	0.1522	0	0	C.	[29]
Ca$_{13}$Y$_6$F$_{43}$	Ca3	8	2.341	0.1444	0.021	0.0142	distorted C.	[30]
	Ca1*	9	2.617	0.0296	0.0627	0.0628	T.t.p.	
	Ca2	10	2.491	0.0478	0.0407	0.0576	T.c.	

The crystal structure data for the table are compiled from Reference [31]. As can be seen, the coordination number (CN) of Ca (coordinated with F atoms only) varies from 6 to 8 or even, rarely, 9 or 10. In the crystal structures with CN6, the coordination is in the form of a nearly perfect octahedron (in colquirite [19]) or subsequently more distorted octahedra in Na$_4$Ca$_4$Al$_7$F$_{33}$ [20], β-NaCaAlF$_6$ [20], and KCaAl$_2$F$_9$ [21]. Ca with CN7 appears in moderately distorted pentagonal bipyramids in jakobssonite [4,22] and Ca$_2$AlF$_7$ (Ca1) [23]. A regular pentagonal bipyramid is the maximum-volume polyhedron for CN7 and the υ values are calculated in comparison with it. Ca2 in Ca$_2$AlF$_7$ and the Ca coordinations in α-NaCaAlF$_6$ [24] have this parameter significantly different from zero because they represent a different type of coordination geometry, close to a "split octahedron", the ideal form of which would have υ = 0.1333 [17]. From values in Table 3, we can see that Ca2 in Ca$_2$AlF$_7$ closely approaches this form, whereas the coordinations in α-NaCaAlF$_6$ are transitional between a pentagonal bipyramid and a "split octahedron". For CN8, the ideal types of coordination polyhedra are, in addition to bisdisphenoid, the cube (0.1522), the hexagonal bipyramid (0.0462), the square antiprism (0.0351), and the bicapped trigonal prism (0.0733). The numbers in parentheses are the υ values in comparison with the maximum-volume bisdisphenoid. The forms observed in Ca fluorides from Table 3 are a regular cube in fluorite and distorted cubes in tveitite (Ca3) [30] and coulsellite [28]. The coordinations in calcioaravaipaite [27] are transitional between the cube and square antiprism, whereas the coordinations in usovite [26] and BaCaAlF$_7$ [25] are moderately distorted square antiprisms. In tveitite, the CN9 and CN10 Ca sites are also present. Ca1 with a tricapped trigonal prism coordination actually occupies a larger CN12 icosahedral void and is a split site, whereas the coordination of Ca2 can be described as a cube where one corner has been substituted by an F$_3$ triangle [30].

Na atoms have a seven-fold coordination with a capped octahedron as the coordination polyhedron (Figure 3). The capped octahedral face is broader than the opposite one (Figure 3).

The average Na–F bond distance is 2.431 Å [13], whereas the polyhedron distortion parameters are eccentricity = 0.0043, asphericity = 0.0615, and υ = 0.0849 (compared to regular pentagonal bipyramid, which is the maximum-volume polyhedron for CN7). There is a considerably larger number of Na containing fluorides than those containing Ca, with a larger variability of coordination types; a comparative study of Na coordinations in fluorides will not be presented here, but in a special following article. We only mention that the Na coordination polyhedron type in verneite seems to be unique compared to other mineral fluorides, similar to the case of the Ca coordination polyhedron.

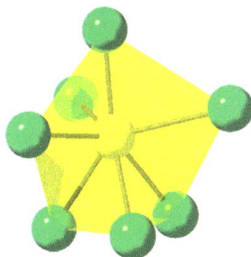

Figure 3. Na coordination in verneite. The three-fold axis is vertical. Projection based on crystal structure data from Reference [13].

The Ca coordination polyhedra build sinuous chains by sharing edges. They are interlaced by the sinuous chains of equally edge-sharing Na coordination polyhedra, and together these two types of chains build layers parallel to {100} (Figure 4). The layers, parallel to the three equivalent crystallographic planes, form a three-dimensional mesh that houses Al coordination octahedra in its interstices (Figure 5). The Al coordination octahedra share one face with a Na coordination polyhedron and three edges with Ca coordination polyhedra. Three F atoms at the corners that belong to the shared face are characterized by longer F–Al bonds (1.824 Å) and are each bonded to one Al, one Ca, and two Na. The three F atoms in the corners of the opposite face with shorter F–Al bonds (1.784 Å) make bonds to one Al and two Ca each. The directions of the sinuous chains of Ca and Na coordination polyhedra on the adjacent {100} faces, e.g., on the (100) and (010) faces, are mutually perpendicular (in this case [010] and [001], respectively), as required by the space group symmetry (Figure 5).

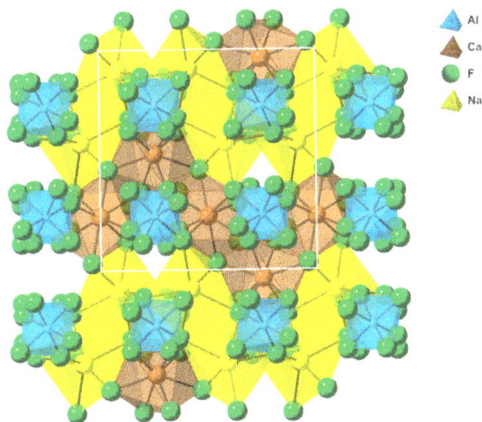

Figure 4. A {100} layer from the crystal structure of verneite, formed by sinuous chains of Ca coordinations and Na coordinations, and overlaid/underlain by the [AlF$_6$] octahedra. Projection based on crystal structure data from Reference [13].

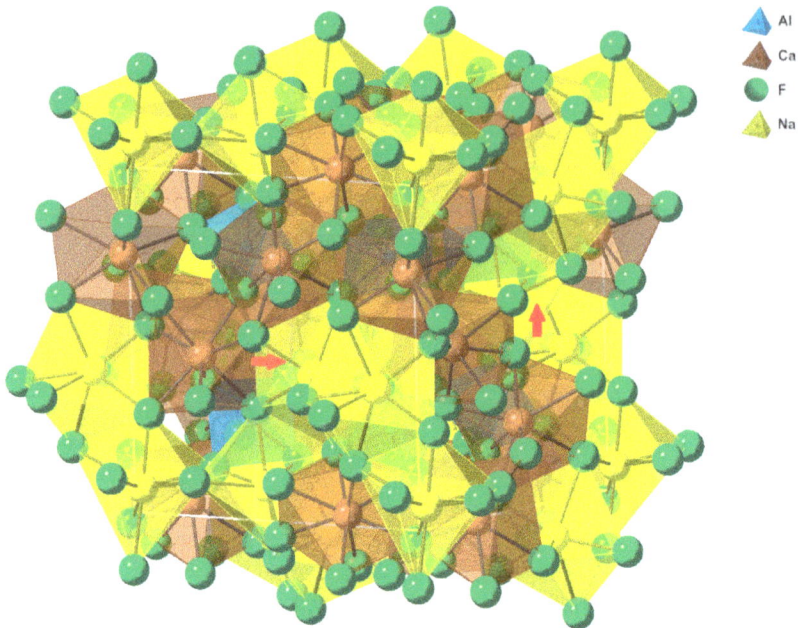

Figure 5. A general direction view of coordination polyhedra in one unit cell of verneite. Red arrows indicate the directions of sinuous chains of coordination polyhedra on the (100) and (010) surfaces of the unit cell. Projection based on crystal structure data from Reference [13].

4.2. NaF–CaF$_2$–AlF$_3$ System and the Natural Occurrences of Phases

Only three ternary phases have been originally reported in the phase system NaF–CaF$_2$–AlF$_3$ [32], the low-temperature form of NaCaAlF$_6$, its high-temperature polymorph, and NaCaAl$_2$F$_9$. Later investigations showed that the material investigated by Craig and Brown [32] and reported as the low-temperature form of NaCaAlF$_6$ actually has the composition Na$_2$Ca$_3$Al$_2$F$_{14}$ [13]. In addition, the composition of NaCaAl$_2$F$_9$ was corrected to the structural formula Na$_4$Ca$_4$Al$_7$F$_{33}$ [20]. NaCaAlF$_6$ was confirmed to have a high-temperature (stable above 620 °C) and a low-temperature form, based on determinations of their structures [20,24]. The crystal structure data for the four confirmed phases in the phase system are represented in Table 4. To the best of our knowledge, none of the phases from this phase system have been observed before in nature, and Na$_2$Ca$_3$Al$_2$F$_{14}$ reported here is the first example. The synthetic work showed that this compound is stable up to 719 °C at atmospheric pressure and decomposes at higher temperatures to a mixture of the high-temperature NaCaAlF$_6$ and fluorite [13]. It forms readily at temperatures lower than 600 °C in the part of the phase system comprising approximately equal molar amounts of Na, Ca, and Al, and it crystallizes rather than the low-temperature form of NaCaAlF$_6$, which is metastable and could be synthesized only through hydrothermal synthesis [20]. The fourth ternary phase in this phase system, Na$_4$Ca$_4$Al$_7$F$_{33}$, is reported to form through a sluggish reaction between Na$_2$Ca$_3$Al$_2$F$_{14}$ and AlF$_3$ [32]. The metastability of NaCaAlF$_6$ below 600 °C and the slow-forming reaction of Na$_4$Ca$_4$Al$_7$F$_{33}$ most probably explain the appearance of verneite as the only ternary Na–Ca–Al fluoride in fumaroles.

Table 4. Crystal structure data for the ternary phases from the NaF–CaF$_2$–AlF$_3$ system.

Formula	Space Group	Crystal lattice Parameters	Structure Type	Aluminofluoride Part	Ref.
β-NaCaAlF$_6$	P321	8.9295(9), 5.0642(2) Å	Na$_2$SiF$_6$	isolated [AlF$_6$]	[20]
α-NaCaAlF$_6$	P2$_1$/c	8.7423(3), 5.1927(2), 20.3514(9) Å, 91.499(2)°	unique	isolated [AlF$_6$]	[24]
Na$_2$Ca$_3$Al$_2$F$_{14}$	I2$_1$3	10.257(1) Å	unique	isolated [AlF$_6$] plus additional F	[13]
Na$_4$Ca$_4$Al$_7$F$_{33}$	Im3m	10.781(3) Å	unique	[Al$_7$F$_{33}$] 3D framework	[20]

Author Contributions: Investigation, T.B.-Ž., A.G., D.P. and D.M.; Methodology, T.B.-Ž., A.G., D.P. and D.M.; Software, T.B.-Ž.; Writing—original draft, T.B.-Ž.; Writing—review & editing, A.G. and D.P.

Funding: This research received no external funding.

Acknowledgments: Authors are indebted to the late S.P. Jakobsson for the collection of the Hekla and Eldfell samples and the initiation of this work, and to E. Leonardsen for the first XRD measurements of verneite. We thank J. Bailey for the correction of the English text and the three anonymous referees for suggestions that improved the manuscript.

Conflicts of Interest: The authors declare no conflict of interest.

References

1. Jakobsson, S.P.; Leonardsen, E.S.; Balić-Žunić, T.; Jónsson, S.S. Encrustations from three recent volcanic eruptions in Iceland: The 1963–1967 Surtsey, the 1973 Eldfell and the 1991 Hekla eruptions. *Fjölrit Náttúrufraedistofnunar* **2008**, *52*, 65.

2. Balić-Žunić, T.; Garavelli, A.; Acquafredda, P.; Leonardsen, E.; Jakobsson, S.P. Eldfellite, NaFe(SO$_4$)$_2$, a new fumarolic mineral from Eldfell volcano, Iceland. *Mineral. Mag.* **2009**, *73*, 51–57. [CrossRef]

3. Garavelli, A.; Balić-Žunić, T.; Mitolo, D.; Acquafredda, P.; Leonadsen, E.; Jakobsson, S.P. Heklaite, KNaSiF$_6$, a new fumarolic mineral from Hekla volcano, Iceland. *Mineral. Mag.* **2010**, *74*, 147–157. [CrossRef]

4. Balić-Žunić, T.; Garavelli, A.; Mitolo, D.; Acquafredda, P.; Leonardsen, E. Jakobssonite, CaAlF$_5$, a new mineral from fumaroles at the Eldfell and Hekla volcanoes, Iceland. *Mineral. Mag.* **2012**, *76*, 751–760. [CrossRef]

5. Mitolo, D.; Garavelli, A.; Balić-Žunić, T.; Acquafredda, P.; Jakobsson, S.P. Leonardsenite, MgAlF$_5$(H$_2$O)$_2$, a new mineral species from Eldfell volcano, Heimaey Island, Iceland. *Can. Mineral.* **2013**, *51*, 377–386. [CrossRef]

6. Jacobsen, M.J.; Balić-Žunić, T.; Mitolo, D.; Katerinopoulou, A.; Garavelli, A.; Jakobsson, S.P. Oskarssonite, AlF$_3$, a new fumarolic mineral from Eldfell volcano, Heimaey, Iceland. *Mineral. Mag.* **2014**, *78*, 215–222. [CrossRef]

7. Balić-Žunić, T.; Garavelli, A.; Mitolo, D. Topsøeite, FeF$_3$(H$_2$O)$_3$, a new fumarolic mineral from the Hekla volcano, Iceland. *Eur. J. Mineral.* **2018**, *30*, 841–848. [CrossRef]

8. Balić-Žunić, T. (University of Copenhagen, Copenhagen, Denmark); Jonasson, K. (Icelandic Institute of Natural History, Gardabaer, Iceland). Personal communication, 2013.

9. Ruste, J. X-Ray spectrometry. In *Microanalysis and Scanning Electron Microscopy*; Maurice, F., Meny, L., Tixier, R., Eds.; Les Editions de Physique: Orsay, France, 1979; pp. 215–267.

10. Acquafredda, P.; Paglionico, A. SEM-EDS microanalyses of microphenocrysts of Mediterranean obsidians: A preliminary approach to source discrimination. *Eur. J. Mineral.* **2004**, *16*, 419–429. [CrossRef]

11. Balić Žunić, T.; Vicković, I. IVTON—Program for the Calculation of Geometrical Aspects of Crystal Structures and Some Crystal Chemical Applications. *J. Appl. Cryst.* **1996**, *29*, 305–306. [CrossRef]

12. Mandarino, J.A. The Gladstone-Dale relationship. I. Derivation of new constants. *Can. Mineral.* **1976**, *14*, 498–502.

13. Courbion, G.; Ferrey, G. Na$_2$Ca$_3$Al$_2$F$_{14}$: A new example of a structure with "independent F$^-$"—A new method of comparison between fluorides and oxides of different formula. *J. Solid State Chem.* **1988**, *76*, 426–431. [CrossRef]

14. Effenberger, H.; Kluger, F. Ralstonit: Ein Beitrag zur Kenntnis von Zusammensetzung und Kristallstruktur. *N. Jahrb. Miner. Monat.* **1984**, *1984*, 97–108.

15. Sawada, H. An electron density residual study of alpha-ferric oxide. *Mater. Res. Bull.* **1996**, *31*, 141–146. [CrossRef]

16. Balić Žunić, T.; Makovicky, E. Determination of the centroid or "the best centre" of a coordination polyhedron. *Acta Cryst.* **1996**, *B52*, 78–81. [CrossRef]

17. Makovicky, E.; Balić-Žunić, T. New measure of distortion for coordination polyhedra. *Acta Cryst.* **1998**, *B54*, 766–773. [CrossRef]

18. Balić-Žunić, T. Use of three-dimensional parameters in the analysis of crystal structures under compression. In *Pressure-Induced Phase Transitions*; Grzechnik, A., Ed.; Transworld Research Network: Kerala, India, 2007; pp. 157–184, ISBN 81-7895-272-6.

19. Bolotina, N.B.; Maximov, B.A.; Simonov, V.I.; Derzhavin, S.I.; Uvarova, T.V.; Apollonov, V.V. Crystal structure and spectral characteristics of $LiCaAlF_6$:Cr^{3+} single crystals. *Crystallogr. Rep.* **1993**, *38*, 446–450.

20. Hemon, A.; Courbion, G. The NaF-CaF_2-AlF_3 system: Structures of beta-$NaCaAlF_6$ and $Na_4Ca_4Al_7F_{33}$. *J. Solid State Chem.* **1990**, *84*, 153–164. [CrossRef]

21. Hemon, A.; Le Bail, A.; Courbion, G. Crystal structure approach of $KCaAl_2F_9$. A new hexagonal tungsten-bronze related structure. *Eur. J. Solid State Inor. Chem.* **1993**, *30*, 415–426. [CrossRef]

22. Hemon, A.; Courbion, G. Refinement of the room-temperature structure of alpha-$CaAlF_5$. *Acta Cryst. C* **1991**, *47*, 1302–1303. [CrossRef]

23. Domesle, R.; Hoppe, R. The crystal structure of Ca_2AlF_7. *Z. Kristallogr.* **1980**, *153*, 317–328. [CrossRef]

24. Le Bail, A.; Hemon-Ribaud, A.; Courbion, G. Structure of alpha-(Na Ca Al F6) determined ab initio from conventional powder diffraction data. *Eur. J. Solid State Inor. Chem.* **1998**, *35*, 265–272. [CrossRef]

25. Werner, F.; Weil, M. Alpha-($BaCaAlF_7$). *Acta Cryst. E* **2003**, *59*, 17–19. [CrossRef]

26. Litvin, A.L.; Petrunina, A.A.; Ostapenko, S.S.; Povarennykh, A.S. The crystal structure of usovite. *Dopov. Akad. Nauk Ukr. RSR Ser. B* **1980**, *3*, 47–80. (In Ukrainian)

27. Kampf, A.R.; Yang, H.; Downs, R.T.; Pinch, W.W. The crystal structures and Raman spectra of aravaipaite and calcioaravaipaite. *Am. Mineral.* **2011**, *96*, 402–407. [CrossRef]

28. Mumme, W.G.; Grey, I.E.; Birch, W.D.; Pring, A.; Bougerol, C.; Wilson, N.C. Coulsellite, $CaNa_3AlMg_3F_{14}$, a rhombohedral pyrochlore with 1:3 ordering in both A and B sites, from the Cleveland mine, Tasmania, Australia. *Am. Mineral.* **2010**, *95*, 736–740. [CrossRef]

29. Swanson, H.E.; Tatge, E. Standard X-ray diffraction powder patterns. *Nat. Bur. Stand. Circ.* **1953**, *539*, 69–70.

30. Bevan, D.J.M.; Straehle, J.; Greis, O. The crystal structure of tveitite, an ordered yttrofluorite mineral. *J. Solid State Chem.* **1982**, *44*, 75–81. [CrossRef]

31. The American Mineralogist Crystal Structure Database. Available online: http://rruff.geo.arizona.edu/AMS/amcsd.php (accessed on 28 November 2018).

32. Craig, D.F.; Brown, J.J. Phase equilibria in the system CaF_2-AlF_3-Na_3AlF_6 and part of the system CaF_2-AlF_3-Na_3AlF_6-Al_2O_3. *J. Am. Ceram. Soc.* **1977**, *63*, 254–261. [CrossRef]

minerals

MDPI

Article

Copper in Natural Oxide Spinels: The New Mineral Thermaerogenite $CuAl_2O_4$, Cuprospinel and Cu-Enriched Varieties of Other Spinel-Group Members from Fumaroles of the Tolbachik Volcano, Kamchatka, Russia

Igor V. Pekov [1,*], Fedor D. Sandalov [1], Natalia N. Koshlyakova [1], Marina F. Vigasina [1], Yury S. Polekhovsky [2,†], Sergey N. Britvin [2,3], Evgeny G. Sidorov [4] and Anna G. Turchkova [1]

[1] Faculty of Geology, Moscow State University, Vorobievy Gory, 119991 Moscow, Russia; fyodor.sandalov@yandex.ru (F.D.S.); nkoshlyakova@gmail.com (N.N.K.); vigasina@geol.msu.ru (M.F.V.); annaturchkova@rambler.ru (A.G.T.)

[2] Institute of Earth Sciences, St. Petersburg State University, Universitetskaya Nab. 7/9, 199034 St Petersburg, Russia; sbritvin@gmail.com (S.N.B.)

[3] Nanomaterials Research Center, Kola Science Centre, Russian Academy of Sciences, Fersman str. 14, 184209 Apatity, Murmansk Region, Russia

[4] Institute of Volcanology and Seismology, Far Eastern Branch of Russian Academy of Sciences, Piip Boulevard 9, 683006 Petropavlovsk-Kamchatsky, Russia; mineral@kscnet.ru

* Correspondence: igorpekov@mail.ru; Tel.: +7-495-939-4676

† Deceased 28 September 2018.

Received: 11 October 2018; Accepted: 29 October 2018; Published: 1 November 2018

Abstract: This paper is the first description of natural copper-rich oxide spinels. They were found in deposits of oxidizing-type fumaroles related to the Tolbachik volcano, Kamchatka, Russia. This mineralization is represented by nine species with the following maximum contents of CuO (wt.%, given in parentheses): a new mineral thermaerogenite, ideally $CuAl_2O_4$ (26.9), cuprospinel, ideally $CuFe^{3+}_2O_4$ (28.6), gahnite (21.4), magnesioferrite (14.7), spinel (10.9), magnesiochromite (9.0), franklinite (7.9), chromite (5.9), and zincochromite (4.8). Cuprospinel, formerly known only as a phase of anthropogenic origin, turned out to be the Cu-richest natural spinel-type oxide [sample with the composition $(Cu_{0.831}Zn_{0.100}Mg_{0.043}Ni_{0.022})_{\Sigma0.996}(Fe^{3+}_{1.725}Al_{0.219}Mn^{3+}_{0.048}Ti_{0.008})_{\Sigma2.000}O_4$ from Tolbachik]. Aluminum and Fe^{3+}-dominant spinels (thermaerogenite, gahnite, spinel, cuprospinel, franklinite, and magnesioferrite) were deposited directly from hot gas as volcanic sublimates. The most probable temperature interval of their crystallization is 600–800 °C. They are associated with each other and with tenorite, hematite, orthoclase, fluorophlogopite, langbeinite, calciolangbeinite, aphthitalite, anhydrite, fluoborite, sylvite, halite, pseudobrookite, urusovite, johillerite, ericlaxmanite, tilasite, etc. Cu-bearing spinels are among the latest minerals of this assemblage: they occur in cavities and overgrow even alkaline sulfates. Cu-enriched varieties of chrome-spinels (magnesiochromite, chromite, and zincochromite) were likely formed in the course of the metasomatic replacement of a magmatic chrome-spinel in micro-xenoliths of ultrabasic rock under the influence of volcanic gases. The new mineral thermaerogenite, ideally $CuAl_2O_4$, was found in the Arsenatnaya fumarole at the Second scoria cone of the Northern Breakthrough of the Great Tolbachik Fissure Eruption. It forms octahedral crystals up to 0.02 mm typically combined in open-work clusters up to 1 mm across. Thermaerogenite is semitransparent to transparent, with a strong vitreous lustre. Its colour is brown, yellow-brown, red-brown, brown-yellow or brown-red. The mineral is brittle, with the conchoidal fracture, cleavage is none observed. D(calc.) is 4.87 g/cm^3. The chemical composition of the holotype (wt.%, electron microprobe) is: CuO 25.01, ZnO 17.45, Al_2O_3 39.43, Cr_2O_3 0.27, Fe_2O_3 17.96, total 100.12 wt.%. The empirical formula calculated on the basis of 4 O *apfu* is: $(Cu_{0.619}Zn_{0.422})_{\Sigma1.041}(Al_{1.523}Fe^{3+}_{0.443}Cr_{0.007})_{\Sigma1.973}O_4$. The mineral is cubic, *Fd-3m*, *a* = 8.093(9) Å,

$V = 530.1(10)$ Å^3. Thermaerogenite forms a continuous isomorphous series with gahnite. The strongest lines of the powder X-ray diffraction pattern of thermaerogenite [d, Å (I, %) (hkl)] are: 2.873 (65) (220), 2.451 (100) (311), 2.033 (10) (400), 1.660 (16) (422), 1.565 (28) (511) and 1.438 (30) (440).

Keywords: thermaerogenite; cuprospinel; gahnite; magnesioferrite; $CuAl_2O_4$; $CuFe_2O_4$; copper oxide; new mineral; spinel supergroup; fumarole sublimate; Tolbachik volcano; Kamchatka

---

1. Introduction

Oxide spinels compose one of the most studied mineral families and have numerous implications in the geosciences, chemistry and materials science [1]. In spite of the apparent simplicity, the spinel-type crystal structure exhibits a remarkable flexibility towards cation and anion substitutions, which results in the appearance of minerals accommodating more than two dozens of chemical elements [2]. Copper–bearing oxide spinels, being well studied as synthetic phases in materials science, are however virtually unknown in nature. The only reported Cu-rich spinel-type oxide mineral, cuprospinel, ideally $CuFe_2^{3+}O_4$, is in fact an anthropogenic phase as it has never been found in the natural environments unaffected by anthropogenic influence factors. Cuprospinel was described as a new mineral species from burnt dumps on the property of Consolidated Rambler Mines Limited near Baie Verte, Newfoundland, Canada, where, together with hematite and a Cu-rich (13.9 wt.% CuO) variety of magnesioferrite, it was formed in the result of a spontaneous fire of the mined copper-zinc ore [3]. Thus, these oxides have definite anthropogenic origin, as well as cuprospinel formed as an incidental product of ore processing in ancient and modern smelters [4,5]. Cuprospinel was mentioned in a volcanic material from Mahanadi, Orissa, India [6], and in ores of the Chahnaly gold deposit in SE Iran [7], however, no analytical evidences were given and the identification of this mineral from both localities seems doubtful.

In the course of ongoing research of oxidizing-type fumaroles related to the Tolbachik volcano at Kamchatka, Russia, we have encountered more than twenty oxide minerals. Among those, ten mineral species belonging to the spinel supergroup [2] were identified, including the new mineral deltalumite, $(Al_{0.67}\square_{0.33})Al_2O_4$, the delta-alumina dimorphous with corundum [8]. A remarkable chemical feature of Tolbachik spinels is the common presence of a significant amount of copper. CuO contents ranging from 1 to 18 wt.% are typical for spinel, magnesioferrite, gahnite, franklinite, magnesiochromite, chromite and zincochromite. Besides these Cu-rich mineral varieties, two minerals with species-defining copper were therein discovered: the genuine natural cuprospinel and a new mineral first described in the present paper, thermaerogenite, ideally $CuAl_2O_4$. The latter was recently approved by the IMA Commission on New Minerals, Nomenclature and Classification (IMA no. 2018-021). The name thermaerogenite (Cyrillic: термаэрогенит) is constructed based on the combination of Greek words θερμός, hot, αέριον, gas, and γενής that means "born by". Thus, in whole it means *born by hot gas*, that reflects the fumarolic origin of the mineral. This name also contains the allusion that Cu-rich spinel-type oxides form in volcanic fumaroles and their anthropogenic counterparts, unlike Cu-free and Cu-poor oxide spinels, are numerous and widespread in other geological formations. The type specimen of thermaerogenite is deposited in the collections of the Fersman Mineralogical Museum of the Russian Academy of Sciences, Moscow, Russia, with the registration number 5192/1.

The Mg-Al-Fe oxide spinels, namely magnesioferrite, magnetite and spinel were previously reported from fumaroles of several active volcanoes [9–13]. However, no information on Cu contents in these minerals was published. Moreover, we did not find any references on Cu-enriched (with CuO content higher than 1 wt.%) natural oxide spinels. Thus, this work is the first report on natural copper-rich oxide members of the spinel supergroup.

2. Occurrence and Mineral Associations

The studied material was collected by us during fieldwork in the period of 2012–2018. The majority of studied samples originate from the Arsenatnaya fumarole, one of the brightest in the mineralogical aspect examples in the world of fumaroles belonging to the oxidizing type (in fumaroles of this type, the increase of oxygen fugacity is a result of the mixing of volcanic gases with the atmospheric air [14,15]). Arsenatnaya is located at the apical part of the Second scoria cone of the Northern Breakthrough of the Great Tolbachik Fissure Eruption (below–NB GTFE), Tolbachik volcano, Kamchatka Peninsula, Far-Eastern Region, Russia (55°41′ N 160°14′ E, 1200 m asl). This scoria cone, formed in 1975, is a monogenetic volcano about 300 m high and approximately 0.1 km^3 in volume [16]. Now, more than 40 years after the eruption, its fumarole fields remain active: numerous gas vents with temperatures up to 490 °C were observed by us in 2012–2018. Now the fumarolic gases at the Second scoria cone are compositionally close to atmospheric air, with the contents of <1 vol.% water vapour and <0.1 vol.% acid species, mainly CO_2, HF and HCl [17], while in 1976–1977 these gases were significantly more enriched in H_2O, CO_2, SO_2, HCl and, in some fumaroles, HF [14].

The Arsenatnaya fumarole was uncovered and first studied by us during fieldworks in July 2012. This active fumarole described in papers [18,19] is a near-meridional linear system of mineralized pockets (up to 10–15 cm wide) and cracks situated between blocks of basalt scoria and volcanic bombs in the near-surface part of the scoria cone. The length of the hot area belonging to Arsenatnaya is about 15 m and its width varies from 1–1.5 m in the southern end to 3–4 m in the northern part. Numerous strongly mineralized pockets occur at depths from 0.3 to 4 m. The sublimate minerals form incrustations in the open space of the pockets, fill cracks and pores or replace basalt. Arsenatnaya is one of the hottest fumaroles at the Second scoria cone of the NB GTFE: the temperature measured by us using chromel-alumel thermocouple in 2012–2018 in different pockets immediately after their partial uncovering varies from 360 to 490 °C and, in general, increases with depth. About 160 valid minerals (including 46 new species first discovered here) and >30 insufficiently studied mineral phases have been identified in this unique mineralogical site.

The Cu-bearing oxide spinels in the Arsenatnaya fumarole were found in two mineral assemblages.

The minerals with Al or Fe^{3+} as species-defining components (spinel, gahnite, thermaerogenite, magnesioferrite, franklinite, and cuprospinel) occur in the intermediate in depth, in the polymineralic zone of the fumarole [19] and are associated with tenorite, hematite, orthoclase (As-bearing variety), fluorophlogopite, langbeinite, calciolangbeinite, aphthitalite-type sulfates, anhydrite, krasheninnikovite, vanthoffite, fluoborite, sylvite, halite, pseudobrookite, rutile, corundum and various arsenates: urusovite, johillerite, ericlaxmanite, kozyrevskite, popovite, lammerite, lammerite-β, tilasite, svabite, nickenichite, bradaczekite, dmisokolovite, shchurovskyite, etc. Cu-bearing spinels are among the latest minerals of this assemblage: they occur in cavities and overgrow not only earlier oxides (hematite, tenorite) and silicates but also arsenates and even "saline" sulfates, such as langbeinite-calciolangbeinite series and aphthitalite group minerals (Figures 1a–c and 2a,b). Overgrowing of hematite by Cu-rich oxide spinels is very common; these minerals form epitactic intergrowths (the crystal face {111} of a spinel-group member is coplanar to the face {0001} of hematite crystal) or clusters of randomly oriented crystals (Figure 3a–c). Sometimes, spinels cover hematite as massive crusts (Figure 4a,b). Some associations contain two or more spinel-group minerals. In such cases, the Cu-richest species are typically the latest (Figure 4b).

Figure 1. (a) Numerous small brown crystals of thermaerogenite on colourless to white langbeinite with light blue urusovite, iron-black tenorite and minor green ericlaxmanite. Field width: 2.2 mm. (b) Abundant small brown crystals of thermaerogenite on white aphthitalite and iron-black hematite and tenorite. Field width: 3.5 mm. (c) Small brown crystals of thermaerogenite on colourless calciolangbeinite iron-black tenorite, light blue urusovite and green kozyrevskite. Field width: 0.8 mm.

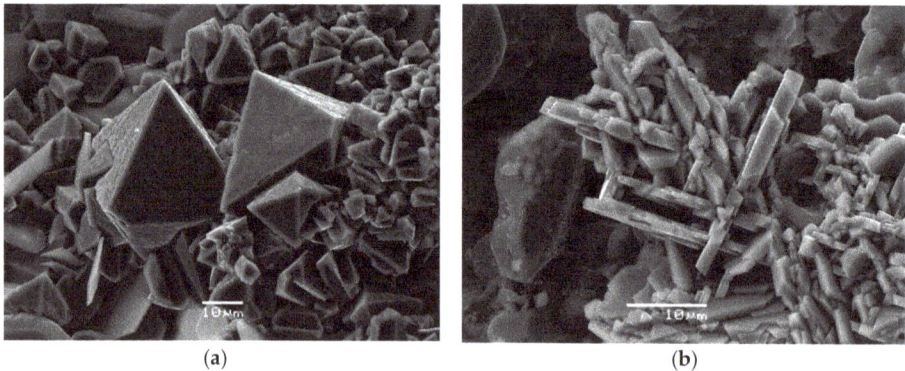

Figure 2. (a) Octahedral crystals of Cu-bearing spinel on crystal crust of tilasite. (b) Flattened spinel-law twins on {111}, additionally combined in complex twinned lattice-like aggregates, of Cu-bearing gahnite on crystal crust of anhydrite. Scanning electron microscopic (SEM)-secondary electron (SE) images.

(a)

(b)

(c)

Figure 3. (**a**) Distorted (flattened on {111}) octahedral crystals of cuprospinel epitactically overgrowing the {0001} face of hematite crystal. (**b,c**) Octahedral crystals of Cu-bearing gahnite overgrowing columnar hematite crystals. Both epitaxy and random orientation of gahnite crystals on hematite are observed. SEM (SE) images.

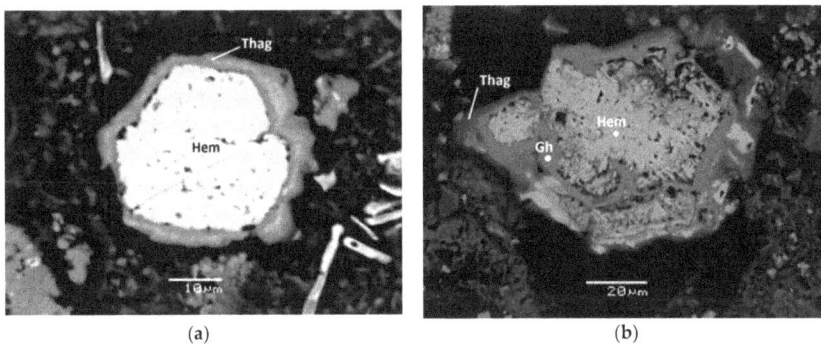

(a)

(b)

Figure 4. (**a**) Solid crust of thermaerogenite (Thag) overgrowing hematite (Hem) crystal. (**b**) Crust consisting of Cu-bearing gahnite (Gh) and later thermaerogenite (Thag) overgrowing skeletal crystal of hematite (Hem). Polished section, SEM-back-scattered electron (BSE) images.

Cu-enriched chrome-spinels were found in Arsenatnaya in several micro-xenoliths (up to 1 mm across) of ultrabasic rock mainly consisting of olivine (Fo_{83-85}). These micro-xenoliths embedded in basalt scoria were strongly altered by fumarolic gases. As a result of gas influence, the peripheral part of olivine grains was significantly replaced by hematite (Figure 5a) and primary, magmatic chrome-spinel (crystals up to 0.06 mm) was altered to Zn- and Cu-bearing species. Figure 5b demonstrates such pseudomorphosed crystal with a core represented by twinned lattice-like aggregates of the newly formed Cu-enriched chromite and a rim consisting of Cu-enriched zincochromite.

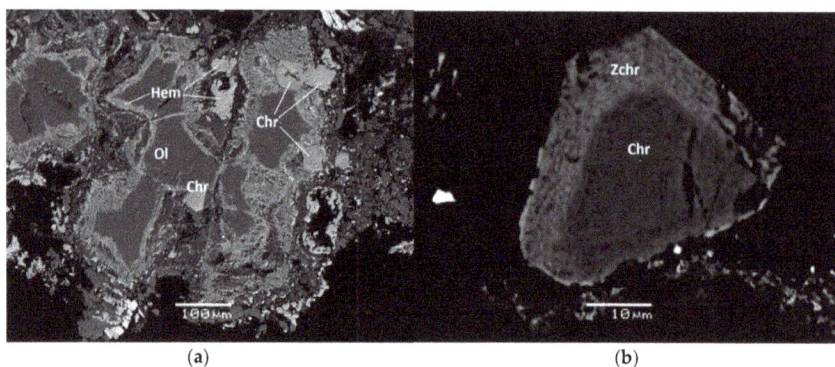

(a)   (b)

Figure 5. (**a**) Altered by fumarolic gases, micro-xenolith of ultrabasic rock mainly consisting of olivine (Ol) and containing Cu-bearing chrome-spinels (Chr); Hem—hematite. (**b**) Crystal of primary, magmatic chrome-spinel replaced by Cu-bearing varieties of chromite (Chr, core) and zincochromite (Zchr, rim). Polished section, SEM (BSE) images.

Besides Arsenatnaya, Cu-bearing varieties of gahnite, magnesioferrite and spinel were found in deposits of extinct fumaroles belonging to the Western paleo-fumarole field at Mountain 1004, a scoria cone located 2 km south of the Second scoria cone of the NB GTFE. Mountain 1004 was formed as a result of an ancient eruption of Tolbachik; the age of this monogenetic volcano and its fumarole fields is evaluated as ca. 2000 years [16]. Spinels occur here in cavities of basalt scoria altered by fumarolic gas. The associated sublimate minerals are diopside, fluorophlogopite, potassic feldspar, indialite, hematite, tenorite, fluorite, sellaite, anglesite, and baryte.

3. Methods

Reflectance spectra for thermaerogenite were obtained in air using a MSF-21 micro-spectrophotometer (LOMO company, St. Petersburg, Russia) with the monochromator slit width of 0.4 mm and beam diameter of 0.1 mm; SiC (Reflectionstandard 474251, No. 545, Germany) was used as a standard.

The Raman spectrum of thermaerogenite was obtained using an EnSpectr R532 Raman microscope (Department of Mineralogy, Moscow State University, Moscow, Russia) with a solid-state laser diode with green radiation (532 nm) at room temperature. The spectrum was processed in the range from 100 to 4000 cm^{-1} with the use of a holographic diffraction grating with 1800 mm^{-1} and a resolution equal to 6 cm^{-1}. The microscope was focused onto the sample using a PLNC 40X objective (NA = 0.65). Backscattered Raman signals were collected at 1 s exposure time with 400 spectra accumulations. The spectrum was obtained for a randomly oriented crystal with the diameter of the focal spot on the sample about 1 μm. The output laser power was about 12 mW. The power of laser radiation on the surface of the mineral was significantly less. No thermal damage of the sample was observed. Before conducting the experiments, the instrument was calibrated with Raman line of crystalline silicon (520 cm^{-1}).

Scanning electron microscopic (SEM) studies in secondary electron (SE) and back-scattered electron (BSE) modes were carried out and chemical composition was determined for all studied samples using a Jeol JSM-6480LV scanning electron microscope equipped with an INCA-Wave 500 wavelength-dispersive spectrometer (Laboratory of Analytical Techniques of High Spatial Resolution, Department of Petrology, Moscow State University, Moscow, Russia), with an acceleration voltage of 20 kV, a beam current of 20 nA, and a 3 μm beam diameter. The standards used are: $MgAl_2O_4$ (Mg,Al), Ni (Ni), Cu (Cu), ZnS (Zn), V (V), $FeCr_2O_4$ (Cr), FeS_2 (Fe), and $MnTiO_3$ (Mn,Ti).

Powder X-ray diffraction data were collected using a Rigaku RAXIS Rapid II diffractometer with a curved image plate detector, a rotating anode with VariMAX microfocus optics, using CoKα radiation, in Debye-Scherrer geometry, at an accelerating voltage of 40 kV, a current of 15 mA and an exposure time 15 min. The distance between the sample and the detector was 127.4 mm. Data processing was carried out using osc2xrd software [20].

Single-crystal X-ray diffraction studies were carried out using an Xcalibur S diffractometer equipped with a CCD detector (MoKα radiation).

4. Results

4.1. General Appearance and Physical Properties of Thermaerogenite and Other Cu-Rich Oxide Spinels from Tolbachik

Thermaerogenite forms octahedral crystals up to 0.02 mm across, sometimes skeletal, typically combined in open-work clusters (Figure 6a) up to 1 mm across. Areas "sprinkled" by crystals of the new mineral (Figure 1b) are up to 0.5 cm × 0.5 cm. Besides the major form {111}, narrow {110} faces were observed on some crystals of thermaerogenite. Being visually perfect (Figure 6a), crystals of thermaerogenite have blocky inner structure and these micro-blocks are slightly misoriented, which hampers the single-crystal X-ray diffraction study of the mineral.

Figure 6. (**a**) Cluster of octahedral, with minor faces {110}, crystals of thermaerogenite. (**b**) Crust of octahedral crystals of cuprospinel with minor faces {100}. (**c**) Aggregate of octahedral crystals of Cu-bearing magnesioferrite partially covered by thin "fluffy" crust of supergene opal from paleo-fumarole (Mountain 1004). (**d**) Crystal of Cu-bearing magnesioferrite of rhombic dodecahedral shape. SEM (SE) images.

Other Cu-rich Al- and Fe^{3+}-dominant spinels in Tolbachik fumaroles occur as octahedral crystals (Figures 2a, 3 and 6a–c), sometimes with minor {110} or/and {100} faces, up to 0.1 mm across, only magnesioferrite was observed as larger (up to 1 mm) crystals, typically rhombic dodecahedra (Figure 6d). Gahnite was also found as flattened spinel-law twins on {111} additionally combined in a complex lattice-like aggregates (Figure 2b). A similar lattice-like shape was observed for Cu-bearing chromite (Figure 5b). Spinel, magnesioferrite and gahnite sometimes form crusts up to 1 cm across and up to 0.5 mm thick. Solid crusts of cuprospinel (up to 2 mm across) were observed on hematite.

Thermaerogenite is semitransparent to transparent, with a yellowish streak and strong vitreous lustre. Its colour is brown, yellow-brown, red-brown, brown-yellow or brown-red (Figure 1).

Colour, transparency and lustre of other Cu-rich Al- and Fe^{3+}-dominant spinels from Tolbachik are in general correlated with the Al:Fe ratio. The Fe-poor varieties are very light (some samples of spinel and gahnite from Mountain 1004 are colourless or greyish-white) with vitreous lustre. Spinel, gahnite and Al-enriched cuprospinel in typical samples from the Arsenatnaya fumarole are visually indistinguishable from the above-described thermaerogenite. Fe-rich cuprospinel is dark reddish-brown with strong vitreous to adamantine lustre. Magnesioferrite has dark brown to iron-black colour and submetallic lustre.

Thermaerogenite is brittle, with conchoidal fracture (observed under the scanning electron microscope); cleavage or parting is not observed. Its Mohs hardness is ca. 7. Density calculated using the empirical formula of the holotype specimen is 4.870 g/cm^3.

4.2. Optical Data for Thermaerogenite

In reflected light, thermaerogenite is optically isotropic, grey, with yellowish internal reflections. The reflectance values are given in Table 1.

Table 1. Reflectance data for thermaerogenite. Reflectance values for four wavelengths recommended by the IMA Commission on Ore Microscopy are given in bold type.

λ (nm)	R	λ (nm)	R
400	16.4	560	14.0
420	16.0	580	13.7
440	15.7	**589**	**13.6**
460	15.4	600	13.4
470	**15.2**	620	13.2
480	15.1	640	13.0
500	14.8	**650**	**12.9**
520	14.5	660	12.8
540	14.2	680	12.5
546	**14.2**	700	12.3

4.3. Raman Spectroscopy of Thermaerogenite

The Raman spectrum of thermaerogenite (Figure 7) is typical for spinel-type oxides [21]. It contains four distinct bands with following wavenumbers of maxima (cm^{-1}, s—strong band): 762s, 590, 284, and 125s. They could be assigned based on the data reported for different oxide spinels in [21]. The strongest band at 762 cm^{-1} could correspond to A_{1g} mode, stretching vibrations of O atoms surrounding Al in tetrahedral coordination. The presence of part of Al in the tetrahedral position (8a) is typical for synthetic Cu-rich members of the series $Zn_{1-x}Cu_xAl_2O_4$ [22]. Broad band with maximum at 590 cm^{-1} probably corresponds to $F_{2g}(2)$ or $F_{2g}(3)$ mode involving divalent cations, (Cu,Zn)–O, whereas weak band at 284 cm^{-1} corresponds to $F_{2g}(1)$ mode. Strong band at 125 cm^{-1} can be assigned as corresponding to lattice modes.

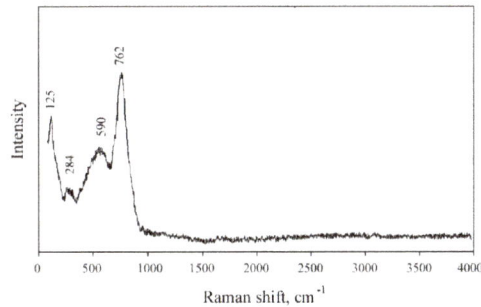

Figure 7. The Raman spectrum of thermaerogenite.

4.4. Chemical Composition of Cu-Rich Oxide Spinels from Tolbachik

Representative electron-microprobe analyses for the studied minerals are given in Table 2 (Cu-bearing Al- and Fe^{3+}-dominant spinels) and Table 3 (Cu-bearing chrome-spinels). There are 32 spot analyses of their different chemical varieties which demonstrate chemical variability of Cu-enriched oxide spinels from Tolbachik fumaroles.

Valent states of cations were not determined by direct methods. For properly sublimate minerals (thermaerogenite, cuprospinel, gahnite, spinel, magnesioferrite and franklinite) Fe and Mn are considered as Fe^{3+} and Mn^{3+}, respectively, taking into account extremely oxidizing conditions of mineral deposition in Arsenatnaya and similar fumaroles [18]. This is clearly confirmed by the calculation of the empirical formulae using the scheme common for oxide spinels, on the basis of four O atoms per formula unit (*apfu*): Sums of divalent and trivalent cations are close to 2.00 and 3.00 *apfu*, respectively (Tables 2 and 3) that is typical for "2-3 spinels" with the general formula $A^{2+}B^{3+}_2O_4$ [2,23]. For the Cu-bearing chrome-spinels formed in the result of the alteration of a primary, magmatic chrome-spinel, we assume the presence of part of iron in a divalent state (while manganese content is below detection limit in these minerals). Their formulae were calculated on the basis of four O *apfu* with fixing of the cation sum $A^{2+} + B^{3+,4+} = 3.00$ *apfu* and the $Fe^{2+}:Fe^{3+}$ ratio was calculated by charge balance.

Maximum contents of copper detected in different spinel-group minerals from Tolbachik fumaroles are presented in Table 4 and atomic ratios of the major bivalent (*A*) and trivalent (*B*) cations are shown in Figure 8a,b (based on 124 spot electron-microprobe analyses).

Table 2. Chemical composition of Cu-bearing Al- and Fe^{3+}-dominant spinel-group oxides from Tolbachik fumaroles: 1–4: thermaerogenite, 5–11: gahnite, 12–16: spinel, 17–20: magnesioferrite, 21: franklinite, and 22–28: cuprospinel.

No.	1 *	2	3	4	5	6	7
			wt.%				
MgO	-	-	-	0.44	-	5.50	-
CuO	25.01 (23.64–26.86)/1.46	25.47	25.67	20.25	20.07	15.00	17.55
ZnO	17.45 (14.46–18.71)/2.00	18.33	17.43	18.99	23.79	18.29	21.71
Al_2O_3	39.43 (34.59–45.43)/4.60	45.43	28.30	22.79	49.05	53.90	56.08
Cr_2O_3	0.27 (0.17–0.33)/0.07	0.17	0.86	-	0.15	0.03	-
Mn_2O_3	-	-	-	0.36	-	0.10	-
Fe_2O_3	17.96 (11.47–22.21)/4.76	11.47	24.48	34.29	7.72	5.99	4.63
TiO_2	-	-	2.59	2.61	-	0.22	0.24
Total	100.12	100.87	99.33	99.73	100.78	99.03	100.21

Table 2. *Cont.*

No.	1 *	2	3	4	5	6	7
formula calculated on the basis of 4 O *apfu*							
Mg	-	-	-	0.023	-	0.242	-
Cu	0.619	0.609	0.673	0.540	0.472	0.334	0.396
Zn	0.422	0.428	0.447	0.496	0.547	0.398	0.479
Al	1.523	1.695	1.156	0.949	1.801	1.874	1.973
Cr	0.007	0.004	0.017	-	0.003	0.001	-
Mn	-	-	-	0.011	-	0.002	-
Fe^{3+}	0.443	0.273	0.639	0.911	0.181	0.133	0.104
Ti	-	-	0.068	0.069	-	0.005	0.006
ΣA^{2+}	**1.041**	**1.037**	**1.119**	**1.059**	**1.019**	**0.975**	**0.874**
$\Sigma B^{3+,4+}$	**1.973**	**1.972**	**1.881**	**1.940**	**1.984**	**2.015**	**2.082**

No.	8	9	10	11	12	13	14
wt.%							
MgO	0.42	-	0.59	2.66	13.27	17.54	19.61
NiO	-	-	-	-	0.87	-	-
CuO	13.72	13.06	13.02	2.94	10.89	6.86	4.83
ZnO	30.66	31.91	29.25	35.66	5.98	8.42	9.37
Al_2O_3	44.70	54.11	44.30	55.81	32.93	61.12	48.44
Cr_2O_3	0.19	-	0.18	-	-	-	-
Mn_2O_3	0.16	-	0.10	-	1.88	0.37	0.93
Fe_2O_3	8.82	1.75	11.40	2.11	34.29	4.77	13.88
TiO_2	1.06	0.11	0.80	-	-	0.41	2.02
Total	99.73	100.94	99.64	99.18	100.11	99.49	99.08
formula calculated on the basis of 4 O *apfu*							
Mg	0.020	-	0.028	0.119	0.601	0.688	0.807
Ni	-	-	-	-	0.021	-	-
Cu	0.333	0.301	0.314	0.066	0.250	0.136	0.100
Zn	0.726	0.718	0.690	0.789	0.134	0.164	0.191
Al	1.690	1.944	1.669	1.970	1.180	1.897	1.576
Cr	0.004	-	0.004	-	-	-	-
Mn	0.004	-	0.003	-	0.048	0.008	0.022
Fe^{3+}	0.213	0.040	0.274	0.048	0.784	0.094	0.288
Ti	0.025	0.003	0.019	-	-	0.008	0.042
ΣA^{2+}	**1.079**	**1.019**	**1.033**	**0.974**	**1.006**	**0.989**	**1.098**
$\Sigma B^{3+,4+}$	**1.935**	**1.987**	**1.969**	**2.018**	**2.012**	**2.008**	**1.928**

Table 2. *Cont.*

No.	15	16	17	18	19	20	21
			wt.%				
MgO	21.94	16.04	17.43	17.65	10.56	20.81	8.54
NiO	-	-	-	-	0.35	-	-
CuO	2.18	4.34	5.90	3.22	14.73	1.26	7.91
ZnO	5.24	12.28	3.09	0.60	0.94	0.60	17.99
Al_2O_3	59.10	58.36	5.30	0.66	5.51	5.72	20.24
Cr_2O_3	-	0.09	0.33	-	-	-	0.17
Mn_2O_3	1.56	0.26	0.24	1.32	1.01	1.24	2.29
Fe_2O_3	8.76	8.10	66.24	77.29	66.64	69.46	36.51
TiO_2	0.48	0.24	0.92	-	-	-	5.50
Total	99.26	99.70	99.45	100.74	99.74	99.09	99.15
		formula calculated on the basis of 4 O *apfu*					
Mg	0.843	0.640	0.873	0.884	0.552	1.014	0.424
Ni	-	-	-	-	0.010	-	-
Cu	0.042	0.088	0.150	0.082	0.391	0.031	0.199
Zn	0.100	0.243	0.077	0.015	0.024	0.015	0.443
Al	1.796	1.842	0.210	0.026	0.228	0.220	0.795
Cr	-	0.002	0.006	-	-	-	0.005
Mn	0.031	0.005	0.007	0.034	0.027	0.031	0.058
Fe^{3+}	0.170	0.163	1.675	1.953	1.760	1.709	0.915
Ti	0.009	0.005	0.023	-	-	-	0.138
ΣA^{2+}	**0.986**	**0.971**	**1.099**	**0.980**	**0.977**	**1.060**	**1.066**
$\Sigma B^{3+,4+}$	**2.006**	**2.018**	**1.922**	**2.013**	**2.015**	**1.960**	**1.910**
No.	22	23	24	25	26	27	28
			wt.%				
MgO	0.75	0.63	4.31	3.68	2.32	0.48	5.05
NiO	0.70	-	-	-	-	-	-
CuO	28.55	27.13	24.48	24.36	25.92	20.81	23.44
ZnO	3.51	7.76	3.39	2.48	4.53	18.82	1.94
Al_2O_3	4.82	7.52	3.44	1.69	4.12	18.69	3.46
Cr_2O_3	-	0.22	-	0.23	-	0.24	0.27
Mn_2O_3	1.65	0.94	2.15	0.85	1.38	0.33	1.20
Fe_2O_3	59.49	53.49	59.77	65.82	61.51	37.06	63.40
TiO_2	0.28	1.75	1.54	-	0.30	2.63	0.41
Total	99.75	99.44	99.08	99.11	100.08	99.06	99.17

Table 2. *Cont.*

No.	22	23	24	25	26	27	28
		formula calculated on the basis of 4 O *apfu*					
Mg	0.043	0.036	0.244	0.211	0.133	0.026	0.283
Ni	0.022	-	-	-	-	-	-
Cu	0.831	0.781	0.704	0.707	0.748	0.573	0.665
Zn	0.100	0.219	0.096	0.071	0.128	0.506	0.054
Al	0.219	0.338	0.154	0.076	0.185	0.802	0.153
Cr	-	0.005	-	0.006	-	0.005	0.006
Mn	0.048	0.030	0.069	0.027	0.045	0.010	0.038
Fe^{3+}	1.725	1.535	1.711	1.902	1.768	1.016	1.793
Ti	0.008	0.050	0.044	-	0.008	0.072	0.011
ΣA^{2+}	0.996	1.036	1.044	0.988	1.008	1.105	1.002
$\Sigma B^{3+,4+}$	2.000	1.958	1.979	2.012	2.006	1.904	2.002

* The holotype specimen of thermaerogenite: averaged from four spot analyses value (range)/standard deviation; dash means the content below detection limit.

Table 3. Chemical composition of Cu-bearing chrome-spinels from the Arsenatnaya fumarole, Tolbachik: 1–2: magnesiochromite, 3: chromite, 4: zincochromite.

No.	1	2	3	4
		wt.%		
MgO	10.48	7.28	7.52	2.60
FeO	2.30	6.19	16.38	1.07
CuO	8.97	6.25	5.86	3.96
ZnO	8.54	12.83	2.16	27.88
Al_2O_3	11.32	11.52	11.06	10.44
V_2O_3	0.19	0.14	0.16	0.19
Cr_2O_3	47.45	46.06	48.20	43.42
Fe_2O_3	10.73	9.70	7.56	10.61
TiO_2	0.83	0.89	0.88	0.81
Total	100.82	100.86	99.78	100.98
	formula calculated on the basis of 4 O *apfu* with $A + B$ = 3.00 *apfu* *			
Mg	0.524	0.374	0.388	0.140
Fe^{2+}	0.058	0.160	0.427	0.029
Cu	0.227	0.163	0.153	0.108
Zn	0.212	0.326	0.056	0.745
Al	0.448	0.467	0.451	0.445
V	0.005	0.004	0.004	0.006
Cr	1.259	1.254	1.319	1.243
Fe^{3+}	0.246	0.228	0.179	0.263
Ti	0.021	0.023	0.023	0.022
ΣA^{2+}	1.021	1.023	1.023	1.022
$\Sigma B^{3+,4+}$	1.979	1.977	1.977	1.978

* $Fe^{2+}:Fe^{3+}$ ratio is calculated by charge balance.

Table 4. Maximum contents of copper detected in different spinel-group members from Tolbachik.

Mineral Species	Ideal Formula	CuO, wt.%	Cu, *apfu*
Cuprospinel	$CuFe^{3+}_2O_4$	28.6	0.83
Thermaerogenite	$CuAl_2O_4$	26.9	0.69
Gahnite	$ZnAl_2O_4$	21.4	0.51
Magnesioferrite	$MgFe^{3+}_2O_4$	14.7	0.39
Spinel	$MgAl_2O_4$	10.9	0.25
Magnesiochromite	$MgCr_2O_4$	9.0	0.23
Franklinite	$ZnFe^{3+}_2O_4$	7.9	0.20
Chromite	$Fe^{2+}Cr_2O_4$	5.9	0.15
Zincochromite	$ZnCr_2O_4$	4.8	0.13

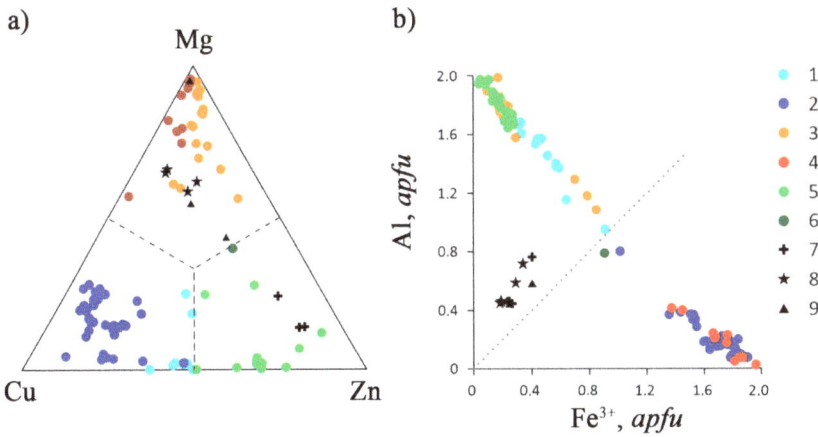

Figure 8. Atomic ratios of the major bivalent (**a**) and trivalent (**b**) cations in Cu-bearing spinel-group oxides from Tolbachik fumaroles. Legend: 1—thermaerogenite, 2—cuprospinel, 3—spinel, 4—magnesioferrite, 5—gahnite, 6—franklinite, 7—zincochromite, 8—chromite, 9—magnesiochromite.

4.5. X-ray Diffraction Data of Cu-Rich Oxide Spinels from Tolbachik

Powder X-ray diffraction (XRD) data of thermaerogenite in comparison with the calculated data for synthetic $CuAl_2O_4$ are given in Table 5. All diffraction reflections are well-indexed in the cubic unit cell, by analogy with common spinel-group minerals [2,23]. Unit-cell dimensions refined from the powder XRD data are presented in Table 6.

Table 5. Powder X-ray diffraction data (*d* in Å) of thermaerogenite and its synthetic end-member analogue.

Thermaerogenite			Synthetic CuAl$_2$O$_4$ *		*h k l*
I_{meas}	d_{meas}	d_{calc}	I_{calc}	d_{calc}	
3	4.659	4.694	2	4.664	111
65	2.873	2.875	51	2.856	220
100	2.451	2.452	100	2.436	311
1	2.329	2.347	0.2	2.332	222
10	2.033	2.033	17	2.020	400
6	1.865	1.865	1	1.853	331
16	1.660	1.660	12	1.649	422
28	1.565	1.565	31	1.555	511
30	1.438	1.437	38	1.428	440
4	1.286	1.286	3	1.277	620
6	1.240	1.240	6	1.232	533
1	1.226	1.226	1	1.218	622
1	1.174	1.174	1	1.166	444

* Calculated from the structure data from [24].

Both single-crystal and powder XRD data for all other studied spinel-group oxides from Tolbachik fumaroles demonstrate their cubic symmetry and systematic absences typical for space group *Fd-3m* characteristic for "2-3 spinels" [2,23]. The obtained unit-cell parameters *a* and *V* are in agreement with the chemical composition of the minerals (Table 6). The major factor which influences these parameters of the studied spinels is the Al:Fe^{3+} ratio. In particular, unit-cell dimensions and the majority of *d*-spacings of thermaerogenite are slightly higher in comparison with its synthetic end-member analogue CuAl$_2$O$_4$ (Tables 5 and 6) that is mainly caused by admixture of Fe^{3+} which partially substitutes Al in the mineral (Table 2).

Single-crystal XRD data of thermaerogenite allowed to obtain the *a* parameter of the cubic unit cell (Table 6), however, the crystal structure of the new mineral was not studied due to the low quality of single-crystal diffraction patterns caused by the imperfectness of all tested crystals which consist of slightly misoriented micro-blocks and in fact can be considered as crystal intergrowths.

Table 6. Unit-cell dimension *a* (Å) and volume (*V*, Å^3) of Cu-rich spinel-group oxides from Tolbachik, cuprospinel from its type locality and synthetic CuAl$_2$O$_4$.

Mineral/Compound	*a*	*V*	Method *	Source
Thermaerogenite	8.093(9)	530.1(10)	SCXRD	this work
Thermaerogenite	8.131(1)	537.6(2)	PXRD	this work
Synthetic CuAl$_2$O$_4$	8.079(3)	527(3)	SCXRD	[24]
Cuprospinel	8.402(11)	593(1)	SCXRD	this work
Cuprospinel **	8.369	586	PXRD	[3]
Gahnite	8.124(4)	536.2(5)	SCXRD	this work
Gahnite	8.1327(6)	537.9(1)	PXRD	this work
Spinel	8.149(1)	541.2(2)	PXRD	this work
Magnesioferrite	8.344(13)	581(1)	SCXRD	this work

* Abbreviations SCXRD and PXRD mean single-crystal and powder X-ray diffraction methods, respectively; ** type specimen from Consolidated Rambler Mines Limited, Baie Verte, Newfoundland, Canada.

Chemical composition of thermaerogenite varies from crystal to crystal in both Cu:Zn and Al:Fe ratios (Figure 8). The latter is a cause of slight discrepancy between unit-cell parameters of the new mineral obtained from single-crystal and powder X-ray diffraction data (Table 6).

5. Discussion

All spinel-group oxides found in the middle zones of the Arsenatnaya fumarole are copper-bearing and the majority of the studied samples contain >1 wt.% CuO (Tables 2 and 3). The maximum contents of copper in these minerals are reported in Table 4. For Tolbachik cuprospinel, the range 20.8–28.6 wt.% CuO (=0.57–0.83 *apfu* Cu) is found and for thermaerogenite 16.7–26.9 wt.% CuO (=0.41–0.69 *apfu* Cu).

Zinc is the most typical admixture in thermaerogenite. Gahnite and this mineral form here the continuous solid-solution series with the main substitution scheme $Cu^{2+} \rightarrow Zn^{2+}$ (Figure 8a). Even the Zn-poorest specimen of thermaerogenite contains 14.5 wt.% ZnO (=0.36 *apfu* Zn), and in the Cu-poorest sample of gahnite from Arsenatnaya, 2.9 wt.% CuO (=0.07 *apfu* Cu) was detected. The continuous isomorphous series with the general formula $Zn_{1-x}Cu_xAl_2O_4$ was reported for synthetic spinels [22].

Some crystals of thermaerogenite contain significant Mg admixture (up to 5.4 wt.% MgO), however, a continuous solid solution between spinel and the new mineral is not observed (Figure 8a), unlike the synthetic system $Mg_{1-x}Cu_xAl_2O_4$ with full isomorphism [25]. The solid-solution series between cuprospinel and thermaerogenite, with the main substitution scheme $Al^{3+} \rightarrow Fe^{3+}$, also demonstrates a significant gap (Figure 8b). Cuprospinel shows the most stable chemical composition among all studied "2-3 spinels" from Tolbachik fumaroles. It is typically characterized by not very high contents of admixtures, including Mg, Zn and Al, and does not form continuous solid-solution series with other oxide spinels (Table 2, Figure 8).

Cuprospinel was described as a new mineral species from burning dumps [3] and its other finds were related to ore smelters [4,5]. These anthropogenic spinel-forming systems are close in conditions to volcanic fumaroles of the oxidizing type in which copper-rich spinel-type oxides crystallize in a completely natural environment at Tolbachik. The similarity of products formed in both systems, as well as data on the synthesis and thermodynamic stability of these phases [26–28] and the absence of data on such minerals in other geological formations, demonstrate that the physical and chemical conditions in fumaroles are optimal for the formation of Cu-enriched oxide spinels: there is the combination of high temperature, atmospheric pressure and high oxygen fugacity.

It is doubtless that Cu-rich Al- and Fe^{3+}-dominant oxide spinels in fumarole systems of Tolbachik where deposited directly from hot gas as volcanic sublimates. The most convincing evidence of their exhalation origin is the location over crusts of very typical sublimate minerals, especially over alkaline sulfates (Figure 1) in the open space of pockets within a recently formed and still active fumarole at the summit part of a volcanic scoria cone. No sign of occurrence of any other process that could result in the formation of spinels (crystallization from melt or solution, solid-state transformation, etc.) is observed there. Hot volcanic gas was a carrier of "ore" constituents, foremost Cu, Zn and Fe. Basalt scoria which composes the walls of fumarole chambers can be a source of elements having low volatilities in such post-volcanic systems, namely Mg, Al and Ti [29,30]. Our temperature measurements in fumaroles of the Second scoria cone of the NB GTFE and fumaroles born by the Ploskiy Tolbachik eruption of 2012–2013 [31] together with data based on the halite–sylvite solid-solution thermometry [19] show that the mineral assemblages with Cu-bearing oxide spinels were formed at temperatures definitely not lower than 500 °C and probably not higher than 800 °C. This assumption is in agreement with data on synthetic $CuAl_2O_4$ prepared under the atmospheric pressure in the temperature range 600–1100 °C [28]. Thus, the most probable temperature interval of crystallization of Cu-rich oxide spinels in Tolbachik fumaroles seems to be 600–800 °C.

For the Cu- and Zn-enriched chrome-spinels found in altered micro-xenoliths of ultrabasic rock (Figure 5), we assume the same physical conditions but rather another mechanism of formation. They could be formed due to gas metasomatism, in terminology by Naboko and Glavatskikh [32], i.e., in the result of the replacement of a primary, magmatic chrome-spinel by chemically different spinel-group species under the influence of hot volcanic gas enriched by "ore" components. Chromium, Mg, Al, V and probably part of Fe (which remained as Fe^{2+}) were inherited from the primary chrome-spinel phase whereas Cu and Zn were taken from gas and Fe^{2+} was partly oxidized to Fe^{3+} in this process.

The distribution of cations between tetrahedral and octahedral sites in Cu-rich oxide spinels from Tolbachik fumaroles was not examined by us. We only can consider, based on the Raman spectrum (see above), that thermaerogenite contains part of Al in tetrahedral coordination. The cation distribution is well-studied for synthetic Cu-bearing spinel-type oxides including the end-member $CuAl_2O_4$ and it was shown that they are "largely normal" spinels (in terminology by [33], i.e., with bivalent cations occupying the tetrahedral site for 2/3 or more) but part of Cu^{2+} typically occurs in the octahedral site and the Cu-B^{3+} disorder in general increases with the temperature increase [22,25,27,28,34]. It should be noted that the distribution of bivalent and trivalent cations between tetrahedral and octahedral sites is not a species-defining sign in the light of the IMA-accepted nomenclature of spinel-supergroup minerals: The classification is based only on chemical formulae as resulting from chemical data only [2].

Copper-rich oxide spinels are extremely rare minerals reliably known in Nature only in Tolbachik fumaroles. Unlike them, the chalcogenide members of the spinel supergroup are not so rare. Nine valid minerals with species-defining Cu belong to the thiospinel and the selenospinel groups [2] and one of them, carrollite $CuCo_2S_4$, is a common sulfide in many ore deposits. Such strong difference in diversity and distribution in nature between oxide and chalcogenide Cu-rich spinels is probably caused by the strong chalcophile character (it is worthy to note that the term *chalcophile* originates from Greek word χαλκός, copper) of copper.

6. Conclusions

Natural copper-rich oxide spinels are characterized for the first time. They were found in the deposits of active and extinct fumaroles of the oxidizing type related to the Tolbachik volcano, Kamchatka, Russia. This mineralization is represented by nine species belonging to the "2-3 spinels" with the general chemical formula $A^{2+}B^{3+}{}_2O_4$, i.e., to the spinel subgroup of the oxyspinel group within the spinel supergroup [2]. There are (content of CuO in wt.% is given in parentheses for each mineral) two minerals with species-defining Cu^{2+}, namely cuprospinel (20.8–28.6) and the new species thermaerogenite (16.7–26.9), and Cu-enriched varieties of gahnite (up to 21.4), magnesioferrite (up to 14.7), spinel (up to 10.9), magnesiochromite (up to 9.0), franklinite (up to 7.9), chromite (up to 5.9), and zincochromite (up to 4.8).

The new mineral species thermaerogenite [ideally $CuAl_2O_4$, cubic, space group *Fd-3m*, $a = 8.093(9)$ Å, $V = 530.1(10)$ Å^3] forms a continuous isomorphous series with gahnite.

Cuprospinel, ideally $CuFe^{3+}{}_2O_4$, earlier reliably known only as a phase of the anthropogenic origin or as a synthetic compound, is a typical oxide mineral in the Arsenatnaya fumarole at Tolbachik. Its sample with composition $(Cu_{0.831}Zn_{0.100}Mg_{0.043}Ni_{0.022})_{\Sigma 0.996}(Fe^{3+}{}_{1.725}Al_{0.219}Mn^{3+}{}_{0.048}Ti_{0.008})_{\Sigma 2.000}O_4$ represents the Cu-richest natural spinel-type oxide so far described.

Cu-bearing oxide spinels at Tolbachik have a properly fumarolic origin. Aluminum- and Fe^{3+}-dominant species (thermaerogenite, gahnite, spinel, cuprospinel, franklinite, and magnesioferrite) were deposited directly from hot gas as volcanic sublimates. The most probable temperature interval of their crystallization seems to be 600–800 °C. Copper-enriched varieties of chrome-spinels (magnesiochromite, chromite and zincochromite) were formed probably under the same physical conditions but in the result of the metasomatic replacement of a magmatic chrome-spinel in micro-xenoliths of ultrabasic rock under the influence of volcanic gas.

The rarity of Cu-rich oxide spinels in nature is probably caused by the very specific character of conditions optimal for their formation: there is the combination of high temperature, atmospheric pressure, high oxygen fugacity, rich source of copper and hot gas as effective carrier of this element. Such combination is mostly realized in volcanic fumaroles of the oxidizing type enriched by "ore" components: there are really rare, exotic geological objects. The strong affinity of Cu-rich oxide spinels to such environments is confirmed by earlier reported [3–5] findings of their anthropogenic counterparts in burning copper mine dumps and ore smelters.

Minerals **2018**, *8*, 498

Author Contributions: I.V.P. and S.N.B. wrote the paper. N.N.K. and F.D.S. obtained, processed and designed chemical data. I.V.P. and S.N.B. obtained and processed X-ray diffraction data. I.V.P., E.G.S., A.G.T. and F.D.S. collected the material and F.D.S. prepared it for laboratory studies. M.F.V. obtained and processed Raman spectrum. Y.S.P. measured reflectance values.

Funding: This work was supported by the Russian Foundation for Basic Research, grant no. 18-05-00051 (in part of mineralogical studies), and the Russian Science Foundation, grant 14-17-00071 (in part of powder XRD studies).

Acknowledgments: We are grateful to Nadezhda V. Shchipalkina for her assistance in the obtaining of powder XRD data for spinel and to Vasiliy O. Yapaskurt for his help with the SEM studies. We thank X-Ray Diffraction Resource Center and Centre for Physical Methods of Surface Investigation of St. Petersburg State University for providing instrumental and computational resources.

Conflicts of Interest: The authors declare no conflict of interest.

References

1. Zhao, Q.; Yan, Z.; Chen, C.; Chen, J. Spinels: Controlled preparation, oxygen reduction/evolution reaction application, and beyond. *Chem. Rev.* **2017**, *117*, 10121–10211. [CrossRef] [PubMed]

2. Bosi, F.; Biagioni, C.; Pasero, M. Nomenclature and classification of the spinel supergroup. *Eur. J. Mineral.* **2018**, in press. [CrossRef]

3. Nickel, E.H. The new mineral cuprospinel {CuFe$_2$O$_4$} and other spinels from an oxidized ore dump at Baie Verte, Newfoundland. *Can. Mineral.* **1973**, *11*, 1003–1007.

4. Lanteigne, S.; Schindler, M.; McDonald, A.M.; Skeries, K.; Abdu, Y.; Mantha, N.M.; Murayama, M.; Hawthorne, F.C.; Hochella, M.F., Jr. Mineralogy and weathering of smelter-derived spherical particles in soils: Implications for the mobility of Ni and Cu in the surficial environment. *Water Air Soil Pollut.* **2012**, *223*, 3619–3641. [CrossRef]

5. Tropper, P.; Krismer, M.; Goldenberg, G. Recent and ancient copper production in the lower inn valley. An overview of prehistoric mining and primary copper metallurgy in the brixlegg mining. *Mitt. Osterr. Miner. Ges.* **2017**, *163*, 97–115.

6. Mallik, B.; Rautray, T.R.; Nayak, P.K. Characterisation of hot material erupted from Mahandi riverbank using EDXRF and XRD techniques. *Indian J. Phys.* **2005**, *79*, 293–296.

7. Sholeh, A.; Rastad, E.; Huston, D.; Gemmell, J.B.; Taylor, R.D. The chahnaly low-sulfidation epithermal gold deposit, Western Makran Volcanic Arc, Southeast Iran. *Econ. Geol.* **2016**, *111*, 619–639. [CrossRef]

8. Pekov, I.V.; Anikin, L.P.; Chukanov, N.V.; Belakovskiy, D.I.; Yapaskurt, V.O.; Sidorov, E.G.; Britvin, S.N.; Zubkova, N.V. Deltalumite, IMA 2016-027. CNMNC Newsletter No. 32, August 2016, page 919. *Mineral. Mag.* **2016**, *80*, 915–922.

9. Rammelsberg, C. Über den sogenannten octaёdrischen Eisenglanz vom Vesuv, und über die Bildung von Magneteisen durch Sublimation. *Ann. Phys. Chem.* **1859**, *107*, 451–454. [CrossRef]

10. Deer, W.A.; Howie, R.A.; Zussman, J. *Rock-Forming Minerals*; Longmans: London, UK, 1962; Volume 5.

11. Stoiber, R.E.; Rose, W.I., Jr. Fumarole incrustations at active Central American volcanoes. *Geochim. Cosmochim. Acta* **1974**, *38*, 495–516. [CrossRef]

12. Yudovskaya, M.A.; Distler, V.V.; Chaplygin, I.V.; Mokhov, A.V.; Trubkin, N.V.; Gorbacheva, S.A. Gaseous transport and deposition of gold in magmatic fluid: Evidence from the active Kudryavy volcano, Kurile Islands. *Miner. Deposita* **2006**, *40*, 828–848. [CrossRef]

13. Balić-Žunić, T.; Garavelli, A.; Jakobsson, S.P.; Jonasson, K.; Katerinopoulos, A.; Kyriakopoulos, K.; Acquafredda, P. Fumarolic minerals: An overview of active European volcanoes. In *Updates in Volcanology—From Volcano Modelling to Volcano Geology*; IntechOpen: London, UK, 2016; pp. 267–322.

14. Meniaylov, I.A.; Nikitina, L.P.; Shapar', V.N. *Geochemical Features of Exhalations of the Great Tolbachik Fissure Eruption*; Nauka Publishing: Moscow, Russia, 1980. (In Russian)

15. Africano, F.; Van Rompaey, G.; Bernard, A.; Le Guern, F. Deposition of trace elements from high temperature gases of Satsuma-Iwojima volcano. *Earth Planets Space* **2002**, *54*, 275–286. [CrossRef]

16. Fedotov, S.A. *The Great Tolbachik Fissure Eruption*; Markhinin, Y.K., Ed.; Cambridge University Press: New York, NY, USA, 1983.

17. Zelenski, M.E.; Zubkova, N.V.; Pekov, I.V.; Boldyreva, M.M.; Pushcharovsky, D.Y.; Nekrasov, A.N. Pseudolyonsite, Cu$_3$(VO$_4$)$_2$, a new mineral species from the Tolbachik volcano, Kamchatka Peninsula, Russia. *Eur. J. Mineral.* **2011**, *23*, 475–481. [CrossRef]

18. Pekov, I.V.; Zubkova, N.V.; Yapaskurt, V.O.; Belakovskiy, D.I.; Lykova, I.S.; Vigasina, M.F.; Sidorov, E.G.; Pushcharovsky, D.Y. New arsenate minerals from the Arsenatnaya fumarole, Tolbachik volcano, Kamchatka, Russia. I. Yurmarinite, $Na_7(Fe^{3+},Mg,Cu)_4(AsO4)_6$. *Mineral. Mag.* **2014**, *78*, 905–917. [CrossRef]

19. Pekov, I.V.; Koshlyakova, N.N.; Zubkova, N.V.; Lykova, I.S.; Britvin, S.N.; Yapaskurt, V.O.; Agakhanov, A.A.; Shchipalkina, N.V.; Turchkova, A.G.; Sidorov, E.G. Fumarolic arsenates—A special type of arsenic mineralization. *Eur. J. Mineral.* **2018**, *30*, 305–322. [CrossRef]

20. Britvin, S.N.; Dolivo-Dobrovolsky, D.V.; Krzhizhanovskaya, M.G. Software for processing the X-ray powder diffraction data obtained from the curved image plate detector of Rigaku RAXIS Rapid II diffractometer. *Zap. Ross. Mineral. Obsh.* **2017**, *146*, 104–107. (In Russian)

21. D'Ippolito, V.; Andreozzi, G.B.; Bersani, D.; Lottici, P.P. Raman fingerprint of chromate, aluminate and ferrite spinels. *J. Raman Spectrosc.* **2015**, *46*, 1255–1264. [CrossRef]

22. Le Nestour, A.; Gaudon, M.; Villeneuve, G.; Andriessen, R.; Demourgues, A. Steric and electronic effects relating to the Cu^{2+} Jahn-Teller distortion in $Zn_{1-x}Cu_xAl_2O_4$ spinels. *Inorg. Chem.* **2007**, *46*, 2645–2658. [CrossRef] [PubMed]

23. Biagioni, C.; Pasero, M. The systematics of the spinel-type minerals: An overview. *Am. Mineral.* **2014**, *99*, 1254–1264. [CrossRef]

24. Arean, C.O.; Vinuela, J.S.D. Structural study of copper-nickel aluminate ($Cu_xNi_{1-x}Al_2O_4$) spinels. *J. Solid State Chem.* **1985**, *60*, 1–5. [CrossRef]

25. Fregola, R.A.; Bosi, F.; Skogby, H.; Hålenius, U. Cation ordering over short-range and long-range scales in the $MgAl_2O_4$-$CuAl_2O_4$ series. *Am. Mineral.* **2012**, *97*, 1821–1827. [CrossRef]

26. Jacob, K.T.; Alcock, C.B. Thermodynamics of $CuAlO_2$ and $CuAl_2O_4$ and phase-equilibria in system Cu_2O-CuO-Al_2O_3. *J. Am. Ceram. Soc.* **1975**, *58*, 192–195. [CrossRef]

27. O'Neill, H.S.C.; Navrotsky, A. Simple spinels: Crystallographic parameters, cation radii, lattice energies, and cation distribution. *Am. Mineral.* **1983**, *68*, 181–194.

28. O'Neill, H.S.C.; James, M.; Dollase, W.A.; Redfern, S.A.T. Temperature dependence of the cation distribution in $CuAl_2O_4$ spinel. *Eur. J. Mineral.* **2005**, *17*, 581–586. [CrossRef]

29. Symonds, R.B.; Reed, M.H. Calculation of multicomponent chemical equilibria in gas-solid-liquid systems: Calculation methods, thermochemical data, and applications to studies of high-temperature volcanic gases with examples from Mount St. Helens. *Am. J. Sci.* **1993**, *293*, 758–864. [CrossRef]

30. Churakov, S.V.; Tkachenko, S.I.; Korzhinskii, M.A.; Bocharnikov, R.E.; Shmulovich, K.I. Evolution of composition of high-temperature fumarolic gases from Kudryavy volcano, Iturup, Kuril Islands: The thermodynamic modeling. *Geochem. Int.* **2000**, *38*, 436–451.

31. Pekov, I.V.; Zubkova, N.V.; Yapaskurt, V.O.; Belakovskiy, D.I.; Chukanov, N.V.; Lykova, I.S.; Saveliev, D.P.; Sidorov, E.G.; Pushcharovsky, D.Y. Wulffite, $K_3NaCu_4O_2(SO_4)_4$, and parawulffite, $K_5Na_3Cu_8O_4(SO_4)_8$, two new minerals from fumarole sublimates of the Tolbachik volcano, Kamchatka, Russia. *Can. Mineral.* **2014**, *52*, 699–716. [CrossRef]

32. Naboko, S.I.; Glavatskikh, S.F. *Post-Eruptive Metasomatism and Ore Genesis: Great Tolbachik Fissure Eruption of 1975–76 at Kamchatka*; Nauka Publishing: Moscow, Russia, 1983. (In Russian)

33. O'Neill, H.S.C.; Navrotsky, A. Cation distributions and thermodynamic properties of binary spinel solid solutions. *Am. Mineral.* **1984**, *69*, 733–753.

34. Cooley, R.F.; Reed, J.S. Equilibrium cation distribution in $NiAl_2O_4$, $CuAl_2O_4$ and $ZnAl_2O_4$ spinels. *J. Am. Ceram. Soc.* **1972**, *55*, 395–398. [CrossRef]

Article

Sharyginite, Ca₃TiFe₂O₈, A New Mineral from the Bellerberg Volcano, Germany

Rafał Juroszek [1,*], **Hannes Krüger** [2], **Irina Galuskina** [1], **Biljana Krüger** [2], **Lidia Jeżak** [3], **Bernd Ternes** [4], **Justyna Wojdyla** [5], **Tomasz Krzykawski** [1], **Leonid Pautov** [6] and **Evgeny Galuskin** [1]

[1] Department of Geochemistry, Mineralogy and Petrography, Faculty of Earth Sciences, University of Silesia, Będzińska 60, 41-200 Sosnowiec, Poland; irina.galuskina@us.edu.pl (I.G.); tomasz.krzykawski@us.edu.pl (T.K.); evgeny.galuskin@us.edu.pl (E.G.)

[2] Institute of Mineralogy and Petrography, University of Innsbruck, Innrain 52, 6020 Innsbruck, Austria; Hannes.Krueger@uibk.ac.at (H.K.); Biljana.Krueger@uibk.ac.at (B.K.)

[3] Institute of Geochemistry, Mineralogy and Petrology, University of Warsaw, Al. Żwirki and Wigury 93, 02-089 Warszawa, Poland; l.jezak@uw.edu.pl

[4] Dienstleistungszentrum Ländlicher Raum (DLR) Westerwald-Osteifel-Aussenstelle Mayen, Bahnhofstrasse 45, DE-56727 Mayen, Germany; Bernd.Ternes@dlr.rlp.de

[5] Swiss Light Source, Paul Scherrer Institute, 5232 Villigen, Switzerland; justyna.wojdyla@psi.ch

[6] Fersman Mineralogical Museum RAS, Leninskiy pr, 18/2, 115162 Moscow, Russia; pla58@mail.ru

* Correspondence: rjuroszek@us.edu.pl; Tel.: +48-516-491-438

Received: 27 June 2018; Accepted: 17 July 2018; Published: 21 July 2018

Abstract: The new mineral sharyginite, $Ca_3TiFe_2O_8$ ($P2_1ma$, Z = 2, a = 5.423(2) Å, b = 11.150(8) Å, c = 5.528(2) Å, V = 334.3(3) Å³), a member of the anion deficient perovskite group, was discovered in metacarbonate xenoliths in alkali basalt from the Caspar quarry, Bellerberg volcano, Eifel, Germany. In the holotype specimen, sharyginite is widespread in the contact zone of xenolith with alkali basalt. Sharyginite is associated with fluorellestadite, cuspidine, brownmillerite, rondorfite, larnite and minerals of the chlormayenite-wadalite series. The mineral usually forms flat crystals up to 100 μm in length, which are formed by pinacoids {100}, {010} and {001}. Crystals are flattened on (010). Sharyginite is dark brown, opaque with a brown streak and has a sub-metallic lustre. In reflected light, it is light grey and exhibits rare yellowish-brown internal reflections. The calculated density of sharyginite is 3.943 g·cm⁻³. The empirical formula calculated on the basis of 8 O *apfu* is $Ca_{3.00}(Fe^{3+}_{1.00}Ti^{4+}_{0.86}Mn^{4+}_{0.11}Zr_{0.01}Cr^{3+}_{0.01}Mg_{0.01})_{\Sigma2}(Fe^{3+}_{0.76}Al_{0.20}Si_{0.04})_{\Sigma1.00}O_8$. The crystal structure of sharyginite, closely related to shulamitite $Ca_3TiFeAlO_8$ structure, consists of double layers of corner-sharing (Ti, Fe³⁺) O₆ octahedra, which are separated by single layers of (Fe³⁺O₄) tetrahedra. We suggest that sharyginite formed after perovskite at high-temperature conditions >1000°C.

Keywords: sharyginite; new mineral; crystal structure; Raman spectroscopy; Bellerberg volcano; Germany

1. Introduction

Sharyginite, $Ca_3TiFe_2O_8$, is a new mineral which was found in thermally-metamorphosed limestone xenoliths in alkali basalts of the Bellerberg volcano lava field, Caspar quarry, Eastern Eifel region, Rhineland-Palatinate, Germany (50°21′6″ N, 7°14′2″ E). Sharyginite ($P2_1ma$, Z = 2, a = 5.423(2) Å, b = 11.150(8) Å, c = 5.528(2) Å, V = 334.3(3) Å³) is a Fe³⁺-analogue of shulamitite, $Ca_3TiFeAlO_8$ [1], a member of the anion deficient perovskite group [2] and also an intermediate member of the pseudobinary perovskite $CaTiO_3$—brownmillerite $Ca_2(Fe,Al)_2O_5$ series [3].

"Grenier phase" is the name of the $Ca_3TiFe^{3+}_2O_8$ compound synthesized in 1976 [4]. This phase was intensively studied due to its ionic and electronic conductivity [4,5]. The crystal structure of $Ca_3TiFe^{3+}_2O_8$ was first determined by Rodriguez-Carvajal and co-authors [6]. Some information on magnetic properties of this phase were reported by Causa et al. as result of an electron paramagnetic resonance (EPR) study [7].

Previously, sharyginite was documented as a mineral in ye'elimite-larnite pyrometamorphic rocks of the Hatrurim Complex [3] and in high-temperature skarns occurring in sedimentary carbonate xenoliths within ignimbrite confined to the Upper Chegem volcanic structure of the North Caucasus, Kabardino-Balkaria, Russia [8]. Some sharyginite crystals were recognized in mineral association of Ca-rich xenoliths in Klöch Basalt quarry, Bad Radkersburg, Styria, Austria [9]. This phase was also described in metacarbonate rocks from some burned dumps of the Donetsk [10] and Chelyabinsk [11] coal basins. Moreover, it was identified in xenoliths from the Eifel region [12], but because of the small size of the crystals, this mineral was not studied in detail, until now.

Sharyginite (IMA 2017-14) was approved by the Commission of New Minerals, Nomenclature and Classification of IMA. Type material was deposited in the mineralogical collection of Fersman Mineralogical Museum RAS, Leninskiy pr., 18/2, 115162 Moscow, Russia, catalogue numbers: 4958/1.

The name sharyginite was given in honour of Victor Victorovich Sharygin (b.1964) from the Sobolev Institute of Geology and Mineralogy, Novosibirsk, Russia. Victor Sharygin is a mineralogist, author and co-author of many publications concerning the mineralogy and petrology of different alkaline rocks (alkali basalts, lamproites, kimberlites, carbonatites, etc.) and pyrometamorphic rocks. He is the author of new mineral descriptions; for example, shulamitite $Ca_3TiFeAlO_8$—Al-analogue of sharyginite [1]. He has also worked on xenolith specimens with sharyginite from Eifel and published some preliminary data on this mineral [12].

In the present paper, we report the detailed description of the new mineral sharyginite from the type locality—Caspar quarry, Bellerberg volcano, Eifel, Germany. In addition, we discuss the composition and Raman spectroscopic data of sharyginite from the other two localities: Upper Chegem volcanic Caldera, Great Caucasus, Russia and Jabel Harmun (Hatrurim Complex), Palestinian Autonomy.

2. Occurrence, Geological Settings, Physical and Optical Properties of Sharyginite

The type locality of sharyginite, is a part of a quaternary volcano region in the Eastern Eifel area, Rhineland-Palatinate, Germany [13]. The Bellerberg volcano is a well-preserved cinder cone: the western part of the crater wall is the actual Ettringer Bellerberg and the eastern one is the Kottenheimer Büden [14]. More than 20 new mineral species were discovered in the area of the Bellerberg volcano [13,15–17]. The active quarry is characterized by the presence of various thermally-metamorphosed silicate and carbonate-silicate xenoliths within a leucite tephrite lava [13,14]. The significant variety of xenoliths is based on different protolith composition and metamorphic transformations. The calcium-rich xenoliths in the crater area of the Bellerberg are mainly composed of gehlenite, pyroxene (diopside), garnet (grossular-andradite), anorthite, and wollastonite. As accessory minerals, chalcopyrite, hematite, magnetite, or spinel are noted. Some of these xenoliths contain larnite, jasmundite, brownmillerite, mayenite, spurrite, and monticellite. The appearance of such mineral assemblages is a result of metamorphism at very high-temperature conditions (pyrometamorphism). The altered limestone xenoliths are known for their diversity of secondary, low-temperature minerals. Ettringite is the most abundant. Portlandite, thaumasite, hydrocalumite, afwillite, tobermorite, $CaCO_3$ polymorphs, and zeolites belong to the low-temperature mineral association [14].

The new mineral sharyginite occurs in altered Ca-rich xenolith, which consists of a few multi-coloured zones characterized by variable mineral associations (Figure 1). The first zone (light brown and pink) is composed of cuspidine, andradite, and minerals of the brownmillerite-srebrodolskite series. Lakargiite, sphalerite, and some Ni-phosphides occur as accessory minerals. Secondary phases are represented by hydrocalumite, ettringite, calcite, brucite, hydrogarnets, and unidentified Ca-hydrosilicates. The second zone (white and creamy) consists of cuspidine and minerals of the brownmillerite-srebrodolskite and chlormayenite-wadalite series.

Rankinite, fluorellestadite, vorlanite, lakargiite, and bismoclite appear in this zone as accessory minerals. The low-temperature mineral association is the same as in the first zone.

Figure 1. Xenolith specimen from Bellerberg with characteristic multi-coloured zones: I-III—xenolith, IV—altered alkali basalt (description in text).

Sharyginite is common in the next zone (Figure 1, zone III). It is associated with fluorellestadite, cuspidine, brownmillerite, rondorfite, larnite, and minerals of the chlormayenite-wadalite series (Figure 2a). Rankinite, magnesioferrite, perovskite, and fluorite are less common in this zone. Low-temperature mineral assemblages are represented by calcite, ettringite-thaumasite, brucite, gypsum and Ca-hydrosillicates. Sharyginite forms euhedral crystals and their intergrowths, which can be up to 200 μm in size (Figure 2b). Abundant inclusions of cuspidine, fluorellestadite, and also chlormayenite are characteristic for sharyginite crystals (Figure 2c). The next zone is represented by completely altered basalt (Figure 1, zone IV). This zone, about 2 cm thick, consists of cuspidine, perovskite, minerals of the chlormayenite-wadalite series, and magnesioferrite. Sharyginite is also noted. In some cavities, clinoenstatite and spinel relicts occur. The low-temperature association is represented by calcite, gypsum, gibbsite, and minerals of ettringite-thaumasite series.

Sharyginite crystals are formed by pinacoids {100}, {010}, {001}, and the rhombic pyramid is observed rarely. Crystals are flattened on (010) (Figure 3). Sharyginite exhibits a dark brown colour and a brown streak. The mineral is opaque in transmitted light and exhibits a sub-metallic lustre. Sharyginite shows good cleavage on (010) and imperfect along (001) and (100). Parting was not observed, tenacity is brittle and fracture is uneven.

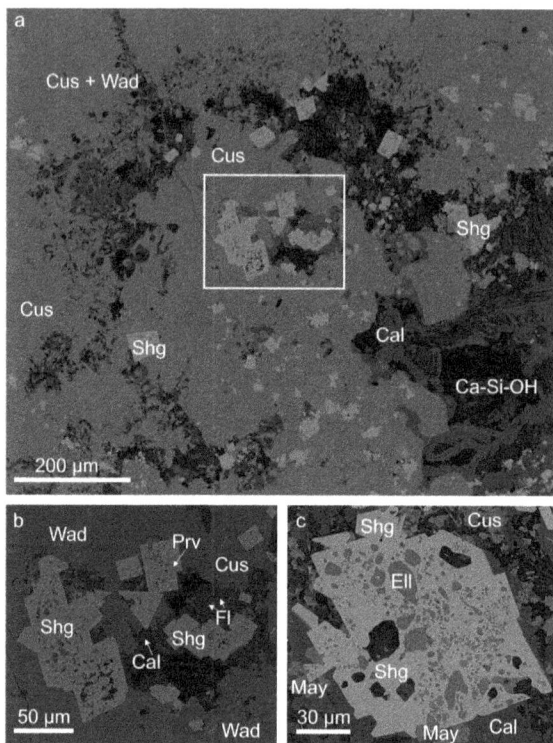

Figure 2. (**a**) General view of a metacarbonate xenolith with sharyginite from the Caspar quarry, Bellerberg volcano; fragment in the frame is magnified in Figure 2b; (**b**) Euhedral crystals and intergrowths of sharyginite; (**c**) Poikilitic crystal of sharyginite with inclusions of fluorellestadite and chlormayenite. BSE (backscattered electron) images. Cal—calcite; Ca-Si-OH—Ca-hydrosilicates; Cus—cuspidine; Ell—fluorellestadite; Fl—fluorite; May—chlormayenite; Prv—perovskite; Shg—sharyginite; Wad—wadalite.

Figure 3. Sharyginite crystals formed by pinacoids {100}, {010}, and {001}, and rhombic pyramid and flattened on (010) at the part of the strongly altered rock. BSE image.

The density could not be measured because of abundant fluorellestadite, chlormayenite, and cuspidine inclusions. The density calculated using the average composition and unit-cell parameters as obtained from single crystal XRD (X-ray Powder Diffraction) analysis is 3.943 g·cm^{-3}. Micro-hardness measurements were carried out using a VHN (Vickers Hardness Number) load of 25 g which gave a mean value of 635 kg/mm^3 (range from 621 to 649 kg/mm^3) based on 35 measurements. This value corresponds to a hardness of ~5.5–6 in the Mohs scale.

In reflected light, sharyginite was characterized by light grey colour and weak pleochroism from grey to very light grey. Under crossed polarizers, weak anisotropy was noted. Rare yellowish brown internal reflections were observed in sharyginite crystals in polarized light (Figure 4).

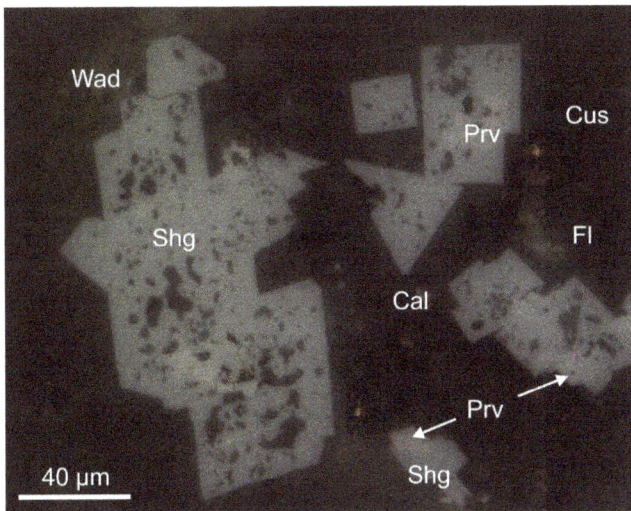

Figure 4. Sharyginite crystals with characteristic yellowish brown internal reflections from the Bellerberg volcano. Reflected light, crossed polarizers. Cal—calcite; Cus—cuspidine; Fl—fluorite; Prv—perovskite; Shg—sharyginite; Wad—wadalite.

Quantitative reflectance measurements for sharyginite from the Bellerberg volcano were measured in air relative to a SiC standard using a UMSP 50D Opton microscope-spectrophotometer (Table 1). Reflectance percentages for the four R_{max} and R_{min} COM (Commission on Ore Mineralogy) wavelengths were: 16.1, 15.5 (470 nm); 14.9, 14.2 (546 nm); 14.6, 14.1 (589 nm), and 14.5, 13.9 (650 nm).

Table 1. Reflectance values for sharyginite.

R_{max}/R_{min} (%)	λ (nm)	R_{max}/R_{min} (%)	λ (nm)
18.7/17.6	400	14.8/14.2	560
18.3/17.4	420	14.7/14.1	580
17.0/16.0	440	14.6/14.1	589 (COM)
16.4/15.6	460	14.6/14.1	600
16.1/15.5	470 (COM)	14.6/14.0	620
15.9/15.4	480	14.6/14.0	640
15.5/14.9	500	14.5/13.9	650 (COM)
15.2/14.5	520	14.4/13.7	660
15.0/14.3	540	14.3/13.5	680
14.9/14.2	546 (COM)	14.1/13.4	700

Sharyginite was also detected in other localities. It was found in the xenolith No.1—the largest of the 11 described altered carbonate xenoliths within ignimbrites of the Upper Chegem volcanic Caldera, North Caucasus, Kabardino-Balkaria, Russia. This unique xenolith was discovered by Victor Gazeev and Alexandr Zadov [18] and specified as a pyrometamorphic rock, formed at high temperature and low-pressure conditions [19]. The Upper Chegem Caldera situated in the eastern part of the Elbrus-Kyugen volcanic region is a type locality for more than 20 new mineral species e.g.: lakargiite, $CaZrO_3$ [8], chegemite, $Ca_7(SiO_4)_3(OH)_2$ [20], elbrusite, $Ca_3(Zr_{1.5}U^{6+}_{0.5})Fe^{3+}_3O_{12}$ [21], rusinovite, $Ca_{10}(Si_2O_7)_3Cl_2$ [22], pavlovskyite, $Ca_8(SiO_4)_2(Si_3O_{10})$ [23], edgrewite, $Ca_9(SiO_4)_4F_2$ [24], dzhuluite, $Ca_3SbSnFe^{3+}_3O_{12}$ [25], etc. More geological information concerning this locality can be found in Gazeev et al. [18], Galuskin et al. [8,20] and Galuskina et al. [26].

In the analysed specimen, sharyginite occurs as an accessory mineral. Cuspidine, fluorellestadite, wadalite, and larnite are widespread rock-forming minerals. Less commonly sharyginite is associated with perovskite, lakargiite, periclase, fluorite, and baryte. Secondary minerals are represented by low-temperature phases belonging to the ettringite-thaumasite series and unidentified Ca-hydrosilicates. Sharyginite from this locality occurs as euhedral and subhedral crystals up to 200 µm in size (Figure 5a). Most of them are partially or totally altered and replaced by phases exhibiting compositions similar to iron oxides. Unaltered sharyginite up to 50 µm in size was observed rarely as single crystals in a matrix of ellestadite-wadalite. Mostly, they are characterized by the presence of wadalite inclusions (Figure 5b).

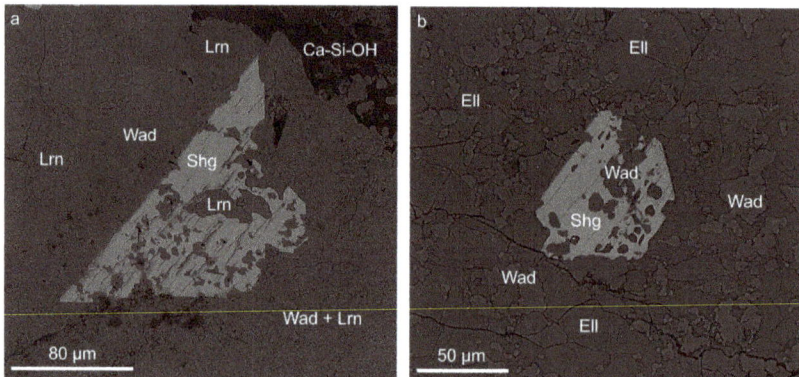

Figure 5. (**a**) Sharyginite crystal in larnite-wadalite rock from the Upper Chegem volcanic Caldera; (**b**) Poikilitic euhedral crystal of sharyginite with wadalite inclusions. BSE images. Ca-Si-OH—Ca-hydrosilicates, Ell—fluorellestadite, Lrn—larnite, Shg—sharyginite, Wad—wadalite.

Sharyginite was also found in larnite-ye'elimite nodules from pseudoconglomerates of the pyrometamorphic rocks of the Hatrurim Complex, Jabel Harmun locality, Judean Mountains, Palestinian Autonomy. Jabel Harmun, as well as other Hatrurim Complex localities, consists of high-temperature rocks represented, mainly, by larnite-, gehlenite- and spurrite-bearing rocks [27]. It is also the type locality for a few new minerals such as: vapnikite, Ca_3UO_6 [27], harmunite, $CaFe_2O_4$ [28], fluormayenite, $Ca_{12}Al_{14}O_{32}F_2$ [29], nabimusaite, $KCa_{12}(SiO_4)_4(SO_4)_2O_2F$ [30], gazeevite, $BaCa_6(SiO_4)_2(SO_4)_2O$ [31] and dzierżanowskite, $CaCu_2S_2$ [32]. Larnite-bearing rocks, which were formed during spontaneous combustion metamorphism of bituminous chalks and marls from the Ghareb and Taqiye Formations [33–36], were collected during fieldworks in 2015. Detailed geological description of the Jabel Harmun locality and the main hypotheses of the Hatrurim Complex formation were described by Galuskina et al. [28].

Sharyginite from Jabel Harmun was recognized in dark-brown homogenous larnite-ye'elimite rock samples, wherein magnesioferrite and minerals of the fluorellestadite-fluorapatite series occurred also as main phases. Accessory and secondary minerals are represented by vorlanite, cuprite, tenorite, baryte, and hydrocalumite. Most frequently, sharyginite forms xenomorphic crystals from20 to 140 μm in size. Poikilitic crystals of sharyginite contain numerous inclusions of larnite, ye'elimite, and fluorellestadite. Homogenous grains occur rarely. Aggregates of sharyginite-magnesioferrite intergrowths up to 70 μm in size were more often noted (Figure 6).

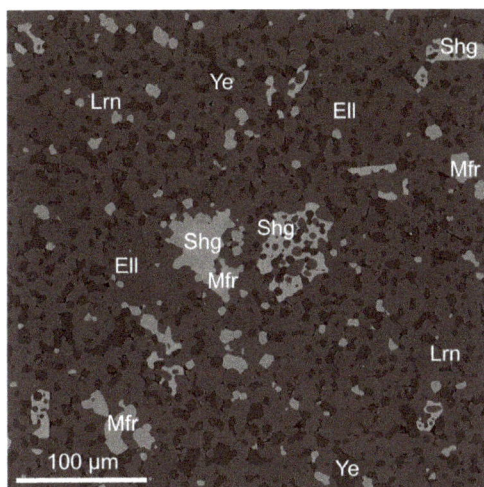

Figure 6. Xenomorphic sharyginite crystals in larnite-ye'elimite rock from Jabel Harmun. BSE image. Ell—fluorellestadite, Lrn—larnite, Mfr—magnesioferrite; Shg—sharyginite, Ye—ye'elimite.

3. Materials and Methods

Mineral association and crystal morphology of sharyginite were examined using a scanning electron microscope Phenom XL, PhenomWorld, ThermoFisher Scientific, Eindhoven, The Netherlands equipped with an EDS (energy-dispersive X-ray spectroscopy) detector (Faculty of Earth Sciences, University of Silesia, Sosnowiec, Poland).

Quantitative chemical analyses of sharyginite were carried out using a CAMECA SX100 electron-microprobe operating in WDS (wavelength dispersive X-ray spectroscopy) mode (Institute of Geochemistry, Mineralogy and Petrology, University of Warsaw, Warsaw, Poland) at 15 kV and 10 nA, beam size ~1 μm. The following lines and standards were used: Ca$K\alpha$, Si$K\alpha$, Mg$K\alpha$—diopside; Al$K\alpha$—orthoclase; Ti$K\alpha$—rutile; Cr$K\beta$—Cr_2O_3; Fe$K\alpha$—Fe_2O_3; Sn$L\alpha$—cassiterite; Sr$L\alpha$—celestine; Mn$K\alpha$—rhodonite; Nb$L\alpha$—Nb; Zr$L\alpha$—zircon.

Powder X-ray diffraction data were obtained using Cu$K\alpha$ radiation (λ = 1.541874 Å) with 40 kV and 40 mA generator settings, on a Panalytical, X'Pert PRO PW3040/60 diffractometer equipped with an X'Celerator strip detector with an active angle of 2.122 °2θ, Malvern PANalytical, Almelo, Netherlands (Faculty of Earth Sciences, University of Silesia, Sosnowiec, Poland). Incident beam optics with divergence slit (1/8°) and anti-scatter slit (1/4°) was used. A Ni filter was used for reduction of the $K\beta$ line intensity. Conditions for recording the XRD pattern were as follow: scan range 5–90 °2θ, step size 0.01 °2θ, the counting time was set to 800 s. The total measurement time was 18 h.

Synchrotron radiation diffraction experiments were carried out at the Swiss Light Source (Paul Scherrer Institute, Villigen, Switzerland) beamline X06DA. The measurement was performed at ambient conditions and the data was collected with the Pilatus 2M-F detector placed at a distance of 120 mm from the crystal. The detector was vertically translated by 67 mm to increase the maximum resolution to 0.67 Å. The X-ray diffraction data were collected from a single crystal mounted onto the multi-axis PRIGo goniometer [37] using the DA+ acquisition software [38]. Data were collected at the

0.7085 Å wavelength with 0.1 s exposure time and 0.1° oscillation range using a focused 80 × 45 μm (h × v) beam. Data reduction, including empirical absorption correction, was performed with the XDS software package [39]. Further details of the intensity data collection and crystal-structure refinement of sharyginite are reported in Table 2.

Table 2. Parameters for X-ray data collection and crystal-structure refinement for sharyginite.

Crystal Data	
Chemical formula	$Ca_3TiFe_{1.68}Al_{0.32}O_8$
Crystal system	orthorhombic
Space group	$P2_1ma$
Unit-cell dimensions	$a = 5.423(2)$ Å
	$b = 11.150(8)$ Å
	$c = 5.528(2)$ Å
Unit cell volume	$334.3(3)$ Å^3
Formula weight	398.6
Density (calculated)	3.943 g/cm^3
Z	2
Crystal size	$60 \times 40 \times 40$ μm
Data Collection	
Diffractometer	beamline X06DA, Swiss Light Source multi-axis goniometer PRIGo, PILATUS 2M-F detector
Radiation wavelength	0.7085 Å
Detector to sample distance	120 mm
Oscillation range	0.1°
No. of frames measured	1800
Time of exposure	0.1 s
Reflection ranges	$-7 \leq h \leq 7;$ $-16 \leq k \leq 10;$ $-7 \leq l \leq 7$
Reflection measured	1810
R_{int}	0.026
Refinement of Structure	**Full Matrix Least-squares on f**
No. of unique reflections	951
No. of observed unique refl. [$I > 3\sigma(I)$]	943
Final R values [$I > 3\sigma(I)$]	$R = 0.024; wR = 0.034$
Final R values (all data)	$R = 0.024; wR = 0.034$
S (all data)	1.62
Refined parameters	74
Weighting scheme	$w = 1/(\sigma^2(F) + 0.0001F^2)$
$\Delta\rho_{min}$ [e Å^{-3}]	-0.62
$\Delta\rho_{max}$ [e Å^{-3}]	0.49

The Raman spectra of sharyginite from Bellerberg, Jabel Harmun and Upper Chegem Caldera were recorded on a WITec alpha 300R Confocal Raman Microscope, WITec, Ulm, Germany (Faculty of Earth Science, University of Silesia, Poland) equipped with an air-cooled solid laser 532 nm and a CCD (closed circuit display) camera operating at −61°C. The laser radiation was coupled to a microscope through a single-mode optical fibre with a diameter of 3.5 μm. An air Zeiss (LD EC Epiplan-Neofluan DIC–100/0.75NA) objective was used. Raman scattered light was focused by effective Pinhole size about 30 μm and monochromator with a 600 mm^{-1} grating. The power of the laser at the sample position was 20–30 mW. Integration times of 5 s with an accumulation of 15 scans and a resolution of 3 cm^{-1} were chosen. The monochromator was calibrated using the Raman scattering line of a silicon plate (520.7 cm^{-1}). Spectra processing such as baseline correction and smoothing was performed using the Spectracalc software package GRAMS (Galactic Industries Corporation, NH, USA). Bands fitting were done using a Gauss-Lorentz cross-product function, with the minimum number of component bands used for the fitting process.

4. Results

4.1. Chemical Composition

The analytical data for sharyginite from the Bellerberg volcano lava field (holotype locality), Jabel Harmun (Hatrurim Complex, Palestinian Autonomy) and Upper Chegem Volcanic Caldera (Kabardino-Balkaria, Russia) are presented in Table 3. Chemical data of other authors [1,3,12,40] are included in Table 4.

Table 3. The representative chemical composition of sharyginite from different localities.

wt %	Bellerberg (Holotype Locality)			Jabel Harmun (Hatrurim Complex)			Upper Chegem Caldera		
	Mean	S.D.	Range	Mean	S.D.	Range	Mean	S.D.	Range
	n = 9			n = 19			n = 4		
Nb_2O_5	n.d.			n.d.			n.d.		
MnO_2	2.27	0.71	1.04–3.22	n.d.			n.d.		
SiO_2	0.58	0.13	0.40–0.80	1.19	0.09	1.07–1.43	0.17	0.11	0.07–0.32
SnO_2	n.d.			n.d.			0.37	0.26	0.15–0.61
TiO_2	17.04	0.89	16.29–19.34	17.97	0.61	16.65–18.76	17.38	0.38	16.97–17.76
ZrO_2	0.27	0.14	0.07–0.57	0.43	0.11	0.25–0.62	0.39	0.23	0.16–0.67
Al_2O_3	2.49	0.53	2.22–3.89	3.83	0.14	3.62–4.13	1.86	0.45	1.36–2.29
Cr_2O_3	0.20	0.11	0.07–0.42	0.25	0.14	0.00–0.47	n.d.		
Fe_2O_3	34.87	0.85	32.81–35.85	32.80	0.79	31.70–34.54	37.43	1.03	36.40–38.72
CaO	41.59	0.37	40.99–42.09	42.19	0.27	41.64–42.60	40.71	0.14	40.53–40.85
FeO	n.d.			n.d.			n.d.		
MgO	0.13	0.05	0.08–0.24	0.08	0.02	0.06–0.11	0.05	0.01	0.04–0.06
MnO	n.d.			n.d.			0.09	0.06	0.01–0.13
SrO	n.d.			0.18	0.05	0.08–0.32	n.d.		
Total	99.44			98.91			98.45		

Table 3. *Cont.*

	Bellerberg (Holotype Locality)	Jabel Harmun (Hatrurim Complex)	Upper Chegem Caldera
		Calculated on 8 O	
Ca^{2+}	3.00	3.03	3.00
Sr^{2+}		0.01	0.00
Sum A	3.00	3.04	3.00
Nb^{5+}			
Mn^{4+}	0.11		
Sn^{4+}			0.01
Ti^{4+}	0.86	0.91	0.90
Zr^{4+}	0.01	0.01	0.01
Cr^{3+}	0.01	0.01	
Fe^{3+}	1.00	1.06	1.06
Fe^{2+}			
Mg^{2+}	0.01	0.01	0.01
Mn^{2+}			0.01
Sum B	2.00	2.00	2.00
Si^{4+}	0.04	0.08	0.01
Al^{3+}	0.20	0.30	0.15
Fe^{3+}	0.76	0.59	0.87
Sum T	1.00	0.97	1.03

Footnotes: S.D. = 1σ standard deviation; n—number of analyses; n.d.—not detected.

Table 4. The chemical composition of sharyginite from different localities-literature data.

Sample wt %	[A] E-2011 Mean n = 4	[B] M7-184 Mean n = 19	[C] E-2-1 Mean n = 1	[D] H-201 Mean n = 2	[E] YV-411 Mean n = 1	[F] YV-568 Mean n = 2	[G] 42-17g Mean n = 2	[H] M-4 Mean n = 1	[I] M-4 Mean n = 4
Nb_2O_5	0.17	0.18	0.17	n.d.	n.d.	n.d.	0.06	0.04	0.06
SiO_2	0.53	0.71	0.71	0.60	2.67	0.94	0.62	0.67	0.37
TiO_2	19.55	20.98	18.81	18.59	17.85	17.20	19.66	18.67	20.23
ZrO_2	0.19	0.03	0.12	0.35	n.d.	n.d.	0.45	0.31	0.41
Al_2O_3	4.27	3.21	3.65	4.85	4.65	5.20	5.51	4.30	3.55
Cr_2O_3	0.12	0.01	0.01	0.45	n.d.	n.d.	0.03	0.02	0.06
Fe_2O_3	31.74	30.74	34.34	32.27	30.94	31.13	30.35	33.52	31.95
V_2O_3	n.d.	n.d.	n.d.	n.d.	n.d.	n.d.	n.d.	0.18	0.10
La_2O_3	n.d.	n.d.	n.d.	n.d.	n.d.	n.d.	0.13	n.d.	n.d.
Ce_2O_3	n.d.	n.d.	n.d.	n.d.	n.d.	n.d.	0.17	n.d.	n.d.
CaO	41.12	41.64	41.68	42.48	42.76	42.79	42.73	41.21	41.10
FeO	0.06	0.51	0.03	n.d.	n.d.	n.d.	0.01	n.d.	0.48
MgO	n.d.	0.30	n.d.	0.12	0.45	0.06	0.07	n.d.	n.d.
MnO	0.09	0.38	0.22	0.05	0.54	0.95	n.d.	n.d.	n.d.
SrO	1.78	0.55	0.41	0.20	n.d.	0.15	0.16	1.41	1.29
ZnO	n.d.	n.d.	n.d.	0.07	n.d.	n.d.	n.d.	n.d.	n.d.
Total	99.62	99.24	100.15	100.03	99.86	98.42	99.95	100.33	99.60

Table 4. *Cont.*

Sample	[A]	[B]	[C]	[D]	[E]	[F]	[G]	[H]	[I]
	E-2011	M7-184	E-2-1	H-201	YV-411	YV-568	42-17g	M-4	M-4
					Calculated on 8 O				
Ca^{2+}	2.95	2.98	2.97	3.00	3.00	3.08	3.02	2.94	2.96
Sr^{2+}	0.07	0.02	0.02	0.01		0.01	0.01	0.05	0.05
$La^{2+}+Ce^{2+}$							0.01		
Sum A	3.02	3.00	2.99	3.01	3.00	3.09	3.04	2.99	3.01
Nb^{5+}	0.01	0.01	0.01				<0.01	<0.01	<0.01
Ti^{4+}	0.98	1.05	0.94	0.92	0.88	0.87	0.97	0.93	1.02
Zr^{4+}	0.01	<0.01	<0.01	0.01			0.01	0.01	0.01
Cr^{3+}	0.01			0.02			<0.01	<0.01	<0.01
Fe^{3+}	0.98	0.85	1.05	1.02	1.05	1.12	0.97	1.06	0.92
V^{3+}		0.01						0.01	0.01
Fe^{2+}		0.03	<0.01				<0.01		0.03
Mg^{2+}		0.03		0.01	0.04	0.01	0.01		
Mn^{2+}	0.01	0.02	0.01	<0.01	0.03				
Zn^{2+}				<0.01					
Sum B	2.00	2.00	2.01	1.98	2.00	2.00	1.96	2.01	1.99
Si^{4+}	0.04	0.05	0.05	0.04	0.17	0.06	0.04	0.04	0.03
Al^{3+}	0.34	0.25	0.29	0.38	0.36	0.41	0.43	0.34	0.28
Fe^{3+}	0.62	0.69	0.67	0.58	0.47	0.46	0.53	0.62	0.69
Sum T	1.00	0.99	1.01	1.00	1.00	0.93	1.00	1.00	1.00

Footnotes: S.D. = 1σ standard deviation; n—number of analyses; n.d.—not detected; [A–C]—metacarbonate xenoliths in alkali basalt, the Bellerberg volcano, Eastern Eifel, Germany [1,12]; [D]—larnite-ye'elimite rocks from the Hatrurim Basin, Israel [1,3]; [E–F]—larnite bearing rocks from the Hatrurim Basin, Israel [40]; [G]—metacarbonate rocks in the contact with parabasalt, burned dump of mine 42, Kopeisk, Chelyabinsk coal basin, Russia [1,11]; [H–I]—metacarbonate rocks, burned dump of the Kalinin mine, Donetsk coal basin, Ukraine [1,10].

45

The empirical formula of the holotype sharyginite calculated on the basis of 8 O apfu was $Ca_{3,00}(Fe^{3+}_{1.00}Ti^{4+}_{0.86}Mn^{4+}_{0.11}Zr_{0.01}Cr^{3+}_{0.01}Mg_{0.01})_{\Sigma2.00}(Fe^{3+}_{0.76}Al_{0.20}Si_{0.04})_{\Sigma1.00}O_8$, which may be simplified to $Ca_3(Fe^{3+}Ti)$ $Fe^{3+}O_8$, as a reflecting end-member formula: $Ca_3(TiFe)FeO_8$ or $Ca_3TiFe_2O_8$. The holotype sharyginite is represented by a complex solid-solution, which contains the following end members: 64% of sharyginite, 20% of shulamitite and 11% of Mn-analogue of sharyginite. Other components like $Ca_3(Zr^{4+}Fe^{3+})Fe^{3+}O_8$, $Ca_3(Fe^{3+}Fe^{3+})Si^{4+}O_8$, $Ca_3(Cr^{3+}Fe^{3+})Si^{4+}O_8$ and $Ca_3(Mg^{2+}Ti^{4+})Si^{4+}O_8$, were minor and their contents were less than 1–2%. The chemical composition of sharyginite from Bellerberg metacarbonate xenoliths was also measured by Sharygin and Wirth [1,12]. The measurements were made for three different samples: E-2011, M7-184, and E-2-1, which are presented in Table 4 as A, B and C, respectively. The differences between these data and the holotype sharyginite were observed in Ti content, which varied between 0.9–1.1 apfu in Sharygin's analyses. In the holotype specimen, lower Ti content was related to Mn^{4+} substitution. Moreover, the iron content in the tetrahedral site indicated that the holotype sharyginite was more enriched in Fe^{3+}. Sharygin also reported small concentrations of Nb^{5+} and Sr^{2+} in his samples [12].

Sharyginite from Jabel Harmun was presented by a solid-solution with prevalent sharyginite $Ca_3TiFeFeO_8$ (62%) and subordinate $Ca_3TiFeAlO_8$ (30%) and $Ca_3Fe(Fe,Cr)SiO_8$ (8%) end members. The mean formula (average of 19 analysis) was $(Ca_{3.03}Sr_{0.01})_{\Sigma3.04}(Fe^{3+}_{1.06}Ti^{4+}_{0.91}Zr_{0.01}Cr^{3+}_{0.01}Mg_{0.01})$ $_{\Sigma2.00}(Fe^{3+}_{0.59}Al_{0.30}Si_{0.08})_{\Sigma0.97}O_8$. Vapnik et al. [40] described opaque minerals occurring within larnite-bearing rocks from the Hatrurm Basin and published some chemical analyses of sharyginite, which were also included in Table 4 (E–F). Sharyginite from these rocks had a similar Ti content as in the holotype sample from the Bellerberg, but relatively lower in comparison with sharyginite from Jabel Harmun. The increase in silicon content (~0.2 apfu) indicated that the iron content in the tetrahedral position does not exceed 0.5 apfu. Despite this, sharyginite end-member prevailed over shulamitite end-member. For example sample YV-411 (E in Table 4) consisted of following end-members: 47% sharyginite, 36% shulamitite, 12% $Ca_3(Cr^{3+}Fe^{3+})Si^{4+}O_8$ and 5% $Ca_3(Mg^{2+}Ti^{4+})Si^{4+}O_8$. Furthermore, chemical analysis for sharyginite from Hatrurim Basin were performed and calculated by Sharygin et al. (D in Table 4) [1,3]. They are very similar in comparison to analysis from Jabel Harmun, mostly in Fe^{3+} and Ti contents, but contained less SiO_2 in chemical composition.

The chemical composition of sharyginite from the Upper Chegem Volcanic Caldera corresponded to the ideal composition. It contained ≈ 85% of sharyginite, 14% of shulamitite, and 1% of Sn or Zr-bearing end-members. The empirical formula yielded on four analysis was as follows: $Ca_{3.00}(Fe^{3+}_{1.06}Ti^{4+}_{0.90}Sn^{4+}_{0.01}Zr_{0.01}Mn^{2+}_{0.01}Mg_{0.01})_{\Sigma2.00}(Fe^{3+}_{0.87}Al_{0.15}Si_{0.01})_{\Sigma1.03}O_8$.

In literature, there are also chemical data of sharyginite from other metacarbonate rocks, collected on a burned dump of mines in Ukraine [10] and Russia [11]. Sharyginite from Chelyabinsk (G in Table 4) and Donetsk (H-I in Table 4) exhibited similar content of main elements: Fe_2O_3, CaO, TiO_2 and Al_2O_3, like in other localities. Differences were observed in impurities. Sharyginite from Chelyabinsk burned dump contained rare earth elements (REE) such as La and Ce, which were below the detection limit for other mentioned localities. We could observe that samples from Donetsk showed similarities to the Bellerberg samples and in turn, samples from Chelyabinsk were more similar to these from Israel, mainly in the context of tetrahedral position occupation by Al^{3+} and Fe^{3+}.

4.2. X-ray Crystallography

The measured powder data, as well as a calculated pattern, are reported in Table 5. The calculated pattern is based on the structure model obtained from single-crystal data and is generated using *XPow* [41]. Powder data were refined with the Rietveld method using the semi-automatic module in HighScore+ software [42]. Unit-cell parameters refined from the powder data are as follows: $a = 5.4262(4)$ Å, $b = 11.1468(7)$ Å, $c = 5.5308(3)$ Å, $V = 334.5(3)$ Å^3, $Z = 2$.

Table 5. Measured and calculated X-ray powder diffraction data (d in Å) for sharyginite. The strongest diffraction lines are given in bold.

h	k	l	I	d$_{meas.}$	d$_{calc.}$	I	d
				Powder Data		**Calculated from the Results of Single-crystal Structure Refinement**	
0	1	0	9	11.0909	11.1641	7	11.1500
0	2	0	5	5.5991	5.5820	4	5.5750
0	1	1	3	4.9385	4.9544	3	4.9527
0	2	1	3	3.9160	3.9281	3	3.9254
1	0	1	8	3.8701	3.8716	11	3.8712
0	3	0	5	3.7127	3.7214	5	3.7167
1	2	1	4	3.1836	3.1813	4	3.1798
0	3	1	6	3.0783	3.0871	5	3.0844
0	0	2	32	2.7633	2.7643	28	2.7640
2	0	0	27	2.7119	2.7118	36	2.7115
1	3	1	100	2.6791	2.6830	100	2.6811
0	4	1	7	2.4874	2.4915	6	2.4890
1	0	2	4	2.4578	2.4628	4	2.4626
1	4	1	7	2.2612	2.2641	6	2.2621
0	3	2	3	-	2.2191	2	2.2179
2	3	0	3	2.1868	2.1916	3	2.1905
2	3	1	2	2.0386	2.0374	2	2.0365
2	0	2	36	1.9359	1.9358	41	1.9356
1	5	1	4	-	1.9342	4	1.9323
2	1	2	3	1.9080	1.9074	3	1.9071
0	6	0	19	1.8566	1.8607	18	1.8583
1	4	2	2	1.8511	1.8467	2	1.8455
2	2	2	3	1.8286	1.8290	3	1.8285
0	2	3	3	1.7554	1.7500	2	1.7496
1	6	1	2	1.6859	1.6771	2	1.6753
0	3	3	4	1.6507	1.6515	4	1.6509
3	2	1	2	1.6393	1.6423	3	1.6419
1	3	3	18	1.5800	1.5798	17	1.5793
3	3	1	12	1.5592	1.5600	16	1.5596
0	6	2	8	1.5450	1.5436	8	1.5422
2	6	0	8	-	1.5342	9	1.5329
3	4	1	2	-	1.4632	2	1.4626
2	3	3	3	1.4064	1.4105	3	1.4101
0	0	4	3	-	1.3821	3	1.3820
4	0	0	3	1.3559	1.3559	4	1.3557
2	6	2	11	1.3410	1.3415	13	1.3405
3	3	3	4	1.2193	1.2193	5	1.2190
4	0	2	2	1.2176	1.2173	2	1.2172
1	9	1	3	1.1806	1.1813	4	1.1799

4.3. Crystal Structure

The crystal structure was refined with Jana2006 [43], starting from the structural model of the synthetic compound $Ca_3TiFe_2O_8$ reported by Rodriguez-Carvajal et al. [6]. As EPMA (electron probe microanalysis) data indicated the presence of up to 3% of Al_2O_3, aluminium was introduced as an additional species on the octahedral as well as on the tetrahedral position. Half of the octahedral site had to be occupied by titanium. Neutral scattering factors and anisotropic displacement parameters were utilized. Furthermore, inversion twinning and isotropic extinction were taken into account. Results showed that an equal amount of twins were present (0.51(3):0.49(3)). Some other crystals showed another type of twinning; a pseudo four-fold axis along b. This was evident in the violation of the $h \neq 2n$ extinction condition of the a-glide plane. The reciprocal space layer *hk0* showed weak reflections at positions $h = 1, 3, 5$. However, these reflections are slightly displaced with respect to a*. Furthermore, reflections with $h = 4$ and $h = 6$, showed splitting along a*. This was caused by the $90°$-rotation twinning along b, as this superimposed the *0kl* layer onto the layer *hk0*. Figure 7 shows a part of the reconstructed *hk0* layer of a crystal affected by this type of twinning. The reflections belonging to the $90°$ rotation twin are marked with arrows. With increasing index h, their distance from integer h positions increased.

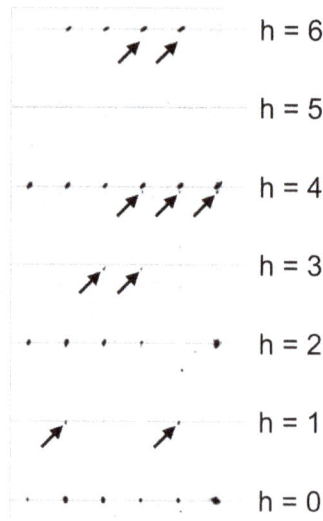

Figure 7. Reconstruction of the reciprocal space layer *hk0* of a crystal affected by 90° rotation twinning. Arrows point to reflections belonging to the smaller twin individual. Synchrotron diffraction data. The curved white bands are caused by gaps between the elements of the Pilatus 2M-F detector and contain no measured data.

The final stage of the refinement included 74 parameters. According to the refinement, the octahedral site is occupied by 3.5% Al, the tetrahedral site exhibits 25% of Al. This corresponds to a composition of x = 0.32 (in $Ca_3TiFe_{2-x}Al_xO_8$). In four additional refinements (using data from further crystals), x was found to vary between 0.26 and 0.38.

Finally, atoms were renumbered, and a cell transformation and origin shift were applied to closely match the model reported for shulamitite [1].

Atom coordinates (x,y,z), occupancies, and equivalent isotropic displacement parameters (U_{iso}, $Å^2$) for sharyginite are given in Table 6. Anisotropic displacement parameters ($Å^2$) were summarized in Table 7. The main interatomic distances ($Å$) are reported in Table 8.

Table 6. Atom coordinates (x,y,z), occupancies and equivalent isotropic displacement parameters (U_{iso}, Å^2) for sharyginite.

Site	Atom	x	y	z	U_{iso}	Occupancy
Ti1	Ti	0.25010(10)	0.16862(4)	0.23966(9)	0.00672(15)	0.5
Fe1	Fe	0.25010(10)	0.16862(4)	0.23966(9)	0.00672(15)	0.465(5)
Al1	Al	0.25010(10)	0.16862(4)	0.23966(9)	0.00672(15)	0.035(5)
Fe2	Fe	0.19825(16)	0.5	0.18318(13)	0.00750(19)	0.751(7)
Al2	Al	0.19825(16)	0.5	0.18318(13)	0.00750(19)	0.249(7)
Ca1	Ca	−0.26824(16)	0.31319(6)	0.27602(11)	0.01048(18)	1
Ca2	Ca	0.2676(3)	0	−0.26624(16)	0.0141(3)	1
O1	O	0.2758(5)	0.35744(18)	0.3210(3)	0.0136(6)	1
O2	O	0.3553(6)	0.5	−0.1222(5)	0.0127(8)	1
O3	O	0.0161(4)	0.15042(14)	0.5135(5)	0.0105(5)	1
O4	O	0.2444(6)	0	0.1700(5)	0.0132(8)	1
O5	O	−0.0146(5)	0.20160(16)	0.0157(5)	0.0127(6)	1

Table 7. Anisotropic displacement parameters (Å^2) for sharyginite.

Site	Atom	U^{11}	U^{22}	U^{33}	U^{23}	U^{13}	U^{12}
Ti1	Ti	0.0055(3)	0.0109(2)	0.0038(3)	−0.0001(3)	−0.00028(18)	0.00037(14)
Fe1	Fe	0.0055(3)	0.0109(2)	0.0038(3)	−0.0001(3)	−0.00028(18)	0.00037(14)
Al1	Al	0.0055(3)	0.0109(2)	0.0038(3)	−0.0001(3)	−0.00028(18)	0.00037(14)
Fe2	Fe	0.0071(4)	0.0094(3)	0.0060(4)	0	0.0003(3)	0
Al2	Al	0.0071(4)	0.0094(3)	0.0060(4)	0	0.0003(3)	0
Ca1	Ca	0.0131(4)	0.0115(3)	0.0069(3)	−0.0012(4)	0.0015(3)	−0.00057(18)
Ca2	Ca	0.0166(6)	0.0142(4)	0.0116(4)	0	−0.0004(4)	0
O1	O	0.0176(12)	0.0131(9)	0.0099(9)	0.0016(8)	−0.0004(9)	0.0011(7)
O2	O	0.0128(15)	0.0130(13)	0.0124(14)	0	0.0032(13)	0
O3	O	0.0105(9)	0.0129(7)	0.0080(9)	0.0001(8)	0.0014(8)	−0.0008(8)
O4	O	0.0159(17)	0.0122(12)	0.0115(13)	0	−0.0009(13)	0
O5	O	0.0134(11)	0.0177(8)	0.0071(9)	0.0007(8)	−0.0022(8)	−0.0005(9)

The crystal structure of synthetic sharyginite, $Ca_3TiFe_2O_8$, has been described before [6]. It is closely related to the structure of shulamitite [1] and consists of double layers of corner-sharing (Ti, Fe^{3+})O_6 octahedra which are separated by single layers of ($Fe^{3+}O_4$) tetrahedra, which form zweier single-chains (Figure 8). These chains are characteristic for sharyginite, shulamitite and the structurally related brownmillerite.

The sharyginite structure exhibited one octahedrally coordinated site, which hosted titanium and iron, as well as minor amounts of aluminium. The occupation factor of titanium was 0.5, whichwas imposed by the stoichiometry. The tetrahedral site was occupied by iron ($\frac{3}{4}$) and aluminium ($\frac{1}{4}$).

The independent calcium cations were located between the two octahedral layers (Ca2) and between octahedral and tetrahedral layers (Ca1).

Table 8. Selected interatomic distances (Å) for sharyginite.

Atom	Atom	distance (Å)
Ti1	O1	2.157(4)
	O3	1.985(4)
	O3	1.996(3)
	O4	1.919(3)
	O5	1.931(3)
	O5	1.938(3)
Mean		1.988
Fe2	O1 × 2	1.812(3)
	O2	1.891(3)
	O2	1.890(3)
Mean		1.851
Ca2	O3 × 2	2.481(3)
	O3 × 2	2.549(3)
	O4	2.887(4)
	O4	2.640(4)
	O4	2.415(3)
	O5 × 2	2.892(3)
Mean		2.643
Ca1	O1	2.534(3)
	O1	2.294(3)
	O2	2.347(3)
	O3	2.719(3)
	O3	2.453(3)
	O5	2.348(3)
	O5	2.436(3)
Mean		2.448

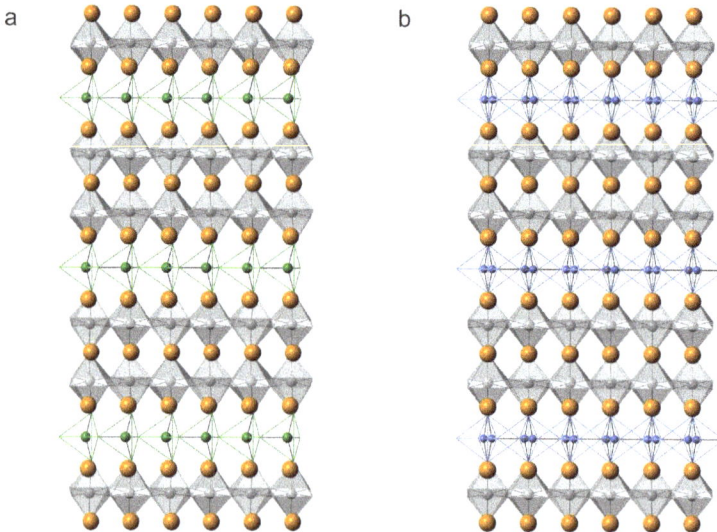

Figure 8. Projections along (001) of the crystal structures of sharyginite (**a**) and shulamitite (**b**). Calcium atoms are shown as orange spheres, $0.5Fe^{3+}/0.5TiO_6$ octahedra are in grey, green, and blue spheres, and bonds corresponding to the tetrahedra centred mainly by Fe^{3+} and Al^{3+} ions, respectively. In shulamitite (**b**), the tetrahedral chains are disordered and both possible configurations are show simultaneously.

4.4. Raman Spectroscopy

In the Raman spectrum of the holotype sharyginite the following main bands were noted (cm^{-1}) (Figure 9a): 114, 145, 190, 248, 307, 389, 486, 560, 710, 752, 785 and 1415, 1475 (overtone). Bands in the OH region were not observed. The spectrum of sharyginite was dominated by an intense Raman band at 710 cm^{-1} which was attributed to ν_1 ($Fe^{3+}O_4$) tetrahedra symmetric stretching vibrations. Two shoulders observed on this band with relative lower intensity at 752 and 785 cm^{-1} were associated with ν_1 (AlO_4) and ν_3 ($Fe^{3+}O_4$) modes, respectively. Bands at 486 and 560 cm^{-1} were related to tetrahedral bending vibrations of ($Fe^{3+}O_4$) group. The low wavenumber region below 400 cm^{-1} was ascribed to the polyhedral CaO_8 and octahedral (Fe^{3+}, Ti)O_6 vibrations [1]. Raman spectra of sharyginite from the other two localities (Figure 9b,c) were similar to the sharyginite spectrum from the Bellerberg. The Raman spectrum was also similar to that of shulamitite [1], with the main distinction in the position of the main band.

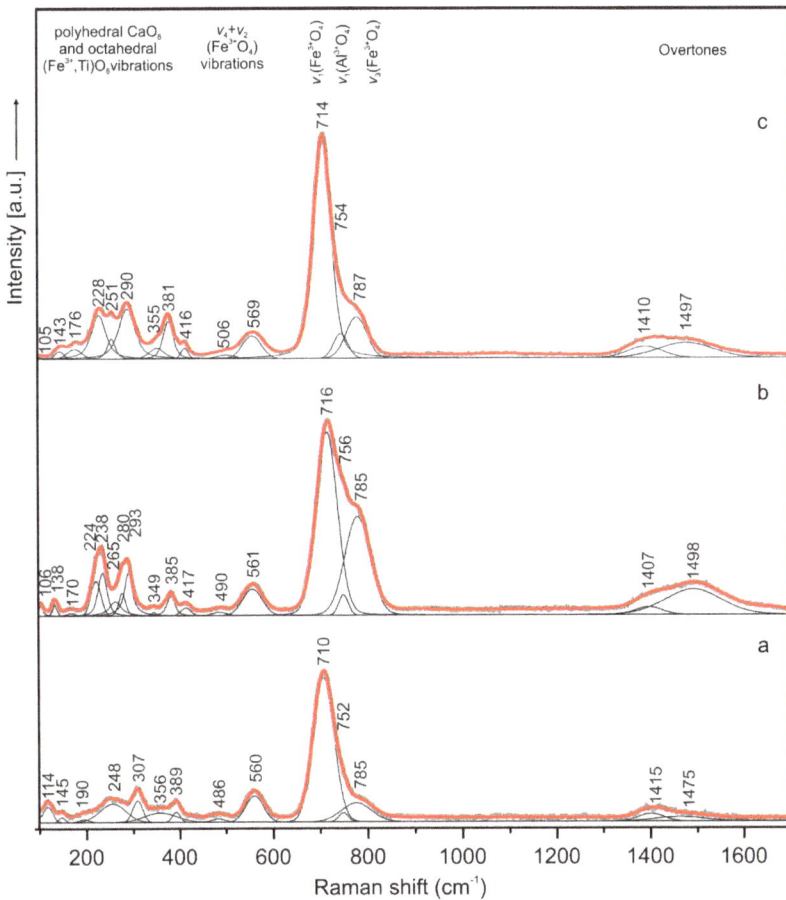

Figure 9. Raman spectra of sharyginite from the Bellerberg volcano (**a**), Jabel Harmun (**b**) and Upper Chegem Volcanic Caldera (**c**) localities.

5. Discussion

The difference between sharyginite, $Ca_3TiFe_2O_8$ and shulamitite, $Ca_3TiFeAlO_8$, is related to the occupancy of the tetrahedral site by Fe^{3+} and Al, respectively and the space group symmetry. The holotype sharyginite from the Bellerberg volcano is close to $Ca_3TiFe(Fe_{0.8} Al_{0.2})O_8$, whereas sharyginite compositions from the Upper Chegem Caldera and the Hatrurim Complex show greater diversity (Tables 3 and 4). The iron content in the holotype specimen, obtained from the refinement calculation, may be slightly overestimated, due to the Mn^{4+} substitution (up to 2.3%, Table 3). Comparing results of chemical composition from the different localities and other authors data we also note the presence of insignificant impurities such as Zr, Sr, Nb, Sn, Mn and REE. The wide variation of trivalent iron and aluminium in the chemical composition allows us to claim that a continuous solid solution exists between sharyginite and shulamitite. This solid solution may be compared with the brownmillerite, $Ca_2(Fe^{3+},Al)O_5$—srebrodolskite, $Ca_2Fe^{3+}_2O_5$ series. All mentioned minerals belong to anion-deficient, perovskites and show differences in TO_4 tetrahedra orientation between end-members which are related to the adopted space group [2].

At a first glance, the main structural differences between shulamitite and sharyginite appear to be in tetrahedral layers: shulamitite exhibits disorder of two possible configurations of the tetrahedral chain, whereas in sharyginite, only one type of chain exists, which also results in the structure being acentric. Looking more closely, another striking difference can be noted in the octahedral layers: Figure 10 shows a view perpendicular to the octahedral layers of both structures. In shulamitite (*Pmma*), the oxygen atoms O3 and O5 reside on the two-fold axis 2[010]. As a consequence, the octahedra cannot be rotated around the b-axis, as their O3-O5 edges are constrained to be parallel to a and c (Figure 10a). The lower symmetry in sharyginite ($P2_1ma$) does not restrict O3 and O5 on a special position, thus allowing the octahedra to rotate around b. The rotation, which can be seen in Figure 10b, is 3.15°. The reason for this rotation is most likely an adaption to the configuration of the tetrahedral chains. As discussed earlier [44], the tetrahedral chains exhibit a dipole moment. In shulamitite, all chains are disordered, resulting in a net dipole moment of zero. Only one type of chain occurs in sharyginite, therefore dipole moments have to be compensated by distortions of the octahedral layers and displacement of Ca cations.

The occurrence of a morphotropic phase boundary (depending on the Al-content) between sharyginite and shulamitite is comparable to the one between srebrodolskite and brownmillerite [45] and will be the subject of another study.

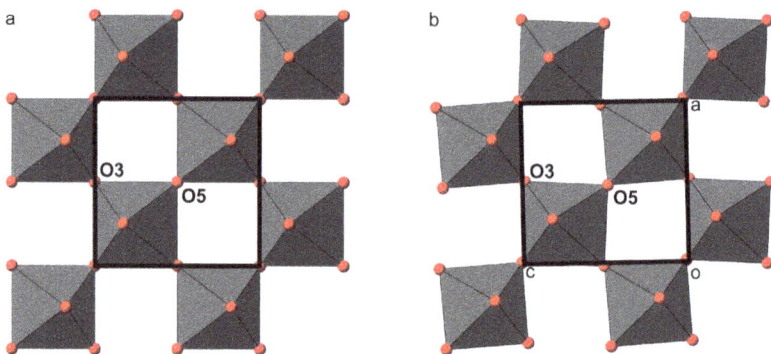

Figure 10. The arrangement of the octahedral layers in shulamitite (**a**) and sharyginite (**b**). Rotation of the octahedra is restricted by symmetry in shulamitite, as O3 and O5 reside on special positions. The lower symmetry of sharyginite allows the octahedra to rotate.

In metacarbonate xenoliths from the Bellerberg locality, the occurrence of sharyginite in association with perovskite is common. The phase relationship indicates that sharyginite is a later phase and overgrows the perovskite or even replaces it, which is very well presented in Figure 4. We suggest that new mineral sharyginite is formed at high-temperature conditions. According to the $CaTiO_3$-$Ca_2Fe_2O_5$ diagram [46], the stoichiometric $Ca_3TiFe_2O_8$ phase and disordered cubic Fe-perovskite are stable at temperatures above 1160°C. Under this temperature in the composition field $0.5 < x < 0.67$, $Ca_3TiFe_2O_8$ coexists with $Ca_4Ti_2Fe_2O_{11}$ phase. In turn in $0.67 < x < 1.00$, two ordered phases: $Ca_3TiFe_2O_8$ and $Ca_2Fe_2O_5$ coexist under 1400°C [3,46]. Moreover, for some larnite rocks of the Hatrurim Complex, where sharyginite was also identified, the minimum temperature of formation was estimated from Fe-perovskite-$Ca_3TiFe_2O_8$ mineral paragenesis. This temperature is in the range of 1170–1200°C [3]. Synthetic data also indicate the formation of $Ca_3TiFe_2O_8$ phase at high temperature (>1000°C) [4,6]. Furthermore, the coexistence of Fe-perovskite and $Ca_3TiFe_2O_8$ phases (sharyginite) may be used for the temperature evaluation of mineral associations in pyrometamorphic systems. However, it is necessary to take into account the specific composition of Fe-perovskite and the modal phase relationship in the association and to be sure that the phases were crystallized under conditions close to equilibrium [10].

Supplementary Materials: The sharyginite_cif.file is available online at http://www.mdpi.com/2075-163X/8/7/308/s1.

Author Contributions: R.J., H.K., B.K., I.G. and E.G. wrote the paper. R.J. performed the petrological investigation, measurements of chemical composition from different localities, and Raman studies. H.K. and B.K. performed SC-XRD investigation and refined the sharyginite structure. J.W. helped in the single-crystal investigation using synchrotron radiation. L.P. carried out optical studies. T.K. performed X-ray powder diffraction data of sharyginite. L.J. helped with chemical measurements as an operator of microprobe analyzer. B.T. collected the samples.

Funding: The investigations were partially supported by the National Science Centre (NCN) of Poland, Grant No. 2016/23/N/ST10/00142.

Acknowledgments: The authors would like to thank Jakub Kaminski for technical assistance at the X06DA beamline, Klaus Zöll and Daniela Schmidmair for assistance during the synchrotron experiments, and Christian Pott for helping with German literature translation. The authors also thank the anonymous reviewers for their useful and constructive comments which helped to improve a previous version of the manuscript.

Conflicts of Interest: The authors declare no conflict of interest.

References

1. Sharygin, V.V.; Lazic, B.; Armbruster, T.M.; Murashko, M.N.; Wirth, R.; Galuskina, I.O.; Galuskin, E.V.; Vapnik, Y.; Britvin, S.N.; Logvinova, A.M. Shulamitite $Ca_3TiFe^{3+}AlO_8$—A new perovskite-related mineral from Hatrurim Basin, Israel. *Eur. J. Miner.* **2013**, *25*, 97–111. [CrossRef]

2. Mitchell, R.H.; Welch, M.D.; Chakhmouradian, A.R.; Mills, S. Nomenclature of the perovskite supergroup: A hierarchical system of classification based on crystal structure and composition. *Miner. Mag.* **2017**, *81*, 411–461. [CrossRef]

3. Sharygin, V.V.; Sokol, E.V.; Vapnik, Y. Minerals of the pseudobinary perovskite-brownmillerite series from combustion metamorphic larnite rocks of the Hatrurim Formation (Israel). *Russ. Geol. Geophys.* **2008**, *49*, 709–726. [CrossRef]

4. Grenier, J.-C.; Darriet, J.; Pouchard, M.; Hagenmuller, P. Mise en evidence d'une nouvelle famille de phases de type perovskite lacunaire ordonnee de formule $A_3M_3O_8$ ($AMO_{2,67}$). *Mater. Res. Bull.* **1976**, *11*, 1219–1225. [CrossRef]

5. Grenier, J.-C.; Pouchard, M.; Hagenmuller, P. Vacancy ordering in oxygen-deficient perovskite-related ferrites. In *Ferrites Transitions Elements Luminescence. Structure and Bonding*; Springer-Verlag: Berlin, Germany, 1981; Volume 47, pp. 1–25.

6. Rodríguez-Carvajal, J.; Vallet-Regí, M.; Calbet, J.M.G. Perovskite threefold superlattices: A structure determination of the $A_3M_3O_8$ phase. *Mater. Res. Bull.* **1989**, *24*, 423–430. [CrossRef]

7. Causa, M.T.; Zysler, R.D.; Tovar, M.; Vallet-Regí, M.; González-Calbet, J.M. Magnetic properties of the $Ca_nFe_2Ti_{n-2}O_{3n-1}$ perovskite related series: An EPR study. *J. Solid State Chem.* **1992**, *98*, 25–32. [CrossRef]

8. Galuskin, E.V.; Gazeev, V.M.; Armbruster, T.; Zadov, A.E.; Galuskina, I.O.; Pertsev, N.N.; Dzierżanowski, P.; Kadiyski, M.; Gurbanov, A.G.; Wrzalik, R.; et al. Lakargiite CaZrO$_3$: A new mineral of the perovskite group from the North Caucasus, Kabardino-Balkaria, Russia. *Am. Miner.* **2008**, *93*, 1903–1910. [CrossRef]

9. Niedermayr, G.; Auer, C.; Bernhard, F.; Brandstätter, F.; Gröbner, J.; Hammer, V.M.F.; Knobloch, G.; Koch, G.; Kolitchs, U.; Konzett, J.; et al. Neue Mineralfunde aus Österreich LX. *Carinth. II* **2011**, *201*, 135–186.

10. Sharygin, V.V. *Lakargiite and Minerals of the Perovskite Brownmillerite Series in Metacarbonate Rocks from Donetsk Burned Dumps*; Scientific works of Donetsk National Technical University, Mining and Geological Series; Donetsk National Technical University: Donetsk, Ukraine, 2011; Volume 15, pp. 114–125.

11. Sharygin, V.V. Minerals of the Ca$_3$TiFeAlO$_8$–Ca$_3$TiFeFeO$_8$ series in natural and technogenic pyrometamorphic systems. In *The Mineralogy of Technogenesis 2012*; Institute of Mineralogy, Uralian Branch of Russian Academy of Sciences: Miass, Russia, 2012; pp. 29–49.

12. Sharygin, V.V.; Wirth, R. Shulamitite and its Fe-analog in metacarbonate xenoliths from alkali basalts, E. Eifel, Germany. In Proceedings of the 29th International Conference Ore Potential of Alkaline, Kimberlite and Carbonatite Magmatism, "Alkaline magmatism of the Earth", Sudak-Moscow, Ukraine-Russia, 14–22 September 2012.

13. Mihajlović, T.; Lengauer, C.L.; Ntaflos, T.; Kolitsch, U.; Tillmanns, E. Two new minerals rondorfite, Ca$_8$Mg[SiO$_4$]$_4$Cl$_2$, and almarudite, K.($\square$,Na)$_2$(Mn,Fe,Mg)$_2$(Be,Al)$_3$[Si$_{12}$O$_{30}$], and a study of iron-rich wadalite, Ca$_{12}$[(Al$_8$Si$_4$Fe$_2$)O$_{32}$]Cl$_6$, from the Bellerberg (Bellberg) volcano, Eifel, Germany. *Neues Jahrb. Miner.* **2004**, *179*, 265–294. [CrossRef]

14. Hentschel, G. *Die Mineralien der Eifelvulkane*, 2nd ed.; Weise Verlag: München, Germany, 1987.

15. Abraham, K.; Gebert, W.; Medenbach, O.; Schreyer, W.; Hentschel, G. Eifelite, KNa$_3$Mg$_4$Si$_{12}$O$_{30}$, a new mineral of the osumilite group with octahedral sodium. *Contrib. Miner. Pet.* **1983**, *82*, 252–258. [CrossRef]

16. Irran, E.; Tillmanns, E.; Hentschel, G. Ternesite, Ca$_5$(SiO$_4$)$_2$(SO$_4$), a new mineral from the Ettringer Bellerberg/Eifel, Germany. *Miner. Petrol.* **1997**, *60*, 121–132. [CrossRef]

17. Galuskin, E.V.; Krüger, B.; Krüger, H.; Blass, G.; Widmer, R.; Galuskina, I.O. Wernerkrauseite, CaFe^{3+}$_2$Mn^{4+}O$_6$: The first nonstoichiometric post-spinel mineral, from Bellerberg volcano, Eifel, Germany. *Eur. J. Miner.* **2016**, *28*, 485–493. [CrossRef]

18. Gazeev, V.M.; Zadov, A.E.; Gurbanov, A.G.; Pertsev, N.N.; Mokhov, A.V.; Dokuchaev, A.Y. Rare minerals from Verkhniechegemskaya caldera (in xenoliths of skarned limestone). *Vestnik. Vladikavkazskogo Nauchnogo Centra.* **2006**, *6*, 18–27.

19. Grapes, R. *Pyrometamorphism*, 2nd ed.; Springer-Verlag: Berlin Heidelberg, Germany, 2010.

20. Galuskin, E.V.; Gazeev, V.M.; Lazic, B.; Armbruster, T.; Galuskina, I.O.; Zadov, A.E.; Pertsev, N.N.; Wrzalik, R.; Dzierżanowski, P.; Gurbanov, A.G.; et al. Chegemite Ca$_7$(SiO$_4$)$_3$(OH)$_2$—A new humite-group calcium mineral from the Northern Caucasus, Kabardino-Balkaria, Russia. *Eur. J. Miner.* **2009**, *21*, 1045–1059. [CrossRef]

21. Galuskina, I.O.; Galuskin, E.V.; Armbruster, T.; Lazic, B.; Kusz, J.; Dzierżanowski, P.; Gazeev, V.M.; Pertsev, N.N.; Prusik, K.; Zadov, A.E.; et al. Elbrusite-(Zr)—A new uranian garnet from the Upper Chegem caldera, Kabardino-Balkaria, Northern Caucasus, Russia. *Am. Miner.* **2010**, *95*, 1172–1181. [CrossRef]

22. Galuskin, E.V.; Galuskina, I.O.; Lazic, B.; Armbruster, T.; Zadov, A.E.; Krzykawski, T.; Banasik, K.; Gazeev, V.M.; Pertsev, N.N. Rusinovite, Ca$_{10}$(Si$_2$O$_7$)$_3$Cl$_2$: A new skarn mineral from the Upper Chegem caldera, Kabardino-Balkaria, Northern Caucasus, Russia. *Eur. J. Miner.* **2011**, 837–844. [CrossRef]

23. Galuskin, E.V.; Gfeller, F.; Savelyeva, V.B.; Armbruster, T.; Lazic, B.; Galuskina, I.O.; Többens, D.M.; Zadov, A.E.; Dzierżanowski, P.; Pertsev, N.N.; et al. Pavlovskyite Ca$_8$(SiO$_4$)$_2$(Si$_3$O$_{10}$): A new mineral of altered silicate-carbonate xenoliths from the two Russian type localities, Birkhin massif, Baikal Lake area and Upper Chegem caldera, North Caucasus. *Am. Miner.* **2012**, *97*, 503–512. [CrossRef]

24. Galuskin, E.V.; Lazic, B.; Armbruster, T.; Galuskina, I.O.; Pertsev, N.N.; Gazeev, V.M.; Włodyka, R.; Dulski, M.; Dzierżanowski, P.; Zadov, A.E.; et al. Edgrewite Ca$_9$(SiO$_4$)$_4$F$_2$-hydroxyledgrewite Ca$_9$(SiO$_4$)$_4$(OH)$_2$, a new series of calcium humite-group minerals from altered xenoliths in the ignimbrite of Upper Chegem caldera, Northern Caucasus, Kabardino-Balkaria, Russia. *Am. Miner.* **2012**, *97*, 1998–2006. [CrossRef]

25. Galuskina, I.O.; Galuskin, E.V.; Kusz, J.; Dzierżanowski, P.; Prusik, K.; Gazeev, V.M.; Pertsev, N.N.; Dubrovinsky, L. Dzhuluite, Ca$_3$SbSnFe^{3+}$_3$O$_{12}$, a new bitikleite-group garnet from the Upper Chegem Caldera, Northern Caucasus, Kabardino-Balkaria, Russia. *Eur. J. Miner.* **2013**, 231–239. [CrossRef]

26. Galuskina, I.O.; Krüger, B.; Galuskin, E.V.; Armbruster, T.; Gazeev, V.M.; Włodyka, R.; Dulski, M.; Dzierżanowski, P. Fluorchegemite, $Ca_7(SiO_4)_3F_2$, a new mineral from the edgrewitw-bearing endoskarn zone of an altered xenolith in ignimbrites from Upper Chegem Caldera, Northern Caucasus, Kabardino-Balkaria, Russia: Occurrence, crystal structure, and new data on the mineral assemblages. *Can. Miner.* **2015**, *53*, 325–344.

27. Galuskin, E.V.; Galuskina, I.O.; Kusz, J.; Armbruster, T.; Marzec, K.M.; Dzierżanowski, P.; Murashko, M. Vapnikite Ca_3UO_6—A new double-perovskite mineral from pyrometamorphic larnite rocks of the Jabel Harmun, Palestinian Autonomy, Israel. *Miner. Mag.* **2014**, *78*, 571–581. [CrossRef]

28. Galuskina, I.O.; Vapnik, Y.; Lazic, B.; Armbruster, T.; Murashko, M.; Galuskin, E.V. Harmunite $CaFe_2O_4$: A new mineral from the Jabel Harmun, West Bank, Palestinian Autonomy, Israel. *Am. Miner.* **2014**, *99*, 965–975. [CrossRef]

29. Galuskin, E.V.; Gfeller, F.; Armbruster, T.; Galuskina, I.O.; Vapnik, Y.; Dulski, M.; Murashko, M.; Dzierżanowski, P.; Sharygin, V.V.; Krivovichev, S.V.; et al. Mayenite supergroup, part III: Fluormayenite, $Ca_{12}Al_{14}O_{32}[\square_4F_2]$, and fluorkyuygenite, $Ca_{12}Al_{14}O_{32}[(H_2O)_4F_2]$, two new minerals from pyrometamorphic rocks of the Hatrurim Complex, South Levant. *Eur. J. Miner.* **2015**, *27*, 123–136. [CrossRef]

30. Galuskin, E.V.; Gfeller, F.; Armbruster, T.; Galuskina, I.O.; Vapnik, Y.; Murashko, M.; Włodyka, R.; Dzierżanowski, P. New minerals with a modular structure derived from hatrurite from the pyrometamorphic Hatrurim Complex. Part I. Nabimusaite, $KCa_{12}(SiO_4)_4(SO_4)_2O_2F$, from larnite rocks of Jabel Harmun, Palestinian Autonomy, Israel. *Miner. Mag.* **2015**, *79*, 1061–1072. [CrossRef]

31. Galuskin, E.V.; Gfeller, F.; Galuskina, I.O.; Armbruster, T.; Krzaˌtała, A.; Vapnik, Y.; Kusz, J.; Dulski, M.; Gardocki, M.; Gurbanov, A.G.; et al. New minerals with a modular structure derived from hatrurite from the pyrometamorphic rocks. Part III. Gazeevite, $BaCa_6(SiO_4)_2(SO_4)_2O$, from Israel and the Palestine Autonomy, South Levant, and from South Ossetia, Greater Caucasus. *Miner. Mag.* **2017**, *81*, 499–513. [CrossRef]

32. Galuskina, I.O.; Galuskin, E.V.; Prusik, K.; Vapnik, Y.; Juroszek, R.; Jeżak, L.; Murashko, M. Dzierżanowskite, $CaCu_2S_2$—A new natural thiocuprate from Jabel Harmun, Judean Desert, Palestine Autonomy, Israel. *Miner. Mag.* **2017**, *81*, 1073–1085. [CrossRef]

33. Gross, S. *The Mineralogy of the Hatrurim Formation Israel*; Geological Survey of Israel: Jerusalem, Israel, 1977.

34. Burg, A.; Starinsky, A.; Bartov, Y.; Kolodny, Y. Geology of the Hatrurim Formation ("Mottled Zone") in the Hatrurim basin. *Isr. J. Earth Sci.* **1991**, *40*, 107–124.

35. Sokol, E.V.; Novikov, I.S.; Vapnik, Y.; Sharygin, V.V. Gas fire from mud volcanoes as a trigger for the appearance of high-temperature pyrometamorphic rocks of the Hatrurim Formation (Dead Sea area). *Dokl. Earth Sci.* **2007**, *413*, 474–480. [CrossRef]

36. Novikov, I.; Vapnik, Y.; Safonova, I. Mud volcano origin of the Mottled Zone, South Levant. *Geosci. Front.* **2013**, *4*, 597–619. [CrossRef]

37. Waltersperger, S.; Olieric, V.; Pradervand, C.; Glettig, W.; Salathe, M.; Fuchs, M.R.; Curtin, A.; Wang, X.; Ebner, S.; Panepucci, E.; et al. PRIGo: A new multi-axis goniometer for macromolecular crystallography. *J. Synchrotron Radiat.* **2015**, *22*, 895–900. [CrossRef] [PubMed]

38. Wojdyla, J.A.; Kaminski, J.W.; Panepucci, E.; Ebner, S.; Wang, X.; Gabadinho, J.; Wang, M. DA+ data acquisition and analysis software at the Swiss Light Source macromolecular crystallography beamlines. *J. Synchrotron Radiat.* **2018**, *25*, 293–303. [CrossRef] [PubMed]

39. Kabsch, W. XDS. *Acta Cryst. D Biol. Cryst.* **2010**, *66*, 125–132. [CrossRef] [PubMed]

40. Vapnik, Y.; Galuskina, I.O.; Palchik, V.; Sokol, E.V.; Galuskin, E.V.; Lindsley-Griffin, N.; Stracher, G. Paralavas in combustion metamorphic complex at the Hatrurim Basin, Israel. In *Coal and Peat Fires: A Global Perspective*, 1st ed.; Elsevier: Amsterdam, Netherlands, 2014; Volume 3, pp. 281–316.

41. Downs, R.T.; Bartelmehs, K.L.; Gibbs, G.V.; Boisen, M.B. Interactive software for calculating and displaying X-ray or neutron powder diffractometer patterns of crystalline materials. *Am. Miner.* **1993**, *78*, 1104–1107.

42. Degen, T.; Sadki, M.; Bron, E.; König, U.; Nénert, G. The HighScore suite. *Powder Diffr.* **2014**, *29*, S13–S18. [CrossRef]

43. Petříček, V.; Dušek, M.; Palatinus, L. Crystallographic Computing System JANA2006: General features. *Z. Kristallogr. Cryst. Mater.* **2014**, *229*, 345–352. [CrossRef]

44. Abakumov, A.M.; Kalyuzhnaya, A.S.; Rozova, M.G.; Antipov, E.V.; Hadermann, J.; Van Tendeloo, G. Compositionally induced phase transition in the $Ca_2MnGa_{1-x}Al_xO_5$ solid solutions: Ordering of tetrahedral chains in brownmillerite structure. *Solid State Sci.* **2005**, *7*, 801–811. [CrossRef]

45. Redhammer, G.J.; Tippelt, G.; Roth, G.; Amthauer, G. Structural variations in the brownmillerite series $Ca_2(Fe_{2-x}Al_x)O_5$: Single-crystal X-ray diffraction at 25 °C and high-temperature X-ray powder diffraction (25 °C ≤ T ≤ 1000 °C). *Am. Miner.* **2004**, *89*, 405–420. [CrossRef]

46. Becerro, A.; McCammon, C.; Langenhorst, F.; Seifert, F.; Angel, R.J. Oxygen-vacancy ordering in $CaTiO_3$-$CaFeO_{2.5}$ perovskites: From isolated defects to infinite sheets. *Phase Transit.* **1999**, *69*, 133–146. [CrossRef]

minerals

MDPI

Article

Nöggerathite-(Ce), $(Ce,Ca)_2Zr_2(Nb,Ti)(Ti,Nb)_2Fe^{2+}O_{14}$, a New Zirconolite-Related Mineral from the Eifel Volcanic Region, Germany

Nikita V. Chukanov [1,*], Natalia V. Zubkova [2], Sergey N. Britvin [3,4], Igor V. Pekov [2,5], Marina F. Vigasina [2], Christof Schäfer [6], Bernd Ternes [7], Willi Schüller [8], Yury S. Polekhovsky [3], Vera N. Ermolaeva [9] and Dmitry Yu. Pushcharovsky [2]

[1] Institute of Problems of Chemical Physics, Russian Academy of Sciences, Chernogolovka, 142432 Moscow, Russia

[2] Faculty of Geology, Moscow State University, Vorobievy Gory, 119991 Moscow, Russia; n.v.zubkova@gmail.com (N.V.Z.); igorpekov@mail.ru (I.V.P.); vigasina55@mail.ru (M.F.V.); dmitp@geol.msu.ru (D.Y.P.)

[3] Institute of Earth Sciences, St Petersburg State University, Universitetskaya Nab. 7/9, 199034 St Petersburg, Russia; sbritvin@gmail.com (S.N.B.); yury1947@mail.ru (Y.S.P.)

[4] Nanomaterials Research Center, Kola Science Centre, Russian Academy of Sciences, Fersman str. 14, 184209 Apatity, Murmansk, Russia

[5] Vernadsky Institute of Geochemistry and Analytical Chemistry, Russian Academy of Sciences, Kosygin str. 19, 119991 Moscow, Russia

[6] Gustav Stresemann-Strasse 34, 74257 Untereisesheim, Germany; mspech612@gmail.com

[7] Bahnhofstrasse 45, 56727 Mayen, Germany; ternes-mayen@web.de

[8] Im Straußenpesch 22, 53518 Adenau, Germany; maschwisch@web.de

[9] Institute of Experimental Mineralogy, Russian Academy of Sciences, Chernogolovka, 142432 Moscow, Russia; cvera@mail.ru

* Correspondence: chukanov@icp.ac.ru; Tel.: +7-4965221556

Received: 7 September 2018; Accepted: 3 October 2018; Published: 12 October 2018

Abstract: The new mineral nöggerathite-(Ce) was discovered in a sanidinite volcanic ejectum from the Laach Lake (Laacher See) paleovolcano in the Eifel region, Rhineland-Palatinate, Germany. Associated minerals are sanidine, dark mica, magnetite, baddeleyite, nosean, and a chevkinite-group mineral. Nöggerathite-(Ce) has a color that ranges from brown to deep brownish red, with adamantine luster; thel streak is brownish red. It occurs in cavities of sanidinite and forms long prismatic crystals measuring up to $0.02 \times 0.03 \times 1.0$ mm, with twins and random intergrowths. Its density calculated using the empirical formula is 5.332 g/cm³. The Vickers hardness number (VHN) is 615 kgf/mm², which corresponds to a Mohs' hardness of 5½. The mean refractive index calculated using the Gladstone–Dale equation is 2.267. The Raman spectrum shows the absence of hydrogen-bearing groups. The chemical composition (electron microprobe holotype/cotype in wt %) is as follows: CaO 5.45/5.29, MnO 4.19/4.16, FeO 7.63/6.62, Al_2O_3 0.27/0.59, Y_2O_3 0.00/0.90, La_2O_3 3.17/3.64, Ce_2O_3 11.48/11.22, Pr_2O_3 1.04/0.92, Nd_2O_3 2.18/2.46, ThO_2 2.32/1.98, TiO_2 17.78/18.69, ZrO_2 27.01/27.69, Nb_2O_5 17.04/15.77, total 99.59/99.82, respectively. The empirical formulae based on 14 O atoms per formula unit (*apfu*) are: $(Ce_{0.59}La_{0.165}Nd_{0.11}Pr_{0.05})_{\Sigma0.915}Ca_{0.82}Th_{0.07}Mn_{0.50}Fe_{0.90}Al_{0.045}Zr_{1.86}Ti_{1.88}Nb_{1.07}O_{14}$ (holotype), and $(Ce_{0.57}La_{0.19}Nd_{0.12}Pr_{0.05}Y_{0.06})_{\Sigma0.99}Ca_{0.79}Th_{0.06}Mn_{0.49}Fe_{0.77}Al_{0.10}Zr_{1.89}Ti_{1.96}Nb_{1.00}O_{14}$ (cotype). The simplified formula is $(Ce,Ca)_2Zr_2(Nb,Ti)(Ti,Nb)_2Fe^{2+}O_{14}$. Nöggerathite-(Ce) is orthorhombic, of the space group *Cmca*. The unit cell parameters are: $a = 7.2985(3)$, $b = 14.1454(4)$, $c = 10.1607(4)$ Å, and $V = 1048.99(7)$ Å³. The crystal structure was solved using single-crystal X-ray diffraction data. Nöggerathite-(Ce) is an analogue of zirconolite-3*O*, ideally $CaZrTi_2O_7$, with Nb dominant over Ti in one of two octahedral sites and *REE* dominant over Ca in the eight-fold coordinated site. The strongest lines of the powder X-ray diffraction pattern (*d*, Å (*I*, %) (*hkl*)) are: 2.963 (91) (202), 2.903 (100) (042), 2.540 (39) (004), 1.823 (15) (400), 1.796 (51) (244), 1.543 (20) (442), and 1.519 (16) (282), respectively. The type material is deposited in the collections of the Fersman Mineralogical Museum of the Russian Academy of Sciences, Moscow, Russia (registration number 5123/1).

Keywords: nöggerathite-(Ce); new mineral; zirconolite; laachite; sanidinite; crystal structure; alkaline volcanic rock; Laacher See; Eifel

---

1. Introduction

Zirconolite-related Ca-*REE*-Zr-Ti-Nb oxides have been described in numerous publications as advanced materials suitable for the immobilization of actinides, which are components of high-level radioactive waste [1–5]. Natural zirconolites and related minerals (zirconolite-3*O* $(Ca,REE)_2Zr_2(Ti,Nb)_3FeO_{14}$, zirconolite-3*T* $(Ca,REE)_2Zr_2(Ti,Nb)_3FeO_{14}$, zirconolite-2*M* $(Ca,REE)_2Zr_2(Ti,Nb)_3FeO_{14}$, laachite $Ca_2Zr_2Nb_2TiFeO_{14}$, stefanweissite $(Ca,REE)_2Zr_2(Nb,Ti)(Ti,Nb)_2Fe^{2+}O_{14}$ (IMA 2018-020) and the here-described nöggerathite-(Ce)) are characterized by a wide compositional diversity and can be considered as prototypes of such materials. These minerals usually contain uranium and thorium, the total content of which can reach 15–20 wt % [6]. As a result of exposure to alpha radiation, most such samples are X-ray amorphous, metamict. Some exceptions are crystalline samples of zirconolite-type minerals from young volcanic rocks [7–10].

This paper describes **nöggerathite-(Ce)**, a new zirconolite-related mineral species from the Laach Lake area situated in the Eifel paleovolcanic region, Germany. This mineral is non-metamict because of the very young geological age of the mother rock [11,12].

The root name of the new mineral is given in honor of Johann Jacob Nöggerath (1788–1877), a prominent German mineralogist and geologist. From 1818 Nöggerath was a professor of mineralogy and geology at the University of Bonn. Among his publications is a geological description of the Laach Lake (Laacher See) paleovolcanic region. The Levinson's modifier -(Ce) in the mineral name reflects the predominance of Ce among rare earth elements, which together are dominant in one structure position. The mineral and its name have been approved by the Commission on New Minerals, Nomenclature and Classificatiob of the International Mineralogical Associatiob (IMA number 2017-107). The type material is deposited in the collections of the Fersman Mineralogical Museum of the Russian Academy of Sciences, Moscow, Russia (registration number 5123/1).

2. Materials and Methods

The new mineral was found in the In den Dellen (Zieglowski) pumice quarry, 1.5 km to the notrheast of Mendig, in the Laach Lake (Laacher See) paleovolcano of the Eifel region, Rhineland-Palatinate, Germany. Two specimens have been investigated and considered as the holotype and the cotype, which are fragments of the same sanidinite volcanic ejectum. Associated minerals are sanidine, dark mica, magnetite, baddeleyite, nosean, and a chevkinite-group mineral.

The origin of sanidinites of the Laacher See area has been discussed earlier [9,12–14]. These rocks are derivatives of different foyaite magmas which are comagmatic with haüyne-bearing rocks (haüyne foyaite, haüyne syenite, haüyne monzonite, etc.), or with noseane–cancrinite–nepheline syenites. Sanidinites of the latter type are enriched in rare element (Nb, Zr, *REE*, U, Th) accessory minerals. These rocks are cogenetic with the phonolitic host magma, and the crystallization took place in an intrusive syenite–carbonatite complex at temperatures below 700 °C in the host rock surrounding the top of the magma chamber 5000 to 20,000 years prior to the eruption of the magma chamber [12]. Most probably, nöggerathite-(Ce) and associated minerals forming crystals on the walls of cavities in sanidinite crystallized from above-critical fluid.

Chemical analyses (five for the holotype and three for the cotype) were carried out using an Oxford INCA Wave 700 electron microprobe (WDS mode, 20 kV, 600 pA, 300-nm beam diameter,

Oxford Instruments plc, London, UK) housed at the Institute of Experimental Mineralogy RAS. The counting time per peak was 100 s.

The Raman spectrum of a randomly oriented nöggerathite-(Ce) crystal was obtained on the cotype sample using an EnSpectr R532 spectrometer based on an OLYMPUS CX 41 microscope coupled with a diode laser (λ = 532 nm) at room temperature (Enhanced Spectrometry, San Jose, USA). The spectrum was recorded in the range from 100 to 4000 cm^{-1} with a diffraction grating (1800 mm^{-1}) and spectral resolution of about 6 cm^{-1}. The output power of the laser beam was about 9 mW. The diameter of the focal spot on the sample was less than 10 μm. The backscattered Raman signal was collected with a 40× objective; thel signal acquisition time for a single scan of the spectral range was 2 s, and the signal was averaged over 100 scans.

The Raman spectrum of an oriented laachite crystal used for comparison was obtained with a HORIBA instrument based on the OLYMPUS BX 41/51 microscope (HORIBA Jobin Yvon, Bensheim Germany) with a diode laser (λ = 532 nm) at room temperature. The power of the laser beam at the sample was about 1.5 mW. The spectrum was recorded in a range from 100 to 3800 cm^{-1}, with a diffraction grating (1800 mm^{-1}) and spectral resolution of about 1 cm^{-1}. The diameter of the focal spot on the sample was less than 7 μm. The backscattered Raman signal was collected with 50× objective; signal acquisition time for a single scan of the spectral range was 10 s and the signal was averaged over five scans.

Both Raman spectrometers were housed at Moscow State University.

Maximal and minimal reflectance values (R_{max}/R_{min}) were measured in air using a MSF-21 micro-spectrophotometer (LOMO company, St. Petersburg, Russia) with a monochromator slit width of 0.4 mm and beam diameter of 0.1 mm. SiC (reflection standard 474251, number 545, Jena, Germany) was used as a standard. The spectrophotometer was housed at St. Petersburg State University.

Powder X-ray diffraction data were collected using a Rigaku RAXIS Rapid II diffractometer (Rigaku Corporation, Tokyo, Japan) with a curved image plate detector and rotating anode in Debye–Scherrer geometry, with an accelerating voltage of 40 kV, current of 15 mA, and exposure time 15 min. The distance between sample and detector was 127.4 mm. Data processing was carried out using osc2xrd software [15]. The diffractometer was housed at St. Petersburg State University.

The single-crystal X-ray diffraction experiment was carried out using a Bruker Kappa APEX DUO CCD diffractometer (Bruker AXS GmbH, Karlsruhe, Germany). The diffractometer was housed at Moscow State University. Experimental details are given in Table 1.

Table 1. Crystal data, data collection information and structure refinement details for the holotype specimen of nöggerathite-(Ce).

Characteristics	Data and Methods
Crystal sizes, mm	$0.01 \times 0.01 \times 0.10$
Temperature, K	293(2)
Radiation and wavelength, Å	Mo$K\alpha$; 0.71073
F_{000}	1530
θ range for data collection, °	2.88–26.98
h, k, l ranges	$-9 \rightarrow 7, -18 \rightarrow 18, -12 \rightarrow 12$
Reflections collected	4962
Independent reflections	617 (R_{int} = 0.0215)
Independent reflections with $I > 2\sigma(I)$	574
Data reduction	Bruker SAINT
Structure solution	Direct methods
Refinement method	Full-matrix least-squares on F^2
Weighting coefficients a, b	0.0251, 7.1886
Number of refined parameters	71
Final R indices (with $I > 2\sigma(I)$)	$R_1 = 0.0198$, w$R_2 = 0.0518$
R indices (with all data)	$R_1 = 0.0224$, w$R_2 = 0.0550$
GoF	1.161
Largest diffraction peak and hole, e/Å^3	1.53 and −0.70

3. Results

3.1. General Appearance and Mechanical Properties

Nöggerathite-(Ce) forms prismatic crystals measuring up to $0.1 \times 0.1 \times 1.0$ mm, elongated along (001), and simple twins, isolated or combined in random aggregates (Figure 1a,b) in cavities in sanidinite. The main observed crystal forms are pinacoids {100} and {010}, as well as prisms {110} and {120}. The other forms are {111} and minor {001}. In most cases, the twinning plane is (130); the angle between the *c* axes of the twin components is 65°. The exception is a growth (possibly a twin) with the angle between the *c* axes of the twin components of 90° (Figure 1c). Some crystals of nöggerathite-(Ce) are embedded in sanidine.

The new mineral is translucent to transparent, with a color ranging from brown to very dark reddish brown, almost black, with adamantine luster. The streak is brownish red.

Nöggerathite-(Ce) is brittle with uneven fractures; no cleavage was observed. Hardness, as determined by the micro-indentation method (Vickers hardness number (VHN) load of 20 g), is equal to 615 kgf/mm^2 which corresponds to a Mohs' hardness of $5\frac{1}{2}$. The density calculated using the empirical formula is 5.332 g/cm^3.

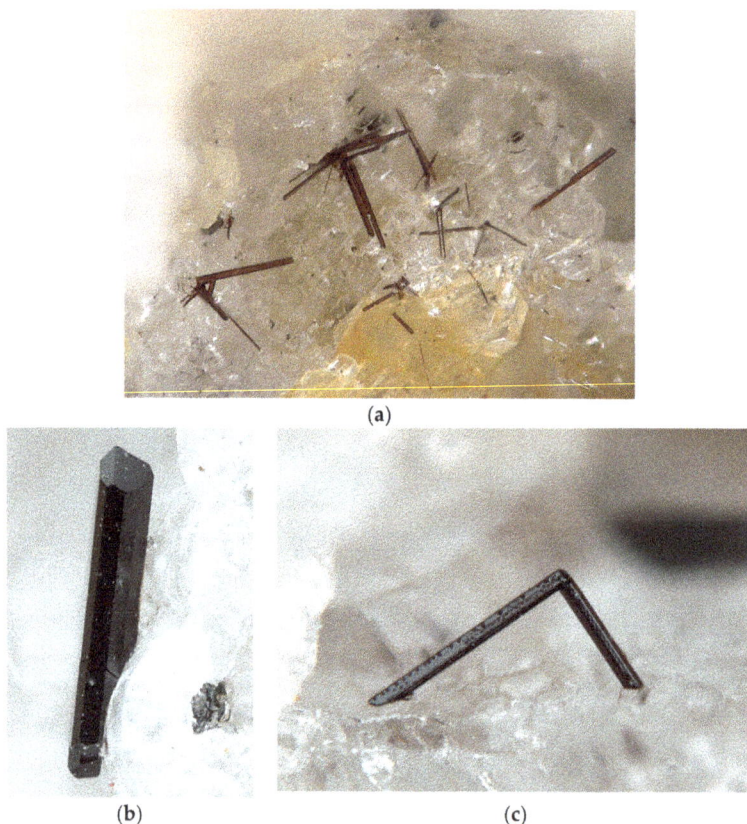

(a)

(b)   (c)

Figure 1. (**a**) Aggregates of brown nöggerathite-(Ce) crystals on sanidine. Photographer: Stefan Wolfsried. Field width: 4 mm; (**b**) Nöggerathite-(Ce) crystal on sanidine. Photographer: Marko Burkhardt. Field width: 0.5 mm; (**c**) Twin of nöggerathite-(Ce) on sanidine. Photographer: Marko Burkhardt. Field width: 1 mm.

3.2. Raman Spectroscopy

The Raman spectrum of nöggerathite-(Ce) (Figure 2) shows the absence of absorption bands of H_2O molecules, OH groups, and CO_3^{2-} anions. The bands in the range of 400–800 cm^{-1} correspond to (Ti,Nb,Zr)–O-stretching vibrations, and the bands in the range of 100–400 cm^{-1} are due to (*REE*,Ca)–O-stretching and O–(Ti,Nb,Zr)–O bending vibrations. Broad features above 900 cm^{-1} in the Raman spectrum of nöggerathite-(Ce) correspond to luminescence due to high amounts of *REE*.

In Raman spectra of laachite (Figure 3), which is a mineral related to nöggerathite-(Ce) but is characterized by much lower *REE*:Ca and Ti:Nb ratios, the bands of (*REE*,Ca)–O- and (Ti,Nb,Zr)–O-stretching vibrations are shifted towards higher and lower values, respectively. This regularity is in accordance with mean masses of corresponding groups of atoms.

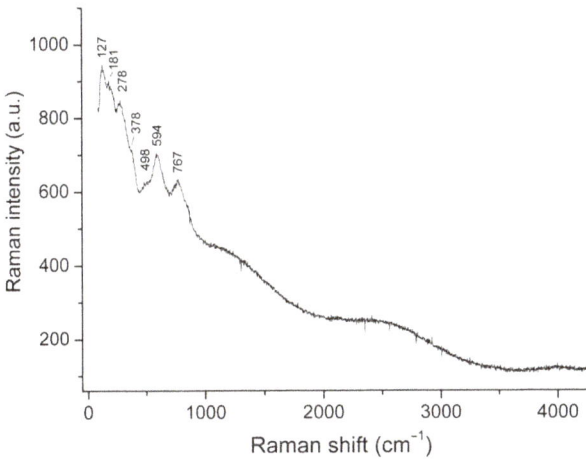

Figure 2. Raman spectrum of nöggerathite-(Ce) (a.u. = arbitrary units).

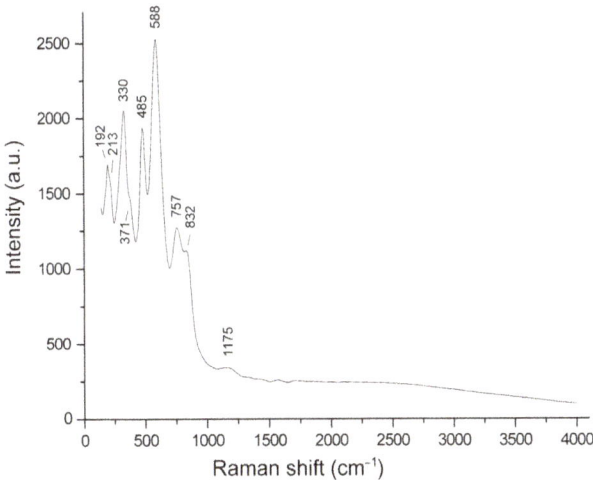

Figure 3. Raman spectrum of laachite $Ca_2Zr_2Nb_2TiFeO_{14}$ obtained with the polarization of the laser beam parallel to the *a* axis of the crystal [9].

3.3. Optical Properties

In reflected light, nöggerathite-(Ce) is optically anisotropic, with $\Delta R_{589} = 1.27\%$. The color is light grey, with reddish brown internal reflections. The reflectance values in the visible range are given in Table 2. Mean refractive index calculated from the Gladstone–Dale equation is 2.267.

Table 2. Reflectance values (R_{max}/R_{min}) for nöggerathite-(Ce). Reflectance values for four wavelengths recommended by the Commission on Ore Microscopy of the International Mineralogical Association are given in bold type.

Wavelength, nm	R_1	R_2
400	17.3	16.8
420	16.8	16.4
440	16.4	16.0
460	16.0	15.5
470	**15.8**	**15.3**
480	15.6	15.2
500	15.3	15.0
520	15.3	14.8
540	15.0	14.7
546	**15.0**	**14.7**
560	15.0	14.6
580	14.9	14.6
589	**14.9**	**14.5**
600	14.8	14.5
620	14.8	14.5
640	14.8	14.4
650	**14.8**	**14.4**
660	14.8	14.4
680	14.7	14.4
700	14.7	14.3

3.4. Chemical Composition

Chemical data for nöggerathite-(Ce) are given in Table 3. Contents of other elements with atomic numbers >8 are below detection limits. Based on structural data (see below) and by analogy with laachite [9], iron and manganese are considered as Fe^{2+} and Mn^{2+}, respectively.

Table 3. Chemical data for nöggerathite-(Ce). Upper and lower values for each constituent correspond to the holotype and the cotype, respectively.

Constituent	wt %	Range	Standard Deviation	Standard Used	X-ray Line Measured
CaO	5.45	5.27–5.55	0.10	Wollastonite	K
	5.29	5.12–5.39	0.34		
MnO	4.19	4.07–4.32	0.09	$MnTiO_3$	K
	4.16	4.06–4.34	0.13		
FeO	7.63	7.46–7.79	0.14	Fe	K
	6.62	6.23–6.83	0.28		
Al_2O_3	0.27	0.18–0.38	0.07	Albite	K
	0.59	0.48–0.78	0.14		
Y_2O_3	0.00	–	–	YPO_4	L
	0.90	0.61–0.99	0.16		
La_2O_3	3.17	3.05–3.28	0.10	$LaPO_4$	L
	3.64	3.47–3.84	0.16		
Ce_2O_3	11.48	11.27–11.73	0.19	$CePO_4$	L
	11.22	10.95–11.69	0.33		
Pr_2O_3	1.04	0.89–1.24	0.12	$PrPO_4$	L
	0.92	0.90–0.97	0.03		

Table 3. Cont.

Constituent	wt %	Range	Standard Deviation	Standard Used	X-ray Line Measured
Nd_2O_3	2.18	2.10–2.34	0.08	$NdPO_4$	L
	2.46	2.28–2.81	0.25		
ThO_2	2.32	2.11–2.50	0.15	ThO_2	M
	1.98	1.79–2.17	0.16		
TiO_2	17.78	17.45–18.12	0.27	TiO_2	M
	18.69	18.49–18.90	0.16		
ZrO_2	27.01	26.82–27.26	0.19	ZrO_2	L
	27.69	27.51–27.86	0.11		
Nb_2O_5	17.04	16.72–17.37	0.28	$LiNbO_3$	L
	15.77	15.53–15.99	0.19		
Total	99.59	-	-	-	-
	99.82				

The empirical formulae (based on 14 O *apfu*) are: $(Ce_{0.59}La_{0.165}Nd_{0.11}Pr_{0.05})_{\Sigma0.915}Ca_{0.82}Th_{0.07}Mn_{0.50}$ $Fe_{0.90}Al_{0.045}Zr_{1.86}Ti_{1.88}Nb_{1.07}O_{14}$ (holotype) and $(Ce_{0.57}La_{0.19}Nd_{0.12}Pr_{0.05}Y_{0.06})_{\Sigma0.99}Ca_{0.79}Th_{0.06}Mn_{0.49}$ $Fe_{0.77}Al_{0.10}Zr_{1.89}Ti_{1.96}Nb_{1.00}O_{14}$ (cotype).

The simplified formula is $(Ce,Ca)_2Zr_2(Nb,Ti)(Ti,Nb)_2Fe^{2+}O_{14}$.

3.5. X-ray Diffraction Data and Crystal Structure

Powder X-ray diffraction data are presented in Table 4. Diffraction peaks are readily indexed in the orthorhombic unit cell, space group *Cmca*. The unit cell parameters calculated from the powder data are: $a = 7.296(1)$, $b = 14.147(2)$, $c = 10.161(1)$ Å, and $V = 1048.9(2)$ Å^3.

Table 4. Powder X-ray diffraction data for the holotype specimen of nöggerathite-(Ce).

I_{obs}, %	d_{obs}, Å	I_{calc}, %	d_{calc}, Å	h	k	l	I_{obs}, %	d_{obs}, Å	I_{calc}, %	d_{calc}, Å	h	k	l
4	7.068	2	7.074	0	2	0	2	1.743	1	1.743	3	1	4
2	5.806	1	5.805	0	2	1	-	-	1	1.742	0	8	1
3	5.463	1	5.466	1	1	1	1	1.735	1	1.734	3	5	2
5	5.085	1	5.081	0	0	2	1	1.721	1	1.722	2	2	5
1	4.131	1	4.126	0	2	2	1	1.710	1	1.709	2	6	3
2	4.001	2	3.999	1	1	2	4	1.690	3	1.688	1	7	3
10	3.689	8	3.690	1	3	1	1	1.667	1	1.668	4	2	2
2	3.536	1	3.537	0	4	0	2	1.646	1	1.647	0	2	6
3	3.343	2	3.340	0	4	1	-	-	1	1.646	3	3	4
2	3.238	1	3.242	2	2	0	-	-	1	1.639	1	1	6
91	**2.963**	100	2.963	2	0	1	1	1.600	1	1.601	4	4	1
100	**2.903**	93	2.903	0	4	2	2	1.571	1	1.572	2	8	1
2	2.731	1	2.733	2	2	2	-	-	1	1.568	0	8	3
2	2.575	1	2.574	1	3	3	1	1.561	1	1.562	2	6	4
39	**2.540**	14	2.540	0	0	4	20	**1.543**	20	1.544	4	4	2
-	-	27	2.539	2	4	0	10	1.532	2	1.536	3	7	1
1	2.393	1	2.391	0	2	4	-	-	10	1.536	2	0	6
1	2.367	1	2.365	1	1	4	-	-	11	1.527	0	4	6
5	2.343	5	2.342	2	2	3	16	**1.519**	17	1.519	2	8	2
-	-	1	2.341	1	5	2	1	1.500	1	1.501	2	2	6
3	2.298	3	2.297	0	6	1	1	1.494	1	1.492	3	5	4
1	2.270	1	2.271	2	4	2	6	1.482	6	1.482	4	0	4
3	2.166	2	2.168	3	1	2	-	-	1	1.481	3	3	5
1	2.139	1	2.138	1	3	4	6	1.451	6	1.451	0	8	4
1	2.113	1	2.114	3	3	1	1	1.441	1	1.440	2	8	3
2	2.082	1	2.081	1	5	3	1	1.427	1	1.428	4	6	1
1	2.062	1	2.063	0	4	4	-	-	1	1.425	1	5	6
1	2.033	1	2.032	2	4	3	3	1.413	2	1.413	3	7	3
1	2.001	1	2.000	2	2	4	2	1.407	1	1.406	1	7	5
3	1.956	2	1.957	3	1	3	1	1.400	1	1.399	1	9	3
-	-	1	1.953	0	2	5	1	1.381	1	1.381	5	3	1

Table 4. *Cont.*

I_{obs}, %	d_{obs}, Å	I_{calc}, %	d_{calc}, Å	h	k	l	I_{obs}, %	d_{obs}, Å	I_{calc}, %	d_{calc}, Å	h	k	l
3	1.940	1	1.944	2	6	1	1	1.365	1	1.367	4	4	4
-	-	1	1.935	0	6	3	-	-	1	1.363	1	3	7
6	1.914	3	1.913	1	7	1	1	1.334	1	1.334	5	1	3
1	1.847	1	1.845	2	6	2	-	-	1	1.334	0	8	5
-	-	1	1.830	1	5	4	-	-	1	1.333	4	2	5
15	1.823	15	1.824	4	0	0	-	-	1	1.333	3	3	6
-	-	1	1.822	3	3	3	1	1.307	1	1.309	3	9	1
51	1.796	56	1.796	2	4	4	-	-	1	1.305	0	10	3
10	1.769	12	1.768	0	8	0	1	1.288	1	1.289	5	3	3

Note: The d_{calc} values are calculated for unit cell parameters obtained from single-crystal data.

The crystal structure of the holotype sample (see Tables 5–7) was solved by direct methods based on single-crystal X-ray diffraction data and refined to R = 0.0198 for 574 unique reflections with $I > 2\sigma(I)$. Nöggerathite-(Ce) is orthorhombic, with space group *Cmca*. The refined unit cell parameters are: a = 7.2985(3), b = 14.1454(4), c = 10.1607(4) Å, V = 1048.99(7) Å^3; and Z = 4.

Solving the crystal structure of nöggerathite-(Ce) reveals the alternation of two types of bent polyhedral layers, namely an octahedral layer (Figure 4A) and a layer of cations with seven- and eight-fold coordination (Figure 4B). The octahedral layer is built by vertex-sharing $M(3)O_6$ and $M(4)O_6$ octahedra forming three- and six-membered rings, whereas $M(5)$ and $M(6)$ sites are located in the centers of six-membered rings. The adjacent sites $M(5)$ and $M(6)$, with coordination numbers 4 and 5, respectively, are statistically occupied and contain Fe^{2+} as the major cation (Figure 4C). The $M(1)$-centered polyhedron is a distorted cube which shares edges with neighboring $M(1)$-centered cubes to form rows along the *a* axis. Similar rows are formed by seven-fold $M(2)$-centered polyhedra (mono-capped octahedra). Adjacent rows of eight- and seven-fold polyhedra are linked with each other via common edges forming a dense layer. A general view of the crystal structure of nöggerathite-(Ce) is shown in Figure 5.

The refined crystal/chemical formula of nöggerathite-(Ce) is as follows (Z = 4, *REE* are modeled by Ce, coordination numbers of cations are indicated with Roman numerals): $^{VIII}(LREE_{0.88}Ca_{0.80}Mn_{0.24}Th_{0.08})$ $^{VII}(Zr_{1.88}Mn_{0.12})^{VI}(Nb_{1.22}Ti_{0.78})^{VI}(Ti_{1.48}Nb_{0.48}Al_{0.04})^{IV}(Fe_{0.48}Mn_{0.08})^{V}(Fe_{0.40}Mn_{0.04})_2O_{14}$.

Table 5. Atom coordinates, equivalent thermal displacement parameters (U_{eq}, Å^2), site populations and site multiplicities (Q) in the structure of nöggerathite-(Ce).

Site	x	y	z	U_{eq}	Site Population	Q
$^{VIII}M(1)$	0.25	0.11753(3)	−0.2500	0.00954(18)	$Ce_{0.44}Ca_{0.40}Mn_{0.12}Th_{0.04}$	8
$^{VII}M(2)$	0.5	0.23411(4)	0.01426(5)	0.0120(2)	$Zr_{0.937(9)}Mn_{0.063(9)}$	8
$^{VI}M(3)$	0.0	0.0	0.0	0.0112(3)	$Nb_{0.608(8)}Ti_{0.392(8)}$	4
$^{VI}M(4)$	0.25	0.13297(5)	0.25	0.0129(2)	$Ti_{0.74}Nb_{0.24}Al_{0.02}$	8
$^{IV}M(5)$	0.4211(8)	0.0	0.0	0.0301(9)	$Fe_{0.24}Mn_{0.04}$	8
$^{V}M(6)$	0.5	0.0140(5)	0.0342(9)	0.0393(19)	$Fe_{0.20}Mn_{0.02}$	8
O(1)	0.1948(4)	0.03150(18)	0.1256(2)	0.0236(6)	O	16
O(2)	0.2131(4)	0.23287(17)	0.1201(2)	0.0218(6)	O	16
O(3)	0.5	0.1081(2)	−0.1007(4)	0.0182(8)	O	8
O(4)	0.0	0.1288(2)	−0.0912(4)	0.0184(8)	O	8
O(5)	0.5	0.1366(3)	0.1756(4)	0.0250(9)	O	8

Table 6. Selected interatomic distances (Å) in the structure of nöggerathite-(Ce).

Cation	Ligand	Distance	Cation	Ligand	Distance
$M(1)$	O(3)	2.376(2) × 2	$M(3)$	O(4)	2.044(3) × 2
$M(1)$	O(4)	2.441(3) × 2	$M(4)$	O(2)	1.952(2) × 2
$M(1)$	O(1)	2.491(3) × 2	$M(4)$	O(1)	1.955(3) × 2
$M(1)$	O(2)	2.508(3) × 2	$M(4)$	O(5)	1.9759(16) × 2
$M(2)$	O(4)	2.091(3)	$M(5)$	O(3)	1.928(4) × 2

Table 6. *Cont.*

Cation	Ligand	Distance	Cation	Ligand	Distance
$M(2)$	O(2)	2.122(3) × 2	$M(5)$	O(1)	2.134(5) × 2
$M(2)$	O(3)	2.132(4)	$M(6)$	O(3)	1.855(8)
$M(2)$	O(5)	2.142(4)	$M(6)$	O(3)	1.910(9)
$M(2)$	O(2)	2.354(3) × 2	$M(6)$	O(5)	2.251(9)
$M(3)$	O(1)	1.961(3) × 4	$M(6)$	O(1)	2.426(5) × 2

Table 7. Cation distribution in structurally investigated samples of zirconolite-3*O*, laachite, and nöggerathite-(Ce).

Site	Zirconolite-3*O*	Laachite	Nöggerathite-(Ce)
$^{VIII}M(1)$	$Ca_{0.53}Ce_{0.41}Na_{0.04}Th_{0.02}$	$Ca_{0.28}Mn_{0.26}Ln_{0.26}Th_{0.14}Y_{0.06}$ $Ca_{0.32}Mn_{0.28}Ln_{0.24}Th_{0.14}Y_{0.02}$	$Ln_{0.44}Ca_{0.40}Mn_{0.12}Th_{0.04}$
$^{VII}M(2)$	Zr	$Zr_{0.78}Mn_{0.22}$	$Zr_{0.937}Mn_{0.063}$
$^{VI}M(3)$	$Ti_{0.52}Nb_{0.47}Ta_{0.01}$	$Nb_{0.82}Ti_{0.18}$	$Nb_{0.608}Ti_{0.392}$
$^{VI}M(4)$	$Ti_{0.88}Nb_{0.12}$	$Ti_{0.72}Nb_{0.28}$ $Nb_{0.44}Ti_{0.40}Al_{0.16}$	$Ti_{0.74}Nb_{0.24}Al_{0.02}$
$^{IV}M(5)$	$Fe_{0.46}$	$Fe_{0.34}Mn_{0.10}Y_{0.06}$	$Fe_{0.24}Mn_{0.04}$
$^{V}M(6)$	$Fe_{0.03}$		$Fe_{0.20}Mn_{0.02}$
Reference	[16]	[9]	This work

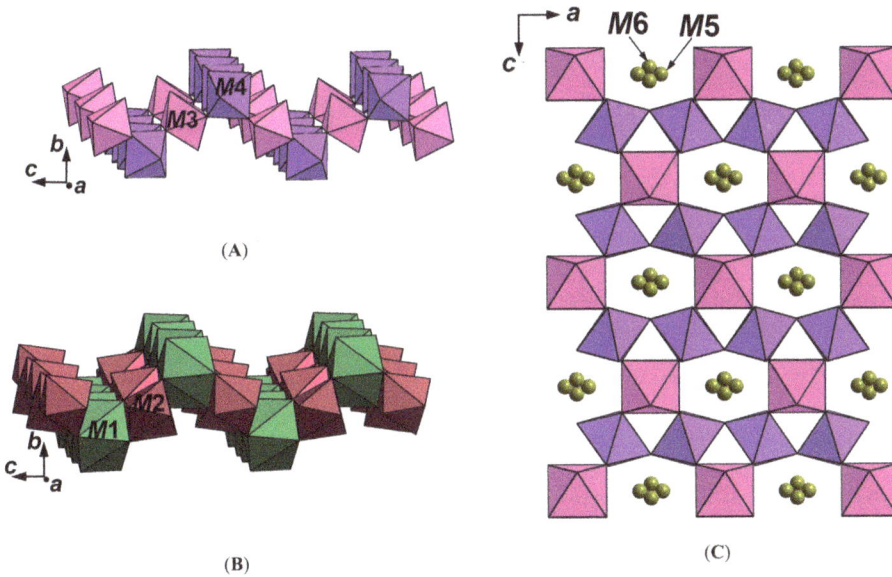

Figure 4. Octahedral (**A**) and large-cation (**B**) layers and arrangement of Fe-dominant sites $M(5)$ and $M(6)$ inside the octahedral layer (**C**) in the structure of nöggerathite-(Ce).

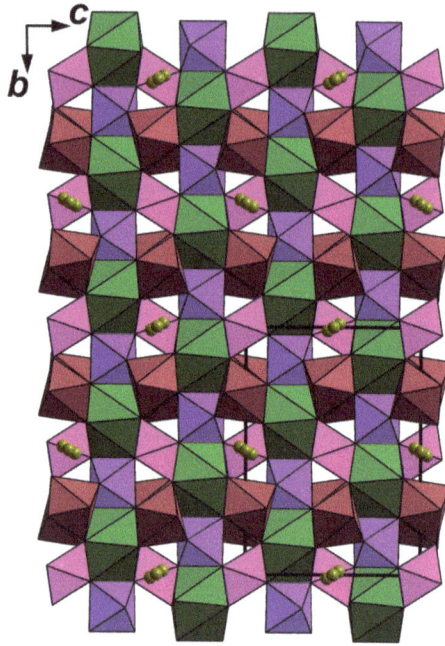

Figure 5. The crystal structure of nöggerathite-(Ce). The unit cell is outlined. For legend see Figure 4.

4. Discussion

Nöggerathite-(Ce) is isostructural with zirconolite-3*O* and is its *REE*-dominant analogue, with Nb prevailing in one of two octahedral sites (see Tables 5–7). Zirconolite-3*O* was originally described as "polymignite" [17,18] and was later redefined and renamed [16,19–21]. However, "polymignite" is usually metamict and, as a rule, its powder X-ray diffraction pattern can be obtained only after calcination.

Unlike most samples of zirconolites, nöggerathite-(Ce) is non-metamict because it was formed recently: the last eruption of the Laach Lake volcano occurred no later than 13,000 years ago. According to Schmitt [14], the age of zircon from sanidinitic ejecta of the Laach Lake volcano is between 17,000 and 30,000 years. Consequently, sanidinite could have formed 4000–17,000 years before the eruption.

A non-metamict *REE*-, Nb-, and Mn-rich variety (or, in accordance with the accepted mineralogical nomenclature rules, an *REE*-analogue) of zirconolite-3*O* from the Laach Lake eruptive center was described by Della Ventura et al. [7] without structural data. Its chemical composition is close to that of nöggerathite-(Ce) and varies within the following ranges (calculated on the basis of 7 O *apfu*): $Ca_{0.312-0.337}Y_{0.079-0.093}La_{0.069-0.086}Ce_{0.269-0.296}Pr_{0.020-0.028}Nd_{0.052-0.062}Sm_{0.004-0.006}Gd_{0.005-0.006}Dy_{0.003-0.006}$ $Er_{0.003-0.005}Th_{0.014-0.073}U_{0.008-0.022}Mn_{0.345-0.397}Mg_{0.003-0.005}Al_{0.020-0.025}Fe_{0.301-0.339}Zr_{0.888-0.910}Hf_{0.006-0.009}$ $Ti_{0.840-0.888}Nb_{0.575-0.613}Ta_{0.007-0.009}Si_{0.000-0.012}O_{7.000}$.

The empirical formula corresponding to spot analysis number 1 from [7], calculated on 14 O *apfu*, is $(Ln_{0.98}Ca_{0.63}Mn_{0.17}Y_{0.16}Th_{0.04}U_{0.02})_{\Sigma2.00}[(Fe_{0.65}Mn_{0.55}Al_{0.05}Mg_{0.01})(Zr_{1.79}Hf_{0.01})(Ti_{1.71}Nb_{1.19}Ta_{0.02})$ $Si_{0.02}]_{\Sigma6.00}O_{14}$. The latter formula may correspond to nöggerathite-(Ce) or its Mn-dominant analogue (with Mn > Fe in the *M*(5) and *M*(6) sites).

Another crystalline mineral related to zirconolite-3*O* and nöggerathite-(Ce) is laachite $(Ca_2Zr_2Nb_2TiFeO_{14})$, which originates from sanidinite of the Laach Lake volcano and demonstrates a perfect crystal structure. Laachite is a monoclinic (pseudo-orthorhombic) analogue of zirconolite-3*O*, with Nb prevailing over Ti in two octahedral sites [9]. Chemical compositions of 24 samples of zirconolite-type minerals from the Laach Lake volcano have been determined by us. Most of them correspond to stefanweissite $(Ca,REE)_2Zr_2(Nb,Ti)(Ti,Nb)_2Fe^{2+}O_{14}$ (IMA 2018-020), which is an analogue of zirconolite-3O with Nb as a species-defining component. Comparative data for nöggerathite-(Ce), zirconolite-3O, and laachite are given in Table 8 and in Figures 6–8. The main substitution scheme following from the compositional data for these minerals is: $Ca^{2+} + Ti^{4+} + Zr^{4+} \leftrightarrow REE^{3+} + Nb^{5+} + Mn^{2+}$. However, as one can see from Figure 7, the substitution Zr^{4+} for Mn^{2+} in the $M(2)$ site is to be completed with other (subordinate) schemes involving zirconium, most probably $Zr^{4+} \leftrightarrow Ti^{4+}$ and/or $REE^{3+} + Zr^{4+} \leftrightarrow Ca^{2+} + Nb^{5+}$.

Table 8. Comparative data for nöggerathite-(Ce) and related minerals.

Mineral	Nöggerathite-(Ce)	Laachite	Zirconolite-3O
Idealized formula	$(Ce,Ca)_2Zr_2(Nb,Ti)$ $(Ti,Nb)_2Fe^{2+}O_{14}$	$Ca_2Zr_2Nb_2TiFeO_{14}$	$CaZrTi_2O_7$
Crystal system	Orthorhombic	Monoclinic	Orthorhombic
Space group	*Cmca*	*C2/c*	*Cmca*
a, Å	7.2985	7.3119	7.278–7.284
b, Å	14.1454	14.1790	14.147–14.18
c, Å	10.1607	10.1700	10.145–10.148
β, °	90	90.072	90
Z	4	4	8
Strong lines of the X-ray powder diffraction pattern: d, Å (I, %)	2.963 (91) 2.903 (100) 2.540 (39) 1.823 (15) 1.796 (51) 1.543 (20) 1.519 (16)	4.298 (22) 2.967 (100) 2.901 (59) 2.551 (32) 1.800 (34) 1.541 (24) 1.535 (23) 1.529 (23)	3.176 (30) 2.914 (100) 2.506 (40) 1.980 (90) 1.792 (90) 1.517 (10)
Refractive index	2.267 (mean, calc.)	2.26 (mean, calc.)	2.215 (meas., metamict); 2.26–2.31 (calc.)
Density, g cm^{-3}	5.332 (calc.)	5.42 (calc.)	4.7 (meas.) 4.9 (calc.)
Sources	This work	[9]	[7,8,16,19–22]

Note: For zirconolite-3O the standard space group *Cmca* is given instead of the space group *Acam* reported in earlier publications.

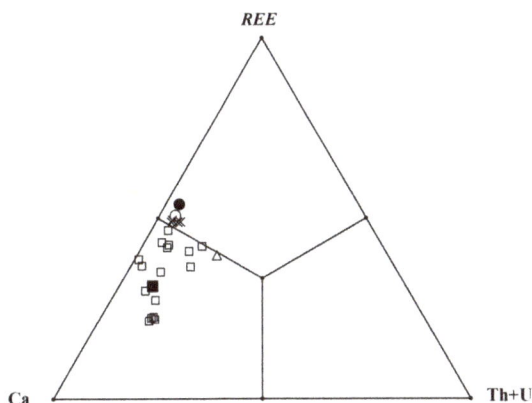

Figure 6. Ratios of large cations with coordination number 8 in zirconolite-type minerals from the Laach Lake volcano: the holotype (○) and cotype (●) nöggerathite-(Ce), other nöggerathite-(Ce) samples (×), the stefanweissite holotype (■), other stefanweissite samples (□), and the holotype laachite (Δ).

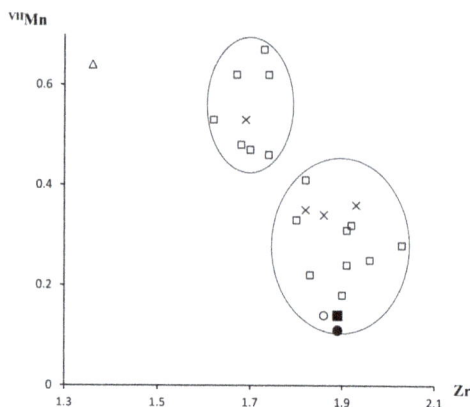

Figure 7. The relationships between the Zr and Mn contents (*apfu*) at the *M*(2) site of zirconolite-type minerals from the Laach Lake volcano. The correlation coefficient is $R = -0.757$ for the whole set of analyses, but almost no correlation is observed within Zr-rich (i.e., Mn-poor) and Zr-poor (i.e., Mn-rich) groups of samples separated with ellipses. For legend see Figure 6.

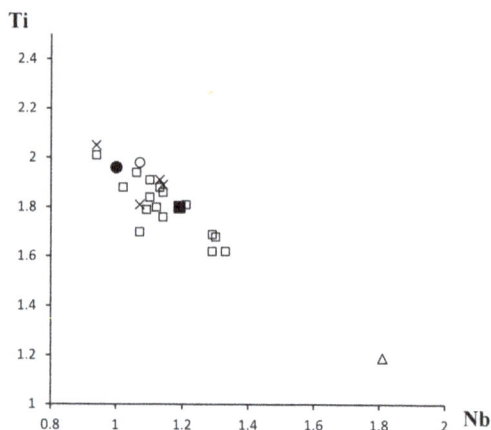

Figure 8. Correlation between contents of Ti and Nb (*apfu*) in zirconolite-type minerals from the Laach Lake volcano. Correlation coefficient $R = -0.932$. For legend see Figure 6.

REE-dominant zirconolite-type minerals (with *REE* > Ca in atomic units and Y prevailing among rare-earth elements) have been described in gabbro pegmatite in St. Kilda, Scotland, United Kingdom, in nepheline syenite from Tchivera, Angola, in metamorphic rocks of the Vestfold Hills, East Antarctica, and in lunar rocks [6], but crystal structures of these minerals have not been studied because of their metamict state.

An important specific feature of nöggerathite-(Ce) is its high content of niobium. A review of localities of zirconolites worldwide and a compilation of chemical compositions of about 300 samples were provided by Williams and Gieré [6]. All analyses show significant predominance of Ti over Nb. For most samples the content of Nb_2O_5 is below 10 wt %. The Nb-richest zirconolite from Vuoriyarvi, Northern Karelia, Russia, was described by Borodin et al. [23] as "niobozirconolite". It contains 0.798 Nb atoms and 1.081 Ti atoms as per a formula calculated on the basis of 7 O atoms. High content of Nb was

Minerals **2018**, *8*, 449

also detected in zirconolite samples from Kovdor, Kola, Russia (0.652 *apfu* Nb vs. 0.857 *apfu* Ti), Kaiserstuhl, Germany (0.687 *apfu* Nb vs. 0.702 *apfu* Ti), and Sokli, Finland (0.608 *apfu* Nb vs. 0.773 *apfu* Ti) [6].

Author Contributions: N.V.C., N.V.Z., and I.V.P. wrote the paper. C.S. collected the type material. B.T. and W.S. collected other samples of zirconolite-type minerals. S.N.B. obtained single-crystal X-ray diffraction data. N.V.Z. and D.Y.P. solved and refined the crystal structure. I.V.P. obtained powder X-ray diffraction data. V.N.E. and N.V.C. conceived and designed chemical data. M.F.V. obtained Raman spectra. N.V.C. analyzed Raman data. Y.S.P. measured reflectance data and hardness.

Funding: This research was funded by the Russian Science Foundation, under grant no. 14-17-00048 (for part of the chemical study of nöggerathite-(Ce) and other zirconolite-type minerals) and the Russian Foundation for Basic Research, under grant no. 18-29-12007 (for parts of the mineralogical study, Raman spectroscopy and X-ray structural analysis). SNB acknowledges St. Petersburg State University for financial support, grant no. 3.42.741.2017 (in part of single-crystal measurements).

Acknowledgments: The authors thank the X-ray Diffraction Centre of St. Petersburg State University for instrumental and computational resources.

Conflicts of Interest: The authors declare no conflict of interest.

References

1. Ringwood, A.E.; Kelly, P.M. Immobilization of high-level waste in ceramic waste forms. *Philos. Trans. R. Soc. A* **1986**, *319*, 63–82. [CrossRef]

2. Donald, I.W.; Metcalfe, B.L.; Taylor, R.N.J. The immobilization of high level radioactive wastes using ceramics and glasses. *J. Mater. Sci.* **1997**, *32*, 5851–5887. [CrossRef]

3. Laverov, N.P.; Yudintsev, S.V.; Stefanovsky, S.V.; Omel'yanenko, B.I.; Nikonov, B.S. Murataite as a universal matrix for immobilization of actinides. *Geol. Ore Depos.* **2006**, *48*, 335–356. [CrossRef]

4. Barinova, T.V.; Borovinskaya, I.P.; Ratnikov, V.I.; Ignat'eva, T.I. Self-propagating high-temperature synthesis for immobilization of high-level waste in mineral-like ceramics: 1. Synthesis and study of titanate ceramics based on perovskite and zirconolite. *Radiochemistry* **2008**, *50*, 316–320. [CrossRef]

5. Zhang, Y.; Gregg, D.J.; Kong, L.; Jovanovich, M.; Triani, G. Zirconolite glass-ceramics for plutonium immobilization: The effects of processing redox conditions on charge compensation and durability. *J. Nucl. Mater.* **2017**, *490*, 238–241. [CrossRef]

6. Williams, C.T.; Gieré, R. Zirconolite: A review of localities worldwide, and a compilation of its chemical compositions. *Bull. Nat. Hist. Mus. Lond.* **1996**, *52*, 1–24.

7. Della Ventura, G.; Bellatreccia, F.; Williams, C.T. Zirconolite with significant REEZrNb(Mn,Fe)O7 from a xenolith of the Laacher See eruptive center, Eifel volcanic region, Germany. *Can. Mineral.* **2000**, *38*, 57–65. [CrossRef]

8. Sinclair, W.; Eggleton, R.A. Structure refinement of zirkelite (zirconolite) from Kaiserstuhl, Germany. *Am. Mineral.* **1982**, *67*, 615–620.

9. Chukanov, N.V.; Krivovichev, S.V.; Pakhomova, A.S.; Pekov, I.V.; Schäfer, Ch.; Vigasina, M.F.; Van, K.V. Laachite, $(Ca,Mn)_2Zr_2Nb_2TiFeO_{14}$, a new zirconolite-related mineral from the Eifel volcanic region, Germany. *Eur. J. Mineral.* **2014**, *26*, 103–111. [CrossRef]

10. Zubkova, N.V.; Chukanov, N.V.; Pekov, I.V.; Ternes, B.; Schüller, W.; Pushcharovsky, D.Y. The crystal structure of natural nonmetamict Nb-rich zirconolite-3T from the Eifel paleovolcanic region, Germany. *Z. Kristallogr.* **2018**, *233*. [CrossRef]

11. Litt, T.; Brauer, A.; Goslar, T.; Merk, J.; Balaga, K.; Mueller, H.; Ralska-Jasiewiczowa, M.; Stebich, M.; Negendank, J.F.W. Correlation and synchronisation of Lateglacial continental sequences in northern Central Europe based on annually laminated lacustrine sediments. *Quat. Sci. Rev.* **2001**, *20*, 1233–1249. [CrossRef]

12. Schmitt, A.K.; Wetzel, F.; Cooper, K.M.; Zou, H.; Wörner, G. Magmatic longevity of Laacher See volcano (Eifel, Germany) indicated by U–Th dating of intrusive carbonatites. *J. Petrol.* **2010**, *51*, 1053–1085. [CrossRef]

13. Frechen, J. Vorgänge der Sanidinit-Bildung im Laacher Seegebiet. *Fortschritte der Mineralogie* **1947**, *26*, 147–166. (In German)

14. Schmitt, A.K. Laacher See revisited: High-spatial-resolution zircon dating indicates rapid formation of a zoned magma chamber. *Geology* **2006**, *34*, 597–600. [CrossRef]

15. Britvin, S.N.; Dolivo-Dobrovolsky, D.V.; Krzhizhanovskaya, M.G. Software for processing the X-ray powder diffraction data obtained from the curved image plate detector of Rigaku RAXIS Rapid II diffractometer. *Zap. Ross. Mineral. Obsh.* **2017**, *146*, 104–107. (In Russian)

16. Mazzi, F.; Munno, R. Calciobetafite (new mineral of the pyrochlore group) and related minerals from Campi Flegrei, Italy; crystal structures of polymignite and zirkelite: Comparison with pyrochlore and zirconolite. *Am. Mineral.* **1983**, *68*, 262–276.

17. Berzelius, J. Undersökning af några Mineralier. 2. Polymignit. *Kongl. Svenska Vetensk.-Acad. Handl.* **1984**, 338–345. (In Swedish)

18. Brøgger, W.C. Die Mineralien der Syenitpegmatitgänge der südnorwegischen Augit und Nephelinsyenite. *Z. Kristallogr. Speziellen Teil* **1890**, *16*, 1–663. (In German)

19. Bayliss, P.; Mazzi, F.; Munno, R.; White, T.J. Mineral nomenclature: Zirconolite. *Mineral. Mag.* **1989**, *53*, 565–569. [CrossRef]

20. Chukhrov, F.V.; Bonshtedt-Kupletskaya, E.M. *Mineraly Vol. II(3)*; Nauka: Moscow, Russia, 1967; pp. 188–190. (In Russian)

21. Pudovkina, Z.V.; Chernitzova, N.M.; Pyatenko, Y.A. Crystallographic study of polymignite. *Zap. Vses. Mineral. Obshch.* **1969**, *98*, 193–199. (In Russian)

22. Borodin, L.S.; Nazarenko, I.I.; Richter, T.L. On a new mineral zirconolite–A complex oxide of AB_3O_7 type. *Dokl. Akad. Nauk SSSR* **1956**, *110*, 845–848. (In Russian)

23. Borodin, L.S.; Bykova, A.V.; Kapitonova, T.A.; Pyatenko, Y.A. New data on zirconolite and its new niobian variety. *Dokl. Akad. Nauk SSSR* **1960**, *134*, 1188–1192. (In Russian)

minerals

MDPI

Article

New Occurrence of Rusinovite, Ca₁₀(Si₂O₇)₃Cl₂: Composition, Structure and Raman Data of Rusinovite from Shadil-Khokh Volcano, South Ossetia and Bellerberg Volcano, Germany

Dorota Środek [1,*], Rafał Juroszek [1], Hannes Krüger [2], Biljana Krüger [2], Irina Galuskina [1] and Viktor Gazeev [3,4]

[1] Department of Geochemistry, Mineralogy and Petrography, Faculty of Earth Sciences, University of Silesia, Będzińska 60, 41-200 Sosnowiec, Poland; rjuroszek@us.edu.pl (R.J.); irina.galuskina@us.edu.pl (I.G.)

[2] Institute of Mineralogy and Petrography, University of Innsbruck, Innrain 52, 6020 Innsbruck, Austria; hannes.krueger@uibk.ac.at (H.K.); biljana.krueger@uibk.ac.at (B.K.)

[3] Insitute of Geology of Ore Deposits, Petrography, Mineralogy and Biochemistry, RAS, Staromonetny 35, 119017 Moscow, Russia; gazeev@igem.ru

[4] Vladikavkaz Scientific Centre of the Russian Academy of Sciences, Markov str. 93a, 362008 Vladikavkaz, Republic of North Ossetia-Alania, Russia

* Correspondence: dsrodek@us.edu.pl; Tel.: +48-693-404-979

Received: 28 August 2018; Accepted: 8 September 2018; Published: 10 September 2018

Abstract: Rusinovite, $Ca_{10}(Si_2O_7)_3Cl_2$, was found at two new localities, including Shadil-Khokh volcano, South Ossetia and Bellerberg volcano, Caspar quarry, Germany. At both of these localities, rusinovite occurs in altered carbonate-silicate xenoliths embedded in volcanic rocks. The occurrence of this mineral is connected to specific zones of the xenolith characterized by a defined Ca:Si < 2 ratio. Chemical compositions, as well as the Raman spectra of the investigated rusinovite samples, correspond to the data from the locality of rusinovite holotype—Upper Chegem Caldera, Northern Caucasus, Russia. The most intense bands of the Raman spectra are related to vibrations of (Si_2O_7) groups. Unit cell parameters of rusinovite from South Ossetia are: $a = 3.76330(4)$ Å, $b = 16.9423(3)$ Å, $c = 17.3325(2)$ Å, $V = 1105.10(4)$ Å^3, $Z = 2$. The performed synchrotron radiation diffraction experiments did not confirm a doubling of c as reported for the synthetic phase, $Ca_{10}(Si_2O_7)_3Cl_2$. However, one-dimensional diffuse scattering parallel to $\mathbf{b^*}$ has been observed. This can be interpreted with an ordered arrangement of Si_2O_7 groups creating layers with a doubled a parameter. Consequently, the two different displacements of neighbouring layers allow random stacking faults to occur.

Keywords: Rusinovite; Raman spectroscopy; pyrometamorphism; stacking faults; Shadil-Khokh volcano; Bellerberg volcano

1. Introduction

Rusinovite, $Ca_{10}(Si_2O_7)_3Cl_2$, was found in 2011 in an altered carbonate-silicate xenolith from the Upper Chegem Caldera located in Kabardino-Balkaria, Northern Caucasus, Russia. These rocks are formed at high-temperature and low-pressure conditions (sanidinite facies) and contain different Cl-bearing minerals, such as rondorfite $MgCa_8(SiO_4)_4Cl_2$, rustumite $Ca_{10}(SiO_4)(Si_2O_7)_2Cl_2(OH)_2$, and minerals of the mayenite supergroup: wadalite $Ca_{12}Al_{10}Si_4O_{32}Cl_6$ and chlorkyuygenite $Ca_{12}Al_{14}O_{32}[(H_2O)_4Cl_2]$ [1–3]. Besides their natural occurrence, rusinovite-type synthetic phases were also described from anthropogenic formations. Firstly, it was reported from the combustion ashes comprised by Cl-bearing Ca-silicates, which were formed after the incineration of waste [4].

Secondly, the $Ca_{10}(Si_2O_7)_3Cl_2$ phase was also described as "chesofiite" from the burned spoil heaps of the coal mine located near the town of Kopeisk, Chelyabinsk Region, Russia [5]. In addition to the geological and anthropogenic occurrences, a phase with rusinovite composition and structure was also synthesized utilizing the flux method [6]. A new green-emitting phosphor was created by doping the synthetic analogue of rusinovite with Europium ($Ca_{10}(Si_2O_7)_3Cl_2$:Eu^{2+}) [7]. Its emission is efficiently excited by the output of near-UV light-emitting diodes (LEDs).

Both rusinovite and the synthetic analogue are reported as strongly elongated grains, which form needle-like single crystals or spherulites composed of fibrous aggregates [1,6]. The structure of the synthetic phase was primarily reported as order-disorder (OD) structure with an orthorhombic unit cell [6], but later its structure was redefined as monoclinic [4]. Galuskin et al. [1] have described the structure of rusinovite as columns of disordered "face-sharing" disilicate units extending parallel to **a**. At a local level, only each second Si_2O_7 unit (along **a**) can be occupied in the structure. Due to the small dimensions of the grains from the holotype locality, some structural issues were still unresolved (such as possible superstructures due to the ordering of the Si_2O_7 groups).

The main goal of this study is to describe rusinovite from two new localities, including chemical composition and Raman spectroscopy. Furthermore, the structure of rusinovite from South Ossetia is to be re-investigated using synchrotron radiation, to check for superstructures, which have been observed in synthetic analogues.

2. The Occurrence and Paragenesis of Rusinovite

Rusinovite was found in pyrometamorphic rock at the two localities. The first one is Shadil-Khokh volcano, belonging to the Kel'sky volcanic plateau located at the southern part of the Greater Caucasus Mountain Range in Southern Ossetia. Small (about 2 m in diameter) altered carbonate-silicate xenolith was revealed at the north-west slope of the volcano within dacite lava.

This xenolith is composed of minerals characteristic for sanidinite facies, such as spurrite, larnite, gehlenite, merwinite, bredigite, rondorfite, and srebrodolskite [8]. Rusinovite occurs as relatively large (100–200 μm in length) elongated crystals with a characteristic macroscopic brown-orange colour in small veins (up to 0.5 cm wide) encountered at the exocontact zone of xenolith (Figure 1A,B). In these zones, another Cl-bearing phase–eltyubyuite, $Ca_{14}Fe^{3+}_{10}Si_4O_{32}Cl_6$, was noted [9]. Rusinovite was found also at the endocontact zone of strongly altered dacite (Figure 2A,B).

The presence of microzonation at both endo- and exocontact zones is reflected in the zonal distribution of minerals with the different Ca:Si ratio. At the endocontact, rock is presented by dacite microbreccia and carbonate-silicate microxenoliths (Figure 2A,B). We distinguish the following zonation (Figure 2C). The first zone (the numbering is according to the Ca:Si ratio increasing)–dacite, consisting of enstatite, albite phenocrysts, and quartz enclosed in the fine-grained aggregate of diopside, plagioclase, and Na–K feldspar. Visible zonation of enstatite (Figure 2C) is caused by variations of minor Fe substitutions in composition. Ilmenite occurs as an accessory mineral. The second zone is composed of massive gehlenite. The third zone is comprised mostly of rusinovite and gehlenite; hydroxylellestadite and wollastonite are noted rarely. The fourth zone is represented by strongly altered rock, mainly composed of secondary Ca–Si hydrosilicates, rusinovite, cuspidine and rondorfite.

Another type of zonation is observed at the dacite-xenolith (exocontact) boundary (Figure 1A,B). The altered dacite is composed of diopside, enstatite, plagioclase phenocrysts, and quartz. The matrix is presented by a fine-grained aggregate of plagioclase. Furthermore, ilmenite, titanite, apatite, and pyrite occur as accessory minerals. The next zone towards xenolith, contains Al–Ca hydrosilicates (Figure 1C). In the second zone, wollastonite and andradite are the main minerals. The accessory phases are represented by cuspidine, gehlenite, and wadalite. In the third zone, rusinovite occurs as a major component together with cuspidine. Wollastonite is present, but not as commonly as in the previous zone. Cl-bearing hydroxylellestadite, gehlenite, magnesioferrite, and rondorfite are also noted. The last zone is a strongly altered xenolith part composed mostly of secondary Ca-hydrosilicates with the relics of rondorfite, larnite, cuspidine, and merwinite.

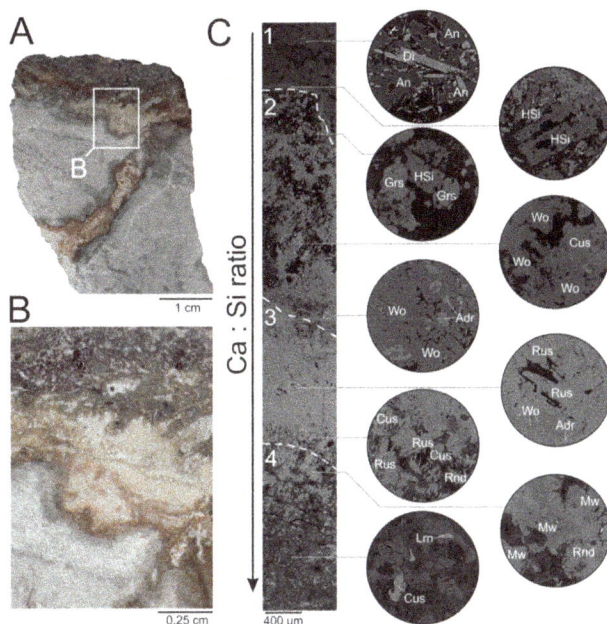

Figure 1. (**A,B**) Macroscopic image of exocontact; (**C**) BSE (backscattered electron) image of the exocontact zone; 1—dacite zone; 2—wollastonite zone; 3—cuspidine—rusinovite zone; 4—altered xenolith. An—anorthite, Adr—andradite, Cus—cuspidine, Di—diopside, Grs—grossular, HSi—Al–Ca hydrosilicates, Lrn—larnite, Mw—merwinite, Rnd—rondorfite, Rus—rusinovite, Wo—wollastonite.

The Bellerberg volcano (Caspar quarry), located near Mayen in the Eifel district in western Germany is the second studied rusinovite locality. In the active quarry, within leucite tephrite, small Ca-rich xenoliths can be found [10]. These xenoliths were formed under similar conditions as the xenoliths from South Ossetia—at high temperatures and low pressure. Bellerberg volcano is a holotype locality for a few Cl-bearing minerals, such as rondorfite $Ca_8Mg[SiO_4]_4Cl_2$ [10], chlormayenite $Ca_{12}Al_{14}O_{32}[\square_4Cl_2]$ [11,12] or vondechenite $Cu_4CaCl_2(OH)_8 \times 4H_2O$ [13].

Macroscopically, the studied xenolith is dark green-grey and contains veins characterized by a colour that is definitely lighter (Figure 3A).

The main part of the xenolith is composed of larnite, gehlenite, rondorfite, Cl-bearing hydroxylellestadite, spurrite, and secondary Ca-hydrosilicates. Perovskite, hematite, cuspidine, Zr-garnet (with a chemical composition close to kerimasite), andradite, and pyrite with chalcopyrite intergrowths are also identified here. The lighter part of the rock is characterized by similar paragenesis, the difference is in the predominance of hydrocalumite among secondary Ca-hydrosilicates. Besides the minerals mentioned above, this zone contains rusinovite, wollastonite, chalcopyrite, baghdadite, and lakargiite (Figure 3B). Aggregates of rusinovite can reach up to 200 μm in size. They contain a lot of mineral inclusions composed of secondary Ca-hydrosilicates, gehlenite, chlorine-bearing ellestadite, perovskite and hematite, mainly, and strongly hydrated zones (Figure 3B).

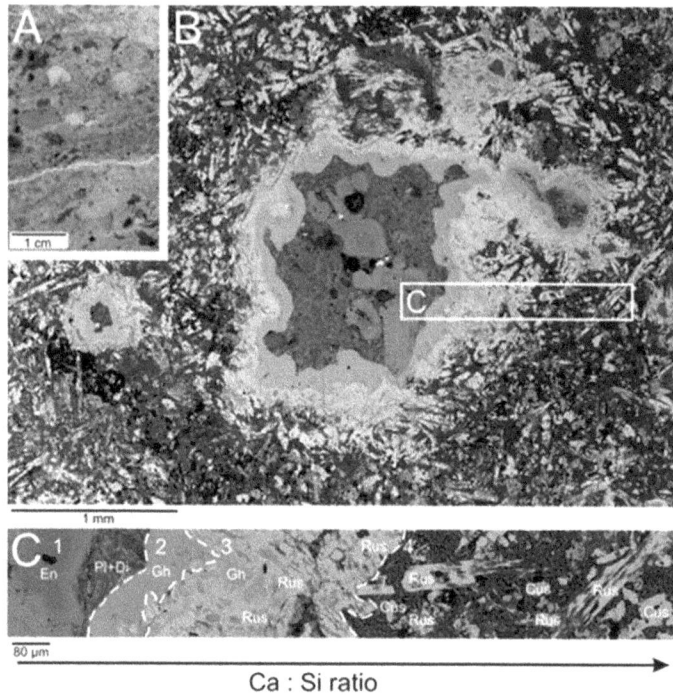

Figure 2. (**A**) Macroscopic image of the endocontact zone; (**B**) BSE image of the endocontact zone; fragment in the frame is magnified in (**C**); Zones: 1—altered dacite zone; 2—gehlenite zone; 3—rusinovite zone; 4—altered xenolith. Cus—cuspidine, Di—diopside, En—enstatite, Gh—gehlenite, Pl—plagioclase, Rus—rusinovite.

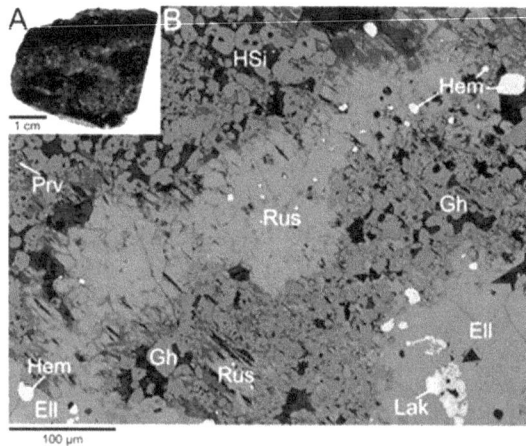

Figure 3. (**A**) Macroscopic photo of vein containing rusinovite from the Bellerberg volcano, Caspar quarry; (**B**) BSE image of rusinovite. Ell—ellestadite, Gh—gehlenite, Hem—hematite, HSi—unidentified secondary Ca-silicates; Lak—lakargiite; Prv—perovskite, Rus—rusinovite.

3. Materials and Methods

The chemical composition and morphology of rusinovite and associated minerals were examined using a scanning electron microscope, the Phenom XL, PhenomWorld (ThermoFisher Scientific, Eindhoven, The Netherlands) equipped with an energy-dispersive X-ray spectroscopy (EDS) detector (Faculty of Earth Sciences, University of Silesia, Sosnowiec, Poland), and a CAMECA SX100 electron microprobe (Institute of Geochemistry, Mineralogy and Petrology, University of Warsaw, Warsaw, Poland). The microprobe analyses were performed at 15 kV and 10 nA using a 1-5 μm beam spot. The following lines and standards were used for rusinovite analyses: Ca$K\alpha$, Si$K\alpha$, Mg$K\alpha$-diopside; Fe$K\alpha$-synthetic Fe$_2$O$_3$; Mn$K\alpha$-rhodonite; Ti$K\alpha$-rutile; Cl$K\alpha$-tugtupite.

The Raman spectra of rusinovite and associated minerals were collected using a WITec confocal Raman microscope CRM alpha 300, WITec, Ulm, Germany (Institute of Physics, University of Silesia, Poland) equipped with an air-cooled solid-state laser (λ = 532 nm) and an electron multiplying charge-coupled device (EMCCD) detector. Raman scattered light was focused onto a multi-mode fiber (50 mm diameter) and monochromator with a 600 line/mm grating. An air Olympus MPLAN 100× objective was used. The spectra were collected with an integration time of 100 ms per spectrum and accumulation of 10 scans, precision of ± 1 cm^{-1} and a resolution of 3 cm^{-1}. The laser radiation power was 10mW at the sample and 1 μm diameter of the incident laser beam. The calibration of the monochromator was held by checking the position of the Si (520.7 cm^{-1}). Spectra processing, such as baseline correction and smoothing, was performed using the Spectralcalc software package GRAMS (Galactic Industries Corporation, Salem, NH, USA). The band fitting was done using a Gauss-Lorentz cross-product function, with the minimum number of component bands used for the fitting process.

Diffraction data were collected at the X06DA beamline at the Swiss Light Source (Paul Scherrer Institute, Villigen, Switzerland) using a Pilatus 2M-F detector at a distance of 80 mm from the sample. A single 180° omega-scan was performed, divided into frames of 0.1° rotation using 0.3 s integration time. The wavelength was tuned to 0.72931 Å. Data reduction, including Lorentz-Polarisation and absorption correction, was performed with X-ray detector software (XDS) [14]. The refinement of the lattice parameters was done with CrysAlis (Rigaku Oxford Diffraction, 2015). For the reconstruction of reciprocal space sections, a modified version of the software Xcavate [15] was utilised. Further details of the intensity data collection and crystal-structure refinement of rusinovite are reported in Table 1. Data about atom coordinates (x,y,z), occupancies, and equivalent isotropic displacement parameters (Uiso, Å2) (Table S1), as well as, anisotropic displacement parameters (Å2) (Table S2), the selected interatomic distances (Å) (Table S3) and cif file (Scheme S1) for rusinovite from Shadil-Khokh volcano, South Ossetia were reported in Supplementary materials.

Table 1. Parameters for X-ray data collection and crystal-structure refinement.

Crystal Data	
Chemical formula	Ca$_{10}$(Si$_2$O$_7$)$_3$Cl$_2$
Crystal system	orthorhombic
Space group	*Cmcm* (No. 63)
	a = 3.76330(4) Å
Unit-cell dimensions	*b* = 16.9423(3) Å
	c = 17.3325(2) Å
Unit cell volume	1105.10(4) Å^3
Formula weight	976.2
Density (calculated)	2.934 g/cm^3
Z	2
Crystal size	28 × 18 × 15 μm
Data collection	
Diffractometer	beamline X06DA, Swiss Light Source single-axis Aerotech goniometer, PILATUS 2M-F detector

Table 1. *Cont.*

Crystal Data	
Radiation wavelength	0.72931 Å
Detector to sample distance	80 mm
Oscillation range	0.1°
No. of frames measured	1800
Time of exposure	0.3 s
Reflection ranges	$-5 \leq h \leq 5$; $-25 \leq k \leq 23$; $-25 \leq l \leq 23$
Reflection measured	5015
R_{int}	0.0274
Refinement of structure	**full matrix least-squares on F**
No. of unique reflections	1087
No. of observed unique refl. [$I > 3s(I)$]	1032
Final R values [$I > 3s(I)$]	$R = 0.021$; $wR = 0.035$
Final R values (all data)	$R = 0.022$; $wR = 0.036$
S (all data)	2.34
Refined parameters	85
Weighting scheme	$w = 1/(s^2(F) + 0.0001F^2)$
$\Delta\rho_{min}$ [e Å^{-3}]	−0.73
$\Delta\rho_{max}$ [e Å^{-3}]	0.32

4. Results

4.1. Chemical Composition and Raman Spectroscopy of Rusinovite

The chemical analysis of rusinovite from the two localities is listed in Table 2. A small loss of the chlorine content in rusinovite from Shadil-Khokh volcano can be caused by slight hydration or methodological problems of chlorine measurements [1].

Table 2. Chemical analysis of rusinovite from the Bellerberg and Shadil-Khokh volcanoes.

Component	Bellerberg Volcano					Shadil-Khokh Volcano				
	Mean 18					Mean 14				
	wt %	S.D.	Range	Atom	pfu	wt %	S.D.	Range	Atom	pfu
SiO_2	36.73	0.23	36.44–37.33	Si	6.00	36.89	0.23	36.12–36.93	Si	5.99
CaO	56.70	0.35	56.24–57.11	Ca	9.93	57.25	0.35	56.79–58.00	Ca	9.95
FeO	0.39	0.13	0.26–0.61	Fe	0.05	0.26	0.13	0.20–0.32	Fe	0.03
MgO	0.06	0.03	0.00–0.09	Mg	0.02	0.09	0.03	0.08–0.11	Mg	0.02
Cl	7.21	0.26	7.11–7.31	Cl	2.00	6.94	0.26	6.76–7.18	Cl	1.91
H_2O *	0.00					0.08			OH	0.08
−O=Cl	1.63					1.57				
Total	99.47					99.95				

* calculated on charge balance; S.D. = 1σ standard deviation.

The Raman spectra of the investigated rusinovite (Figure 4A,B) are similar to the spectrum from the holotype specimen. The most intense bands correspond to the vibrations of (Si_2O_7) group. The stretching motions of this group generally occur between 800 and 1100 cm^{-1}, and it can be differentiated into vibrations of SiO_4 (800–1000 cm^{-1}) and stretching vibrations of Si–O–Si bridges (1000–1100 cm^{-1}) [16]. A sharp, strong band at 905 cm^{-1} is related to the ν_1 symmetric stretching vibrations of Si_2O_7. The bands at 1041, 658 and 641, 365 cm^{-1} are attributed to the ν_3, ν_4, and ν_2, respectively. Galuskin [1] reported, that the rusinovite spectrum is highly similar to the spectrum of rankinite $Ca_3Si_2O_7$ [1]. The presence of a band doublet about 640 and 660 cm^{-1} linked to the symmetric bending ν_4 (Si_2O_7) vibrations is characteristic for rusinovite. Bands occurring below 350 cm^{-1} are connected with Ca–O–Ca, O–Ca–O and librational vibrations of the Si_2O_7 group [1,17].

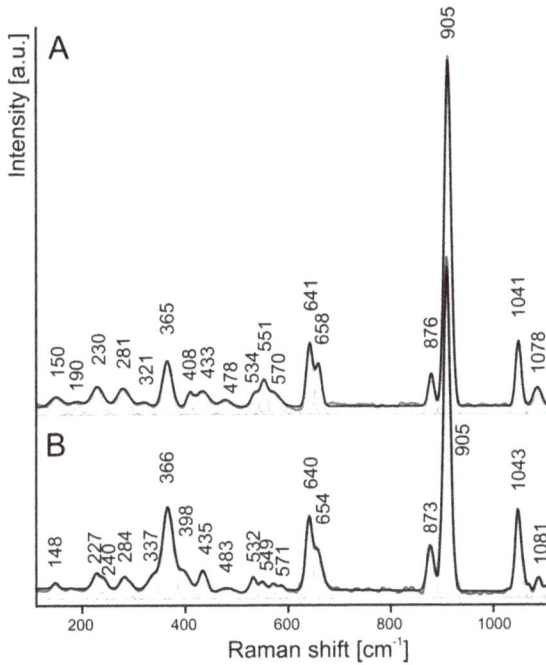

Figure 4. Raman spectra of rusinovite from (**A**) Shadil-Khokh volcano and (**B**) Bellerberg volcano.

4.2. Crystal Structure of Rusinovite

Rusinovite [1], as well as its synthetic analogue [6], exhibits columns of disordered disilicate groups. The first crystal structure refinement of the average structure of rusinovite from single-crystal data was reported by Galuskin et al. [1].

As the rusinovite crystals from the new locality (Shadil-Khokh volcano, South Ossetia) showed almost identical lattice parameters (Table 1) as reported before [1], we started the structure refinement from the known model. The refinement proves that the structures are basically the same. Final atom coordinates, anisotropic displacement parameters, as well as interatomic distances for rusinovite from Shadil-Khokh volcano can be found in a cif file (Supplementary Material).

The average crystal structure of $Ca_{10}(Si_2O_7)_3Cl_2$ consists of the Ca1- and Ca3-polyhedrons and two disilicate units, which form undulating layers perpendicular to **b** (Figure 5). The disilicate units appear to be "face-sharing" (Figure 6), which is an artifact of the disorder. In the real structure, only each second Si_2O_7 unit is occupied. The single layers are connected by the Ca2-site and Cl atoms. Chlorine atoms with characteristic distorted tetrahedral coordination are surrounded only by calcium ions. The bond lengths between Ca1–Cl and Ca2–Cl are within the range from 2.75Å to 2.92Å.

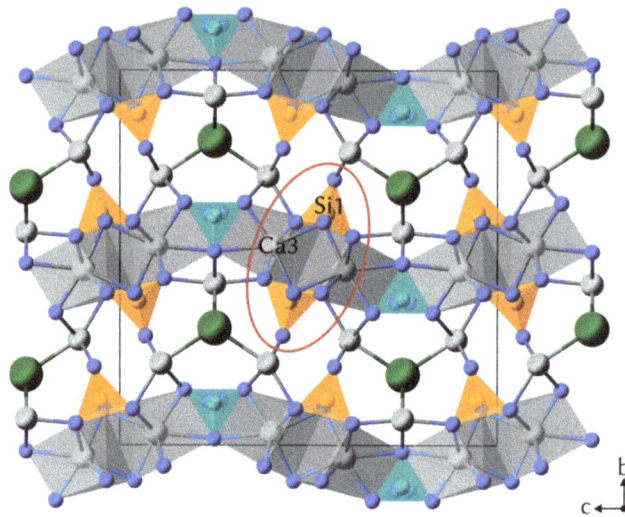

Figure 5. Crystal structure of rusinovite, projected along [100]. There are two types of Si_2O_7 disilicate units: Si1 and Si2 tetrahedra are light brown and greenish blue, respectively. Calcium atoms are shown as grey spheres and in addition, coordination polyhedra of Ca3 are depicted in grey. Oxygen atoms are presented as small blue spheres and Cl as green spheres.

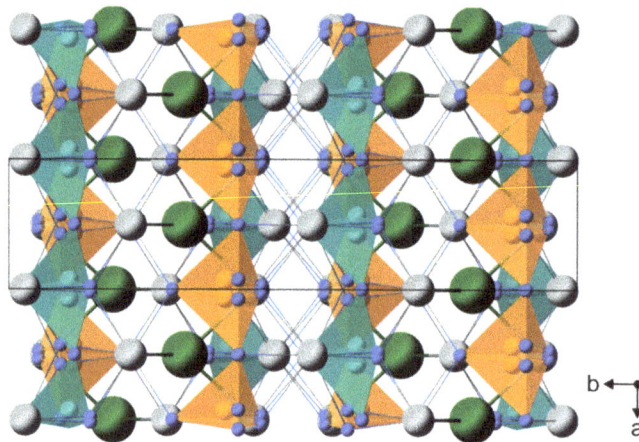

Figure 6. "Face-sharing" disilicates units appear in the average structure of rusinovite. The real structure exhibits ordered columns, where every second Si_2O_7 group is present.

In addition to the lattice formed by Bragg reflections, the diffraction pattern exhibits diffuse streaks which are parallel to **b*** and do not intersect any of the lattice nodes. The streaks are located at $h = 0.5, 1.5, 2.5 \ldots$. Reciprocal space sections (*hkl*) with $l = 0, 1, 2, \ldots$ show Bragg reflections and lines of diffuse scattering. Sections (*hkl*) at $h = 0.5, 1.5, 2.5, \ldots$ exhibit rods of diffuse scattering only (Figure 7).

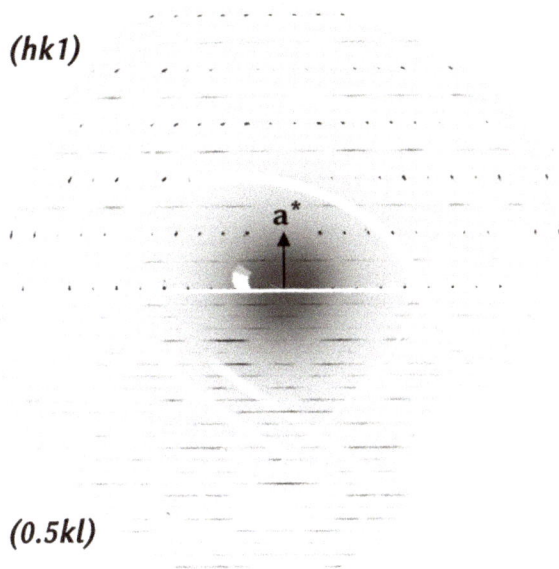

Figure 7. Layers of the reciprocal space of rusinovite as reconstructed from synchrotron diffraction data. Layer *hk1* (top half) exhibits Bragg reflections and diffuse lines parallel to **b***, whereas the layer *(0.5kl)* (lower half) shows only diffuse lines. In both images, **b*** is oriented horizontally parallel to the diffuse lines. The intensity scale (arb. units) was chosen to range from zero (white) to 60 (black) to show details of the weak diffuse intensities. For comparison, the intensities of the Bragg reflections in the upper half of the image range up to 12.5×10^3.

The observed pattern of diffuse lines proves the existence of stacking faults in rusinovite. The positions of the diffuse lines suggest that layers (parallel to the **a**, **c**-plane) are stacked along **b** with random disorder. The most likely mechanism is that within each layer the Si_2O_7 groups are ordered—which results in the doubling of the lattice parameter *a*—and thus creating two possible origins (along **a**) for each layer. The Si_2O_7 groups, which in the average structure appear to be disordered, are indeed ordered. Between neighboring columns of Si_2O_7 groups, strong correlations exist which create ordered layers.

Information on the relative arrangement of $Si_{12}O_7$ groups may be obtained from bond-valence sum (BVS) calculations of the Ca3 atom, which connects to two neighboring $Si_{12}O_7$ units (Figure 5). Ca3 is coordinated by eight oxygen atoms in the average structure. Two of them belong to the Si2 tetrahedra and their BVS contribution is independent of the two possible positions of the tetrahedra. The other six oxygen atoms are connected to two Si1-columns. These six atoms are split-atoms belonging to one of two possible positions of the $Si_{12}O_7$ groups. Looking at the two neighboring $Si_{12}O_7$ columns, there may be two different relative arrangements of the Si_2O_7 groups: Ladder-like (Si_2O_7 groups are side by side along **a**), or zig-zag-like (Figure 8). Taking these two arrangements into account, we must consider four different environments for Ca3: (1) includes bridging oxygen atoms (O3) of both columns (8-fold coordination), (2) no bridging O3 atom is included (6-fold coordination), (3) and (4) only one bridging O3 is coordinating Ca3 (7-fold coordination). In the ladder-like arrangement the environments (1) and (2) occur, whereas in the zig-zag-arrangement only (3) and (4) are possible. The BVS for (1) and (2) are 1.74 and 2.38, respectively. The zig-zag arrangement seems more likely as the corresponding BVS are 1.84 and 2.28 for the environments (3) and (4).

Correlations between the Si1 and Si2-columns are less obvious as neither Ca2 nor Ca1 have bridging oxygen atoms of both columns in their first coordination sphere. Additionally, distortions of the local structure may be involved.

A careful re-investigation of the original data used in [1], revealed that diffuse scattering is detectable, although very weak. The observation of diffuse streaks in X-ray rotation photographs [6] is consistent with our observation in rusinovite crystals. However, the doubled lattice parameter **c** (corresponds to **b** in the setting used by Hermoneit et al. [6]) is not observed in our data. However, this could be explained by a more complex order of disilicate groups within the layers.

A relation to the monoclinic cell (a = 18.66, b = 14.11, and c = 18.14 Å, β = 111.6°) described by Stemmermann et al. [4] seemingly does not exist. It is possible that the monoclinic cell corresponds to a polymorph, which has not yet been further characterized.

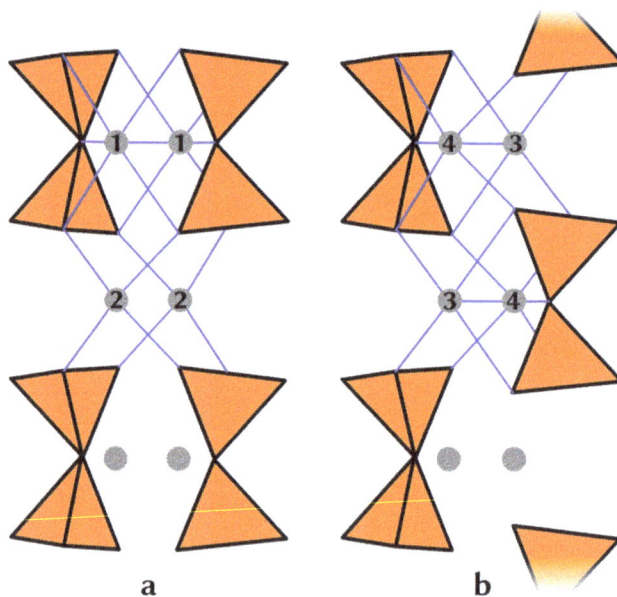

Figure 8. Two possible relative arrangements of neighboring "Si1$_2$O$_7$-columns" and resulting bonds of Ca3, projected along [011], the a-direction is vertical (the shown part of the structure is encircled in red in Figure 5). (**a**) shows a ladder-like arrangement, whereas in (**b**) the zig-zag arrangement is depicted. Bond valence sums for the resulting environments of Ca3 (1–4) are discussed in the text. Bonds from Ca3 to oxygen atoms of the Si2$_2$O$_7$-column are omitted.

5. Discussion

Rusinovite was previously found in South Ossetia [9] and at the Bellerberg volcano [18], but this study is a first mineralogical description of this mineral from these localities. The obtained chemical data are in good agreement with the composition of the holotype specimen from Chegem Caldera, Northern Caucasus, Russia [1]. An insignificant presence of Mg^{2+} and Fe^{2+} impurities at Ca site is characteristic for the studied samples. The Raman spectra of rusinovite from both localities are analogous to the one of the holotype rusinovite with a characteristic doublet about 640–660 cm^{-1} (Figure 4; [1]). In the Raman spectra of disilicates, the split of 640–660 cm^{-1} band was noted before [19]. It is connected with the change Si–O–Si angle [20] and suggests that two types of (Si$_2$O$_7$) groups with different Si–O–Si angles [1] exist in rusinovite.

One of the goals of this study was to check rusinovite for superstructure reflections, which have been reported for synthetic material [6]. Additional Bragg reflections were not observed, however diffuse scattering revealed that stacking faults of layers with doubled *a* are present. Ordered polytypes may exhibit larger periodicity in **b**, the doubling in **c** as suggested by Hermoneit et al. [6] is not explained. The new data does not show evidence for any significant correlation in the stacking sequence. The diffraction pattern can be fully explained by random stacking faults.

For the samples from South Ossetia, we were enabled to distinguish successive stages of mineral formation with a decrease of silica content towards the xenolith from volcanic rock. Minerals occurring in the subsequent zones are characterized by the variable Ca:Si ratio. Provisionally it is possible to distinguish the following successive metasomatic zones in the Shadil-Khokh xenolith: wollastonite/rankinite zone (also as endocontact zone), Ca:Si $\approx$ 1–1.5; larnite/cuspidine/rondorfite zone, Ca:Si $\approx$ 2:1; spurrite/Ca-humite zone, Ca:Si $\approx$ 2.33–2.5:1; and central marble zone (Figure 1C, [21]). In both, xenolith exo- and endocontact, the field of rusinovite prevalence is distinct and corresponds to a Ca:Si ratio between 1.5:1 and 2:1. This ratio is similar to the Ca:Si ratio of the rusinovite-bearing zone from the holotype locality [1,21].

The occurrence of rusinovite in the specific zones with characteristic Ca:Si ratio proves that its formation is a result of metasomatic processes, which require Si diffusion to Ca-bearing xenolith and Ca diffusion through a porous system of volcanic rock to endocontact zones (Figure 2). At the first stage chlorine-free minerals such as wollastonite or rankinite are formed [21]. Later, these minerals were affected by the flow of Cl-bearing fluids of volcanic origin, and as a result rusinovite was formed.

6. Conclusions

The rusinovite from the two new localities shows similar chemical composition and physical properties as the holotype specimen. The single-crystal synchrotron diffraction experiments on rusinovite from the Shadil-Khokh volcano did not confirm the presence of a superstructure as observed in its synthetic analogue. However, the measured one-dimensional diffuse scattering revealed more details of the stacking fault mechanism.

Rusinovite genesis is connected with metasomatic processes, which occurred at the contact zones of the xenolith and volcanic rock.

Supplementary Materials: The following are available online at http://www.mdpi.com/2075-163X/8/9/399/s1, Table S1: Atom coordinates (x,y,z), occupancies and equivalent isotropic displacement parameters (Uiso, Å2) for rusinovite from Shadil-Khokh volcano, South Ossetia, Table S2: Anisotropic displacement parameters (Å2) for rusinovite from Shadil-Khokh volcano, South Ossetia, Table S3: Selected interatomic distances (Å) for rusinovite from Shadil-Khokh volcano, South Ossetia, Scheme S1: The Rusinovite_cif.file.

Author Contributions: D.Ś., R.J., H.K. and I.G. wrote the paper, D.Ś. performed the mineralogical and petrological investigations and Raman studies, R.J., H.K. and B.K. performed investigation of rusinovite structure, V.G. collected the samples.

Funding: This work was supported by the National Science Centre (NCN) of Poland, grant no. 2015/17/N/ST10/03141 (D.Ś.). The investigations were partially supported by the National Science Centre (NCN) of Poland, grant no. 2016/23/N/ST10/00142 (R.J.). The research leading to these results has received funding from the European Union's Horizon 2020 research and innovation programme under grant agreement no. 730872, project CALIPSOplus. The publication has been partially financed from the funds of the Leading National Research Centre (KNOW) received by the Centre for Polar Studies of the University of Silesia, Poland, grant no. 2018/D2/K20-3.

Acknowledgments: The authors would like to thank B.T. for donating the samples from the Bellerberg volcano. H.K. acknowledges help from A.P., V.K. and T.G. during the synchrotron experiments.

Conflicts of Interest: The authors declare no conflict of interest.

References

1. Galuskin, E.V.; Galuskina, I.O.; Lazic, B.; Armbruster, T.; Zadov, A.E.; Krzykawski, T.; Banasik, K.; Gazeev, V.M.; Pertsev, N.N. Rusinovite, $Ca_{10}(Si_2O_7)_3Cl_2$: A new skarn mineral from the Upper Chegem caldera, Kabardino-Balkaria, Northern Caucasus, Russia. *Eur. J. Mineral.* **2011**, *23*, 837–844. [CrossRef]
2. Galuskin, E.V.; Galuskina, I.O.; Kusz, J.; Gfeller, F.; Armbruster, T.; Bailau, R.; Dulski, M.; Gazeev, V.M.; Pertsev, N.N.; Zadov, A.E.; et al. Mayenite supergroup, part II: Chlorkyuygenite from Upper Chegem, Northern Caucasus, Kabardino-Balkaria, Russia, a new microporous mineral with "zeolitic" H_2O. *Eur. J. Mineral.* **2015**, *27*, 113–122. [CrossRef]
3. Gfeller, F.; Armbruster, T.; Galuskin, E.V.; Galuskina, I.O.; Lazic, B.; Savelyeva, V.B.; Zadov, A.E.; Dzierżanowski, P.; Gazeev, V.M. Crystal chemistry and hydrogen bonding of rustumite $Ca_{10}(Si_2O_7)_2(SiO_4)(OH)_2Cl_2$ with variable OH, Cl, F. *Am. Mineral.* **2013**, *98*, 493–500. [CrossRef]
4. Stemmermann, P.; Pöllmann, H. The system $CaO-SiO_2-CaCl_2$-phase equilibria and polymorphs below 1000 °C. An interpretation on garbage combustion ashes. *Neues Jahrb. Mineral. Monatsh.* **1992**, *9*, 409–431.
5. Chesnokov, B.V.; Vilisov, V.; Bushmakin, A.; Kotlyarov, V.; Belogub, E.V. New minerals from a fired dump of the Chelyabinsk coal basin. *Ural Mineral. Zbor.* **1994**, *3*, 3–34.
6. Hermoneit, B.; Ziemer, B.; Malewski, G. Single crystal growth and some properties of the new compound $Ca_3Si_2O_7 \cdot 13CaCl_2$. *J. Cryst. Growth* **1981**, *52*, 660–664. [CrossRef]
7. Ding, W.; Wang, J.; Zhang, M.; Zhang, Q.; Su, Q. Luminescence properties of new $Ca_{10}(Si_2O_7)_3Cl_2:Eu^{2+}$ phosphor. *Chem. Phys. Lett.* **2007**, *435*, 301–305. [CrossRef]
8. Gazeev, V.M.; Gurbanova, O.; Zadov, A.E.; Gurbanov, A.G.; Leksin, A. Mineralogy of skarned carbonate xenoliths from Shadil-Khokh volcano (Kelski Vulcan area of Great Caucasian Range). *Vestn. Vladikavkazskogo Nauchnogo Cent.* **2012**, *2*, 18–27.
9. Gfeller, F.; Środek, D.; Kusz, J.; Dulski, M.; Gazeev, V.; Galuskina, I.; Galuskin, E.; Armbruster, T. Mayenite supergroup, part IV: Crystal structure and Raman investigation of Al-free eltyubyuite from the Shadil-Khokh volcano, Kel' Plateau, Southern Ossetia, Russia. *Eur. J. Mineral.* **2015**, *27*, 137–143. [CrossRef]
10. Mihajlovic, T.; Lengauer, C.L.; Ntaflos, T.; Kolitsch, U.; Tillmanns, E. Two new minerals, rondorfite, $Ca_8Mg[SiO_4]_4Cl_2$, and almarudite, $K(\square,Na)_2(Mn,Fe,Mg)_2(Be,Al)_3[Si_{12}O_{30}]$, and a study of iron-rich wadalite, $Ca_{12}[(Al_8Si_4Fe_2)O_{32}]Cl_6$, from the Bellerberg (Bellberg) volcano, Eifel, Germany. *Neues Jahrb. Mineral. Abh.* **2004**, *179*, 265–294.
11. Hentschel, G. Mayenit, $12CaO \cdot 7Al_2O_3$, und Brownmillerit, $2CaO \cdot (Al, Fe)_2O_3$, zwei neue Minerale in den Kalksteineinschlussen der Lava des Ettringer Bellerberges. *Neues Jahrb. Mineral. Mh.* **1964**, 22–29. (In German)
12. Galuskin, E.V.; Gfeller, F.; Galuskina, I.O.; Armbruster, T.; Bailau, R.; Sharygin, V.V. Mayenite supergroup, part I: Recommended nomenclature. *Eur. J. Mineral.* **2015**, *27*, 99–111. [CrossRef]
13. Schlüter, J.; Malcherek, T.; Pohl, D.; Schäfer, C. Vondechenite, a new hydrous calcium copper chloride hydroxide, from the Bellerberg, East-Eifel volcanic area, Germany. *Neues Jahrb. Mineral.* **2018**, *195*, 79–86. [CrossRef]
14. Kabsch, W. XDS. *Acta Crystallogr. D* **2010**, *66*, 125–132. [CrossRef] [PubMed]
15. Estermann, M.A.; Steurer, W. Diffuse scattering data acquisition techniques. *Phase Transit.* **1998**, *67*, 165–195. [CrossRef]
16. Le Cleac'h, A.; Gillet, P. IR and Raman spectroscopic study of natual lawsonite. *Eur. J. Mineral.* **1990**, *2*, 43–53. [CrossRef]
17. Dulski, M.; Bulou, A.; Marzec, K.M.; Galuskin, E.V.; Wrzalik, R. Structural characterization of rondorfite, calcium silica chlorine mineral containing magnesium in tetrahedral position $[MgO_4]^{6-}$, with the aid of the vibrational spectroscopies and fluorescence. *Spectrochim. Acta. A. Mol. Biomol. Spectrosc.* **2013**, *101*, 382–388. [CrossRef] [PubMed]
18. Galuskin, E.V.; Krüger, B.; Krüger, H.; Blass, G.; Widmer, R.; Galuskina, I.O. Wernerkrauseite, $CaFe^{3+}_2Mn^{4+}O_6$: The first nonstoichiometric post-spinel mineral, from Bellerberg volcano, Eifel, Germany. *Eur. J. Mineral.* **2016**, *28*, 485–493. [CrossRef]
19. Lazarev, A.N. *Vibrational Spectra and Structure of Silicates*; Consultants Bureau: New York, NY, USA, 1972.

20. McMillan, P. Structural studies of silicate glasses and melts—applications and limitations of Raman spectroscopy. *Am. Mineral.* **1984**, *69*, 622–644.
21. Galuskina, I.O.; Krüger, B.; Galuskin, E.V.; Armbruster, T.; Gazeev, V.M.; Włodyka, R.; Dulski, M.; Dzierżanowski, P. Fluorchegemite $Ca_7(SiO_4)_3F_2$, a new mineral from the edgrewite-bearing endoskarn zone of an altered zenolith in ignimbrites from Upper Chegem caldera, Northern Caucasus, Kabardino-Balkaria, Russia: Occurrence, crystal structure and new data on the mineral assemblages. *Can. Mineral.* **2015**, *53*, 325–344. [CrossRef]

Article

New Mineral with Modular Structure Derived from Hatrurite from the Pyrometamorphic Rocks of the Hatrurim Complex: Ariegilatite, $BaCa_{12}(SiO_4)_4(PO_4)_2F_2O$, from Negev Desert, Israel

Evgeny V. Galuskin [1,*], **Biljana Krüger** [2], **Irina O. Galuskina** [1], **Hannes Krüger** [2], **Yevgeny Vapnik** [3], **Justyna A. Wojdyla** [4] and **Mikhail Murashko** [5]

[1] Faculty of Earth Sciences, Department of Geochemistry, Mineralogy and Petrography, University of Silesia, Będzińska 60, 41-200 Sosnowiec, Poland; irina.galuskina@us.edu.pl

[2] Institute of Mineralogy and Petrography, University of Innsbruck, Innrain 52, 6020 Innsbruck, Austria; biljana.kruger@gmail.com (B.K.); hannes.krueger@gmail.com (H.K.)

[3] Department of Geological and Environmental Sciences, Ben-Gurion University of the Negev, P.O. Box 653, Beer-Sheva 84105, Israel; vapnik@bgu.ac.il

[4] Swiss Light Source, Paul Scherrer Institute, 5232 Villigen, Switzerland; justyna.wojdyla@psi.ch

[5] Saint Petersburg State University, Faculty of Geology, 7-9 Universitetskaya nab., St. Petersburg 199034, Russia; mzmurashko@gmail.com

* Correspondence: evgeny.galuskin@us.edu.pl; Tel.: +48-32-3689365

Received: 19 February 2018; Accepted: 5 March 2018; Published: 8 March 2018

Abstract: Ariegilatite, $BaCa_{12}(SiO_4)_4(PO_4)_2F_2O$ ($R\bar{3}m$, a = 7.1551(6) Å, c = 41.303(3) Å, V = 1831.2(3) Å^3, Z = 3), is a new member of the nabimusaite group exhibiting a modular intercalated antiperovskite structure derived from hatrurite. It was found in a few outcrops of pyrometamorphic rocks of the Hatrurim Complex located in the territories of Israel, Palestine and Jordan. The holotype specimen is an altered spurrite marble from the Negev Desert near Arad city, Israel. Ariegilatite is associated with spurrite, calcite, brownmillerite, shulamitite, CO_3-bearing fluorapatite, fluormayenite-fluorkyuygenite and a potentially new mineral, $Ba_2Ca_{18}(SiO_4)_6(PO_4)_3(CO_3)F_3O$. Ariegilatite is overgrown and partially replaced by stracherite, $BaCa_6(SiO_4)_2[(PO_4)(CO_3)]F$. The mineral forms flat disc-shaped crystals up to 0.5 mm in size. It is colorless, transparent, with white steaks and vitreous luster. Optically, ariegilatite is uniaxial, negative: ω = 1.650(2), ε = 1.647(2) (λ = 589 nm). The mean composition of the holotype ariegilatite, $(Ba_{0.98}K_{0.01}Na_{0.01})_{\Sigma1}(Ca_{11.77}Na_{0.08}Fe^{2+}_{0.06}Mn^{2+}_{0.05}Mg_{0.04})_{\Sigma12}(Si_{3.95}Al_{0.03}Ti_{0.02})_{\Sigma4}(P_{1.70}C_{0.16}Si_{0.10}S^{6+}_{0.03}V_{0.01})_{\Sigma2}F_{2.04}O_{0.96}$, is close to the end-member formula. The structure of ariegilatite is described as a stacking of the two modules $\{F_2OCa_{12}(SiO_4)_4\}^{4+}$ and $\{Ba(PO_4)_2\}^{4-}$ along (001). Ariegilatite, as well as associated stracherite, are high-temperature alteration products of minerals of an early clinker-like association. These alterations took place under the influence of pyrometamorphism by-products, such as gases and fluids generated by closely-spaced combustion foci.

Keywords: ariegilatite; nabimusaite group; new mineral; crystal structure; intercalated hexagonal antiperovskite; CO_3-group; Raman; pyrometamorphic rocks; Hatrurim Complex

---

1. Introduction

Ariegilatite, $BaCa_{12}(SiO_4)_4(PO_4)_2F_2O$ ($R\bar{3}m$, a = 7.1551(6) Å, c = 41.303(3) Å, V = 1831.2(3) Å^3, Z = 3), is the first P-bearing mineral in the nabimusaite group, which combines nabimusaite, $KCa_{12}(SiO_4)_4(SO_4)_2O_2F$ ($R\bar{3}m$, a = 7.1905(4), c = 41.251(3) Å, V = 1847.1(2) Å^3, Z = 3) and dargaite, $BaCa_{12}(SiO_4)_4(SO_4)_2O_3$ ($R\bar{3}m$, a = 7.1874(4); c = 41.292(3) Å; V = 1847.32(19) Å^3, Z = 3) [1,2]. Ariegilatite was found in altered spurrite marble belonging to the pyrometamorphic Hatrurim Complex in the Negev Desert near Arad, Israel. Rocks of the Hatrurim Complex mainly consist of spurrite marbles, larnite pseudoconglomerates and gehlenite hornfelses. Furthermore, pyrometamorphic rocks and by-products of their low-temperature alterations commonly occur along the Dead Sea Rift in the territories of Israel, Palestine and Jordan (the "Mottled zone") [3–5].

Minerals of the nabimusaite group are silicates with additional anions and are isostructural with arctite, a phosphate with an ideal crystal chemical formula of $Ba(Ca_7Na_5)(PO_4)_6F_3$ [6]. Minerals of the nabimusaite group and arctite can be described as hexagonal intercalated antiperovskites with the general crystal chemical formula $AB_{12}(TO_4)_4(TO_4)_2W_3$, A = Ba, K, … ; B = Ca, Na, … ; T = Si, P, S^{6+}, V^{5+}, Al, … ; W = O^{2-}, F^- [2,7]. Nabimusaite group minerals show hatrurite-like triple antiperovskite layers $\{[W_3Ca_{12}](SiO_4)_4\}^{e+}$, which alternate with $A(TO_4)_2^{e-}$ single layers [1,2]. The same type of layers is found in arctite. However, the composition of the triple antiperovskite module is $\{[F_3(Ca_7Na_5)](PO_4)_4\}^{4+}$, and the second module is $Ba(PO_4)_2^{4-}$ [6,8].

Minerals of the zadovite group can also be described as intercalated antiperovskites. Their general formula is $AB_6(TO_4)_2[(TO_4)_{2-x}(CO_3)_x]W$ (A, B, T, W = as above; $x \approx 0$ or $x \approx 1$): zadovite, $BaCa_6[(SiO_4)(PO_4)](PO_4)_2F$; aradite, $BaCa_6[(SiO_4)(VO_4)](VO_4)_2F$; gazeevite, $BaCa_6(SiO_4)_2(SO_4)_2O$ and stracherite $BaCa_6(SiO_4)_2[(PO_4)(CO_3)]F$. In these minerals, single antiperovskite layers $\{[WB_6](TO_4)_2\}^{e+}$ intercalate with single $Ba(TO_4)_2^{e-}$ layers [7,9]. In stracherite, about half of the $(PO_4)^{3-}$ tetrahedra are replaced by $(CO_3)^{2-}$ groups [10].

Ariegilatite is named in honor of Dr. Arie Gilat (b. 1939). Arie Gilat is retired from the Geological Survey of Israel, where he was involved in geological mapping, tectonics and geochemical studies for more than 30 years. He is the author and co-author of numerous geological papers. At present, his main interests are related to the study of earthquake physics and processes at the core and mantle interface. His continuous support, consulting and several new unconventional ideas on the genesis of the Hatrurim Complex are greatly appreciated by the authors.

Ariegilatite was approved as a new mineral species by Commission on New Minerals, Nomenclature and Classification, International Mineralogical Association (CNMNC IMA) in March 2017 (IMA2016-100). The material was deposited in the mineralogical collection of the Fersman Mineralogical Museum, Leninskiy pr., 18/k2, 115162 Moscow, Russia, Catalogue Number 4956/1.

2. Methods of Investigation

The crystal morphology and chemical composition of ariegilatite and associated minerals were examined using optical microscopes, as well as the analytical electron scanning microscopes Philips XL30 and Phenom XL, PhenomWorld, ThermoFisher Scientific, Eindhoven, The Netherlands (Faculty of Earth Sciences, University of Silesia, Sosnowiec, Poland). Chemical analyses of ariegilatite were performed with a CAMECA SX100 microprobe analyzer (Institute of Geochemistry, Mineralogy and Petrology, University of Warsaw, Warszawa, Poland) at 15 kV and 10 nA using the following lines and standards: $BaL\alpha$, baryte; $PK\alpha$, apatite; $CaK\alpha$, wollastonite; $MgK\alpha$, $SiK\alpha$, diopside; $FeK\alpha$, hematite, $AlK\alpha$, $KK\alpha$, orthoclase; $TiK\alpha$, rutile; $NaK\alpha$, albite, $SrL\alpha$, $SrTiO_3$, $FK\alpha$, fluorphlogopite.

Raman spectra of ariegilatite were recorded on a WITec alpha 300R Confocal Raman Microscope, WITec, Ulm, Germany (Department of Earth Science, University of Silesia, Sosnowiec, Poland) equipped with an air-cooled solid-state laser (532 nm) and a CCD camera operating at −61 °C. The laser radiation was coupled to a microscope through a single-mode optical fiber with a diameter of 3.5 μm. An air Zeiss LD EC Epiplan-Neofluan DIC-100/0.75NA objective was used. Raman scattered light was focused on a broad band single mode fiber with an effective pinhole size of about 30 μm, and a monochromator with a 600-mm^{-1} grating was used. The power of the laser at the sample position was ca. 40 mW. Integration times of 5 s with an accumulation of 15–20 scans and a resolution 3 cm^{-1} were chosen. The monochromator was calibrated using the Raman scattering line of a silicon plate (520.7 cm^{-1}). Fitting of spectra was performed with the help of the "GRAMS" program using the mixed Lorentz–Gauss function.

Single-crystal X-ray diffraction data were collected from a crystal of ariegilatite ($\sim$38 $\times$ 32 $\times$ 25 μm) using synchrotron radiation at the super-bending magnet beamline X06DA at the Swiss Light Source, Paul Scherrer Institute, Villigen, Switzerland. The multi-axis goniometer PRIGo [11] and a PILATUS 2M-F detector, which was placed at a distance of 120 mm from the crystal with a vertical offset of 67 mm, was used. The wavelength was set to λ = 0.70848 Å. The data were collected in a single 180° omega scan with steps of 0.1° and 0.1-s exposures, controlled by the DA + software [12]. Data evaluation and processing was performed using the CrysAlisPro software package [13]. As a starting model, the structure of the isostructural analogue nabimusaite was used, adapted to the expected composition. With subsequent analyses of difference-Fourier maps, the structure was refined to R_1 = 1.95%. The refinements include anisotropic displacement-parameters and have been carried out with neutral atom scattering-factors, using the program SHELX97 [14].

3. Results

3.1. Occurrence and Description of Holotype Specimen

Ariegilatite was found in a few outcrops of pyrometamorphic rocks of the Hatrurim Complex located in the territories of Israel, Palestine and Jordan. Investigations of a new mineral were performed in samples of spurrite rocks collected in the Negev Desert (Hatrurim Basin, N31°13′ E35°16′) near Arad, Israel. Ariegilatite forms strongly flattened crystals of disc-shaped form. In thin-sections, its pseudo-aciculate morphology is usually observed (Figure 1A). The size of some highly-fractured crystals of ariegilatite reaches 0.5 mm with a thickness of 0.1 mm (Figure 1A). Ariegilatite is associated with spurrite, calcite, brownmillerite, shulamitite, CO_3-bearing fluorapatite, fluormayenite-fluorkyuygenite, periclase, brucite, barytocalcite, baryte, garnets of elbrusite-kerimasite series, unidentified Ca-Fe- and Rb-bearing K-Fe sulfides and a potentially new mineral, $Ba_2Ca_{18}(SiO_4)_6(PO_4)_3(CO_3)F_3O$ [15]. Ariegilatite is often overgrown and replaced by stracherite (Figure 1B). Ariegilatite and stracherite are usually limited to re-crystallization zones of dark-grey fine-grained spurrite rocks, which differ from the surrounding rocks by discoloration, development of thin calcite veins and also by local appearance of large spurrite metacrysts (up to 1 cm in size), as well as the presence of sulfide mineralization.

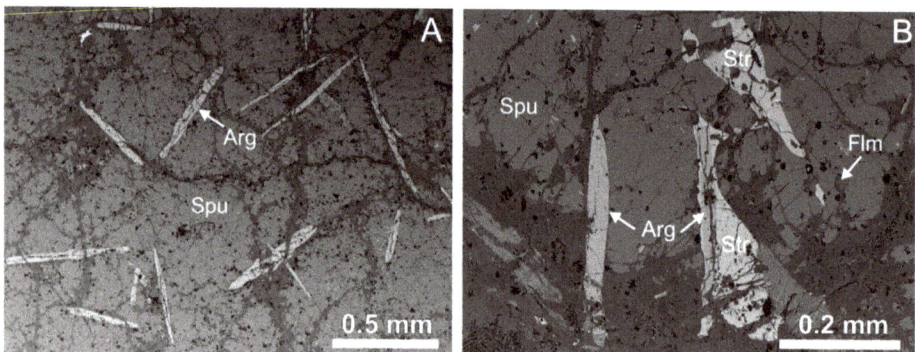

Figure 1. Disc-shaped, flattened ariegilatite crystals in spurrite rock (**A**), holotype specimen L15; overgrowing and replacing of ariegilatite by stracherite (**B**). Arg = ariegilatite, Flm = fluormayenite, Spu = spurrite, Str = stracherite.

Ariegilatite is a colorless, transparent mineral with white streaks and vitreous luster. It does not show any fluorescence. The mineral is uniaxial, negative: ω = 1.650(2), ε = 1.647(2) (λ = 589 nm) and non-pleochroic. The hardness was measured using micro-indentation and determined to VHN$_{50}$ = 356(16) kg·mm^{-2}, as an average of 16 measured spots ranging from 331–378 kg·mm^{-2}, which corresponds to 4–4.5 on the Mohs scale. In contrast to minerals of the nabimusaite-dargaite series [1,2], ariegilatite does not show pronounced cleavage on (001), and the fracture is irregular. The small size of separated crystal fragments does not allow measuring the density. Therefore, the density was calculated on the basis of structural data and the mean composition of ariegilatite: 3.329 g·cm^{-3}. Compatibility index $1 - (K_p/K_c) = -0.017$ (superior) was calculated for the empirical formula of ariegilatite of the holotype specimen (Table 1).

The Raman spectrum of ariegilatite exhibits the following strong bands (Figure 2; cm^{-1}): 129, 179, 229 and 309 (lattice mode, Ba-O, Ca-O vibrations); 403 [$\upsilon_2(SiO_4)^{4-}$]; 427 [$\upsilon_2(PO_4)^{2-}$]; 520 [$\upsilon_4(SiO_4)^{4-}$]; 569 and 591 [$\upsilon_4(PO_4)^{3-}$]; 834 and 874 [$\upsilon_1(SiO_4)^{4-}$]; 947 [$\upsilon_1(PO_4)^{3-}$]; 993 [$\upsilon_1SO_4)^{2-}$]; 1030 [$\upsilon_3(PO_4)^{2-}$]; 1066 [$\upsilon_1(CO_3)^{2-}$]. The main strong Raman bands of ariegilatite are related to vibrations in (SiO$_4$)$^{4-}$ and (PO$_4$)$^{3-}$ (Figure 2). The band at about 1066 cm^{-1} can be attributed to $\upsilon_1(CO_3)^{2-}$. Its position is very close to the $\upsilon_1(CO_3)^{2-}$ mode in CO$_3$-bearing hydroxylapatite and hydroxylellestadite [16,17]. Raman spectroscopy data indicate that H$_2$O is absent in ariegilatite. Direct determination of the CO$_2$ content was not feasible due to abundant microscopic inclusions of other minerals.

The mean composition of the holotype ariegilatite, (Ba$_{0.98}$K$_{0.01}$Na$_{0.01}$)$_{\Sigma1}$(Ca$_{11.77}$Na$_{0.08}$Fe$^{2+}_{0.06}$Mn$^{2+}_{0.05}$ Mg$_{0.04}$)$_{\Sigma12}$(Si$_{3.95}$Al$_{0.03}$Ti$_{0.02}$)$_{\Sigma4}$(P$_{1.70}$C$_{0.16}$Si$_{0.10}$S$^{6+}_{0.03}$V$_{0.01}$)$_{\Sigma2}$F$_{2.04}$O$_{0.96}$, is close to the end-member formula BaCa$_{12}$(SiO$_4$)$_4$(PO$_4$)$_2$F$_2$O. The amount of CO$_3$ in ariegilatite is calculated assuming charge balance and 13 non-tetrahedral cations (Table 1, **L15**).

Experimental details and refinement data of ariegilatite are summarized in Tables 2–5. The structural formula of ariegilatite Ba1Ca1$_6$Ca2$_6$[(Si1O$_4$)$_2$(Si2O$_4$)$_2$](P1O$_4$)$_{~1.8}$F1$_2$O$_2$ (Table 4) is simplified to BaCa$_{12}$(SiO$_4$)$_4$(PO$_4$)$_2$F$_2$O according to the general formula of the nabimusaite group minerals [2]. The ariegilatite structure is most easily described as a 1:1 stacking of the two modules {[F$_2$OCa$_{12}$](SiO$_4$)$_4$}$^{4+}$ and {Ba(PO$_4$)$_2$}$^{4-}$ along (001) (Figure 3). The sites F1 and O7 are coordinated by six Ca atoms in an octahedral arrangement (Figure 3B). The module {Ba(PO$_4$)$_2$}$^{4-}$ is characterized by (P1O$_4$) tetrahedra connected to six-coordinated Ba1 (Figure 3C).

Table 1. Chemical composition of ariegilatite, wt %.

		L15			YV595			SS20b	
Mean	22	s.d.	Range	23	s.d.	Range	14	s.d.	Range
SO$_3$	0.17	0.07	0.05–0.31	0.60	0.22	0.32–1.02	n.d.		
V$_2$O$_5$	0.10	0.06	0–0.17	0.32	0.08	0.13–0.51	0.41	0.07	0.26–0.52
P$_2$O$_5$	9.83	0.45	8.96–10.55	10.39	0.55	9.63–12.53	10.61	0.4	10.05–11.40
TiO$_2$	0.12	0.06	0.05–0.25	0.31	0.04	0.24–0.40	0.26	0.16	0.10–0.56
SiO$_2$	19.87	0.26	19.52–20.42	19.21	0.36	17.70–19.62	18.64	0.27	18.21–19.15
Fe$_2$O$_3$				0.34	0.09		0.17	0.05	0.11–0.24
Al$_2$O$_3$	0.12	0.03	0.07–0.18	0.17	0.11	0.10–0.68	0.24	0.06	0.17–0.38
BaO	12.26	0.06	12.14–12.41	12.01	0.35	10.81–12.49	12.29	0.34	11.74–12.94
FeO	0.32	0.06	0.24–0.46	0.49					
MnO	0.29	0.08	0.09–0.39	0.26	0.07	0.14–0.41	n.d.		
CaO	53.84	0.24	53.19–54.40	53.77	0.30	53.13–54.27	54.24	0.31	54.30–55.62
MgO	0.14	0.03	0.11–0.22	0.38	0.05	0.27–0.53	0.03	0.02	0–0.07
K$_2$O	0.04	0.03	0–0.10	n.d.			n.d.		
Na$_2$O	0.22	0.05	0.16–0.36	0.05	0.02	0.02–0.09	0.31	0.05	0.20–0.38
F	3.17	0.10	2.96–3.34	3.24	0.45	2.27–3.27	3.05	0.11	2.84–3.17
CO$_2$ *	0.57			0.00			0.62		
−O=F	1.33			1.36			1.28		
Total	99.72			100.18			99.58		
Ba	0.98			0.96			0.99		
K	0.01								
Na	0.01			0.02			0.01		
Ca				0.02					
A	**1**			**1**			**1**		
Ca	11.77			11.75			11.88		
Mn^{2+}	0.05			0.05					
Fe^{2+}	0.06			0.08					
Mg	0.04			0.12			0.01		
Na	0.08						0.11		

Table 1. *Cont.*

	L15			YV595			SS20b		
Mean	22	s.d.	Range	23	s.d.	Range	14	s.d.	Range
B	12			12			12		
Si	3.95			3.86			3.81		
Ti^{4+}	0.02			0.05			0.04		
Fe^{3+}				0.05			0.03		
Al	0.03			0.04			0.06		
P							0.07		
T1	4			4			4		
P	1.70			1.80			1.77		
Si	0.10			0.07					
V^{5+}	0.01			0.04			0.06		
S^{6+}	0.03			0.09					
C	0.16						0.17		
T2	2			2			2		
O	0.96			0.91			1.03		
F	2.04			2.09			1.97		
W	2			2			2		

	SS27A			MA5b			IS129			teor
Mean	29	s.d.	Range	11	s.d.	Range	12	s.d.	Range	
SO_3	0.05	0.08	0–0.34	1.35	0.22	1.10–1.92	n.d.			
V_2O_5	0.42	0.07	0.24–0.62	0.18	0.11	0.04–0.41	n.d.			
P_2O_5	10.52	0.52	9.48–11.65	12.91	0.43	11.89–13.60	9.02	0.49	8.31–10.16	11.53
TiO_2	0.27	0.16	0.05–0.59	0.18	0.06	0.09–0.28	n.d.			
SiO_2	18.85	0.42	18.25–20.30	16.46	0.45	15.81–17.16	19.83	0.27	19.44–20.23	19.53
Fe_2O_3	n.d.						n.d.			
Al_2O_3	0.25	0.08	0.14–0.48	0.53	0.10	0.40–0.75	0.15	0.1	0.09–0.47	
BaO	12.26	0.41	11.39–13.15	12.17	0.49	11.42–13.31	12.36	0.17	12.12–12.66	12.46
FeO				0.17	0.11	0.03–0.38				
MnO	n.d.			n.d.			n.d.			
CaO	54.26	0.36	54.00–55.58	54.49	0.13	54.25–54.68	54.78	0.15	54.56–55.07	54.68
MgO	0.08	0.16	0–0.93	0.09	0.02	0.06–0.12	n.d.			
K_2O	n.d.			0.06	0.02	0.02–0.10	0.04	0.02	0.02–0.08	
Na_2O	0.24	0.08	0.06–0.37	0.23	0.04	0.15–0.32	0.66	0.11	0.49–0.82	
F	3.02	0.15	2.68–3.37	2.30	0.07	2.15–2.48	2.92	0.14	2.56–3.22	3.08
CO_2 *	0.55			0.18			1.68			0.00
–O=F	1.27			0.97			1.23			1.29
Total	99.50			100.32			100.21			100.00
Ba	0.98			0.97			0.97			1.000
K				0.02			0.01			
Na	0.02			0.01			0.02			
Ca										
A	1			1			1			
Ca	11.90			11.87			11.76			12.000
Mn^{2+}										
Fe^{2+}				0.03						
Mg	0.02			0.03						
Na	0.08			0.07			0.24			
B	12			12			12			
Si	3.86			3.35			3.97			4.000
Ti^{4+}	0.04			0.02						
Fe^{3+}										
Al	0.06			0.13			0.03			
P	0.04			0.50						
T1	4			4			4			
P	1.78			1.72			1.53			2.000
Si							0.01			
V^{5+}	0.06			0.02						
S^{6+}	0.01			0.21						
C	0.15			0.05			0.46			
T2	2			2			2			
O	1.04			1.52			1.15			1.000
F	1.96			1.48			1.85			2.000
W	2			2			2			2

* Calculated on charge balance, Fe^{2+}: calculated as $13 - (A + B)$ cations, n.d., not detected.

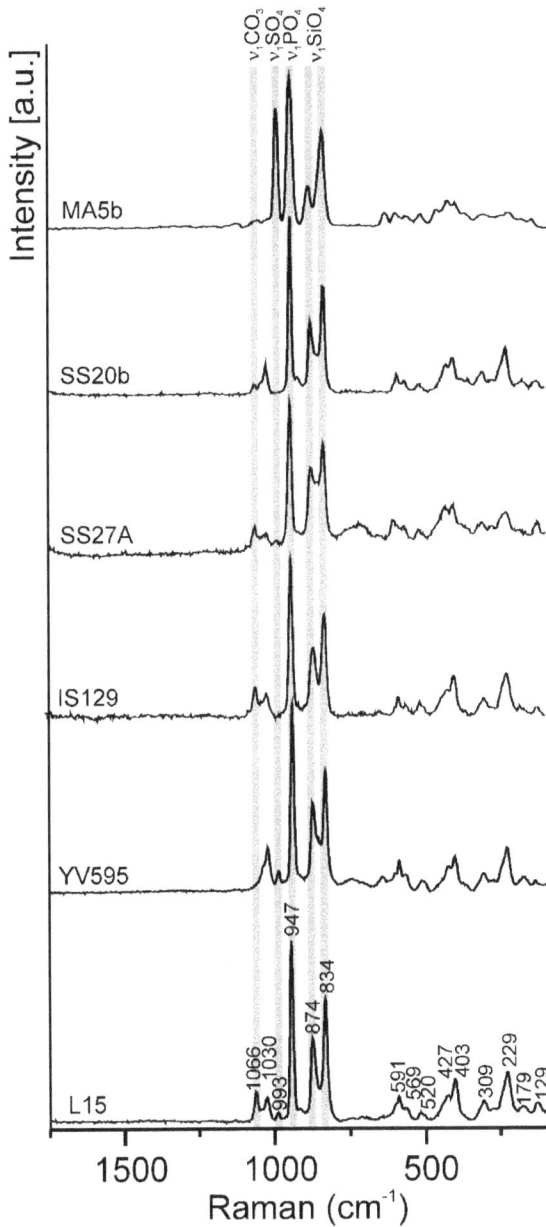

Figure 2. Raman spectra of ariegilatite: holotype specimen, Negev Desert, Israel (Sample Number **L15**); gehlenite-larnite rock, Daba-Siwaqa, Jordan (**YV595**); apatite-spurrite rock, Gurim Anticline, Negev Desert (**IS129**); spurrite rock, Daba-Siwaqa, Jordan (**SS27A** and **SS20b**); exsolution structures after flamite (**MA5b**). Grey vertical bars show the positions of characteristic bands of stretching vibrations (v_1) in tetrahedral anion groups and $(CO_3)^{2-}$.

Table 2. Parameters for X-ray data collection.

Crystal Data	
Crystal system	trigonal
Unit cell dimensions (Å)	$a = 7.1551(6)$, $c = 41.303(3)$ Å, $b = 7.1551(6)$
	$c = 41.303(3)$
	$\alpha, \beta = 90°$ $\gamma = 120°$
Space group	$R\bar{3}m$ (no.166)
Volume (Å^3)	1831.2(3)
Z	3
Density (calculated)	3.329 g·cm^{-3}
Chemical formula	~BaCa$_{12}$(SiO$_4$)$_4$(PO$_4$)$_{1.8}$F$_2$O
Crystal size (mm)	$38 \times 32 \times 25$ μm
Data Collection	
Diffractometer	beamline X06DA, SLS
	multi-axis goniometer PRIGo
	PILATUS 2M-F detector
	$\lambda = 0.70848$ Å
Max. $\theta°$-range for data collection	32.139
Index ranges	$-10 \leq h \leq 6$
	$-8 \leq k \leq 10$
	$-59 \leq l \leq 44$
No. of measured reflections	3445
No. of unique reflections	825
No. of observed reflections (I > 2σ (I))	822
Refinement of the Structure	
No. of parameters used in refinement	58
Rint	0.0199
Rσ	0.0105
R1, I > 2σ(I)	0.0195
R1 all Data	0.0195
wR2 on (F2)	0.0555
GooF	1.128
Δρ min (−e. Å^{-3})	−0.54
Δρ max (e. Å^{-3})	0.82

The antiperovskite module {[F$_2$OCa$_{12}$](SiO$_4$)$_4$]$^{4+}$ consists of columns formed by Ca-triplets Ca$_3$O$_{14}$, rotated relative to each other by 60° (Figure 4A,B). Ca-triplets form four layers with (SiO$_4$)$^{4-}$ tetrahedra in structural cavities (Figure 4C). The structure of hatrurite (Ca$_3$SiO$_5$) shows similar antiperovskite modules [1,18,19].

Raman and electron microprobe data (Figure 2, Table 1, **L15**) indicate the presence of small amounts of (CO$_3$)$^{2-}$ in ariegilatite, replacing (PO$_4$)$^{3-}$ groups. Refinement of the occupancy factor of P1 (Table 3) shows a small reduction from full occupation (89%), which is in agreement with the substitution by CO$_3$. The structural formula of ariegilatite calculated assuming a balanced charge, BaCa$_{12}$(SiO$_4$)$_4$ {[(PO$_4$)$_{0.89}$□$_{0.11}$](CO$_3$)$_{0.11}$}$_2$F$_{1.78}$O$_{1.22}$, is close to the empirical formula obtained from microprobe analyses (Table 1, **L15**). However, the exact location and orientation of the planar CO$_3$ group cannot be determined from diffraction data. Most likely, replacement of (PO$_4$)$^{3-}$ by planar (CO$_3$)$^{2-}$ groups takes place the same way as in stracherite: CO$_3$ triangles are randomly located along one of the faces of the replaced tetrahedra excluding the face parallel to (001) [10]. Replacement of (PO$_4$)$^{3-}$ by (CO$_3$)$^{2-}$ in ariegilatite takes place according to the schemes: Ca^{2+} + (PO$_4$)$^{3-}$ ↔ Na$^+$ + (CO$_3$)$^{2-}$, 2(PO$_4$)$^{3-}$ ↔ (SiO$_4$)$^{4-}$ + (CO$_3$)$^{2-}$ and (PO$_4$)$^{3-}$ + F$^-$ ↔ (CO$_3$)$^{2-}$ + O^{2-}.

As ariegilatite occurs only in tiny amounts, and the crystals contain numerous inclusions of other phases, useful X-ray powder diffraction data could not be collected. However, a powder pattern was calculated (Table S1) using the model derived from the single-crystal structure refinements.

Table 3. Atom coordinates, U_{eq} (Å^2) values for ariegilatite.

Atom	x/a	y/b	z/c	sof	Ueq
Ba1	0	0	0	1	0.01731(9)
Ca1	0.16368(3)	0.83632(3)	0.39407(2)	1	0.01059(12)
Ca2	0.15534(3)	0.84466(3)	0.53179(2)	1	0.01036(12)
P1	0	0	0.67473(2)	0.892(4)	0.0085(2)
Si1	0	0	0.20615(2)	1	0.00650(17)
Si2	0	0	0.08404(2)	1	0.00723(17)
O1	0.55031(11)	0.44969(11)	0.64495(3)	1	0.0148(4)
O2	0.12532(10)	0.87468(10)	0.19359(3)	1	0.0130(4)
O3	0.12681(11)	0.87319(11)	0.07376(3)	1	0.0140(4)
O4	0	0	0.36244(5)	1	0.0125(5)
O5	0	0	0.75407(5)	1	0.0132(5)
O6	0	0	0.12420(5)	1	0.0113(5)
O7	0	0	0.5	1	0.0057(6)
F1	0	0	0.43179(4)	1	0.0134(4)

Table 4. Anisotropic displacement parameters U^{ij} for ariegilatite.

Atom	U^{11}	U^{22}	U^{33}	U^{23}	U^{13}	U^{12}
Ba1	0.01604(12)	0.01604(12)	0.01985(15)	0.00802(6)	0	0
Ca1	0.00850(14)	0.00850(14)	0.01499(17)	0.00441(12)	0.00013(5)	−0.00013(5)
Ca2	0.00855(14)	0.00855(14)	0.01337(17)	0.00382(12)	−0.00066(5)	0.00066(5)
P1	0.0087(3)	0.0087(3)	0.0082(4)	0.00437(15)	0	0
Si1	0.0053(2)	0.0053(2)	0.0089(3)	0.00264(11)	0	0
Si2	0.0063(2)	0.0063(2)	0.0091(3)	0.00316(11)	0	0
O1	0.0156(4)	0.0156(4)	0.0144(5)	0.0088(5)	0.0010(2)	−0.0010(2)
O2	0.0113(4)	0.0113(4)	0.0194(6)	0.0079(5)	0.0009(2)	−0.0009(2)
O3	0.0107(4)	0.0107(4)	0.0220(6)	0.0065(5)	0.0015(2)	−0.0015(2)
O4	0.0136(6)	0.0136(6)	0.0103(8)	0.0068(3)	0	0
O5	0.0147(6)	0.0147(6)	0.0102(8)	0.0074(3)	0	0
O6	0.0114(6)	0.0114(6)	0.0110(8)	0.0057(3)	0	0
O7	0.0059(7)	0.0059(7)	0.0054(10)	0.0030(4)	0	0
F1	0.0131(5)	0.0131(5)	0.0140(8)	0.0066(3)	0	0

Table 5. Selected interatomic distances (Å) for ariegilatite.

Atom	-Atom	Distance		Atom	-Atom	Distance	
Ba1	-O1	2.8346(10)	×6	P1	-O4	1.535(2)	
	mean	2.8346			-O1	1.5484(11)	×3
Ca1	-O5	2.3735(11)			mean	1.5451	
	-O3	2.4006(9)	×2		bvs	4.859(9)	
	-O4	2.4128(12)		Si1	-O2	1.6374(10)	×3
	-O1	2.4628(12)	×2		-O5	1.643(2)	
	F1	2.5577(12)			mean	1.6388	
	mean	2.4387			bvs	3.843(7)	
	bvs	1.921(2)		Si2	-O3	1.6279(10)	×3
Ca2	-O6	2.2495(5)			-O6	1.659(2)	
	-F1	2.4431(12)			mean	1.6356	
	-O2	2.3301(8)	×2		bvs	3.879(7)	
	-O7	2.3303(5)		O7	-Ca2	2.3303(5)	×6
	-O2	2.5114(15)			mean	2.3303	
	-O3	2.6028(15)			bvs	2.248(1)	
	mean	2.3996		F1	-Ca1	2.5577(12)	×3
	bvs	2.196(2)			-Ca2	2.4431(12)	×3
	bvs	1.024(1)			mean	2.5004	

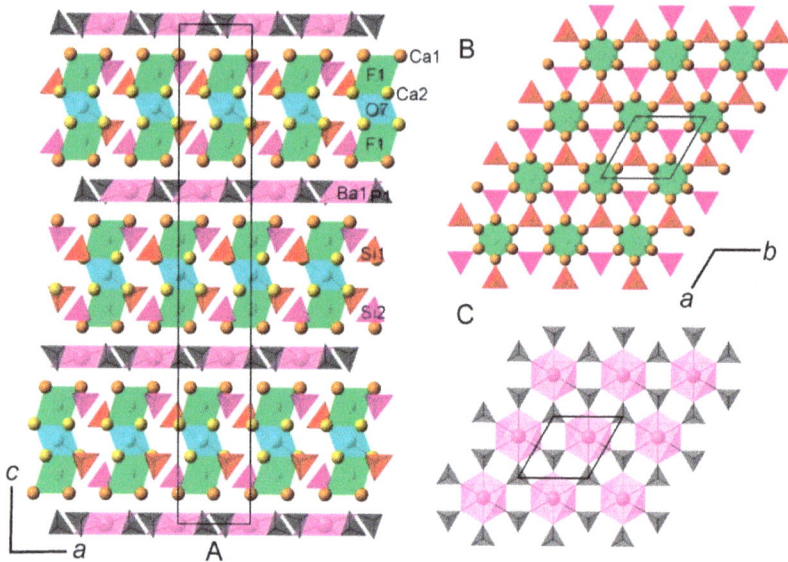

Figure 3. (**A**) The structure of ariegilatite, $BaCa_{12}(SiO_4)_4(PO_4)_2F_2O$, described as a 1:1 stacking of the two modules $\{[F_2OCa_{12}](SiO_4)_4\}^{4+}$ and $\{Ba(PO_4)_2\}^{4-}$ along (001). Atoms F1 and O7 are surrounded by six Ca atoms and form face-sharing octahedra (green and blue). (**B**) Module $\{[F_2OCa_{12}](SiO_4)_4\}^{4+}$ consisting of close-packed seven-fold coordinated Ca (dark yellow filled circles) with (SiO_4) filling the gaps (red and purple tetrahedra). (**C**) The $\{Ba(PO_4)_2\}^{4-}$ module is characterized by (PO_4) tetrahedra (lilac) connected to six-fold coordinated Ba (purple translucent octahedra).

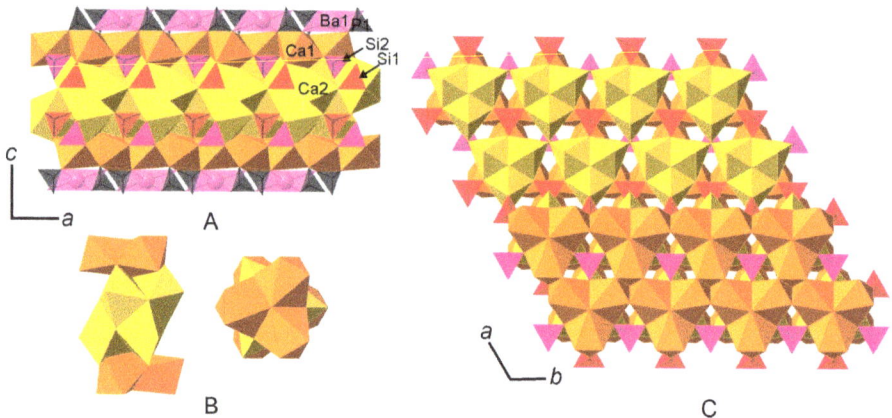

Figure 4. Using a cation-centered representation of the ariegilatite structure (**A**–**C**), the antiperovskite module can be decomposed into four layers assembled of triplets of Ca coordination polyhedra. Structural voids within each layer accommodate the silicate tetrahedra. Perpendicular through the layers, columns of four Ca-triplets can be identified. These are shown in a projection along c and [010] in (**B**). The layers of Ca-triplets are connected by common anions in octahedral coordination: Ca1 and Ca2 layers are sharing F1; Ca2 and Ca2 are connected by O7. In (**C**), two layers of Ca-triplets are removed at the top half of the image.

3.2. Occurrences of Ariegilatite in Other Localities of the Hatrurim Complex Rocks

Ariegilatite occurs both in larnite and spurrite rocks of the Hatrurim Complex, whereas minerals of the nabimusaite-dargaite series are found in larnite rocks only [1,2]. The first samples of ariegilatite were found in 2011 by M. Murashko in larnite pebble, in the northern part of the Siwaqa pyrometamorphic rock area, 80 km south of Amman, Jordan. Daba-Siwaga is the largest area of the Hatrurim Complex within the Dead Sea rift region [20–22]. In this samples, ariegilatite is associated with larnite, gehlenite, spinel, fluormayenite, fluorapatite, perovskite and a bredigite-like Ba-bearing mineral (Figure 5A). It forms poikilitic crystals up to 0.25 mm in size and fine reaction rims on fluorapatite (Figure 5A). The mean ariegilatite composition from larnite rock is $(Ba_{0.96}K_{0.02}Na_{0.02})_{\Sigma 1}(Ca_{11.75}Mg_{0.12}Fe^{2+}_{0.08}Mn^{2+}_{0.05})_{\Sigma 12}(Si_{3.86}Ti^{4+}_{0.05}Fe^{3+}_{0.05}Al_{0.04})_{\Sigma 4}(P_{1.80}S^{6+}_{0.09}Si_{0.07}V^{5+}_{0.04})_{\Sigma 2}F_{2.09}O_{0.91}$ (Table 1, **YV595**); here, Mg is slightly increased in comparison to the holotype specimen (Table 1, **L15**). The Raman spectrum of ariegilatite from larnite rocks in Jordan shows no characteristic bands of $(CO_3)^{2-}$ groups (Figure 2, **YV595**). Furthermore, the calculation of the stoichiometric formula does not indicate the presence of $(CO_3)^{2-}$ (Table 1, **YV595**).

Investigation of spurrite rocks (+calcite, fluorapatite, fluormayenite-fluorkyuygenite, brownmillerite and periclase) occurring within a 1–2-km radius from the place, where the larnite pebble was found at the location in Siwaqa, revealed ariegilatite metacrysts 30–50 μm in size (Figure 3B). These rocks are locally enriched in sphalerite and oldhamite and are interspersed with calcite veins of 1–2 mm in thickness. As well as the holotype material, ariegilatite from spurrite rocks of Jordan contains about 0.15–0.17 $(CO_3)^{2-}$ *pfu* (Table 1, **SS20b** and **SS27A**). The presence of $(CO_3)^{2-}$ is also confirmed by Raman spectroscopy (Figure 2, **SS20b** and **SS27A**).

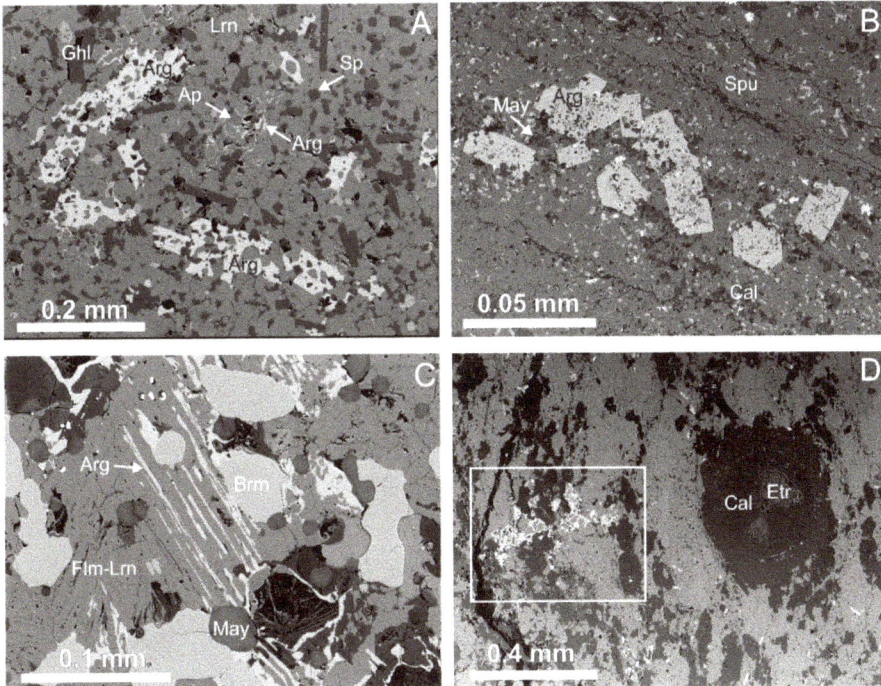

Figure 5. *Cont.*

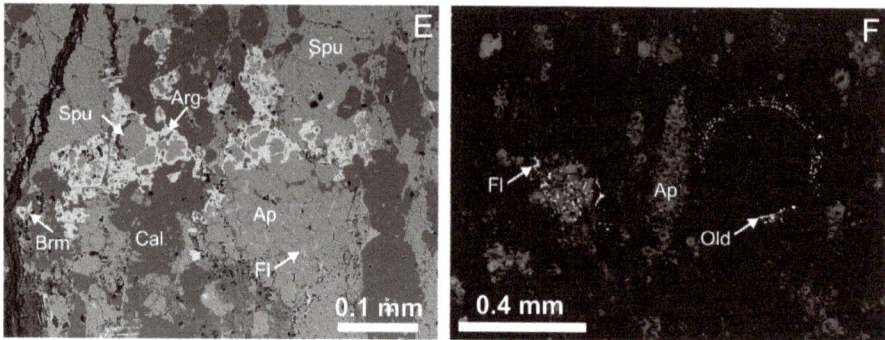

Figure 5. Poikilitic crystals of ariegilatite in gehlenite-larnite rock (**A**), where it also forms reaction rims on fluorapatite, Daba-Siwaqa, Jordan (Sample No. **YV595**); poikilitic crystals of ariegilatite in spurrite rock (**B**), Daba-Siwaqa, Jordan (**SS20b**); replacement of flamite by ariegilatite with exsolution structures (**C**), Ma'ale Adummim, Palestinian Autonomy (**MA5b**); apatite-spurrite rock with ariegilatite and channels filled with calcite (**D**). The fragment magnified in (**E**) is shown in the frame; Gurim Anticline, Negev Desert, Israel (**IS129**); xenomorphic aggregates of ariegilatite with spurrite inclusions (**E**); cathodoluminescence image of the area shown in (**D,F**), round shape of the channel, the walls of which are incrusted by oldhamite, having intensive luminescence. Arg = ariegilatite, Ap = fluorapatite, Brm = brownmillerite, Cal = calcite, Etr = ettringite, Flm = flamite, Fl = fluorite, Ghl = gehlenite, Lrn = larnite; May = fluormayenite-fluorkyuygenite, Old = oldhamite, Sp = spinel, Spu = spurrite.

Ariegilatite was also found in the territory of Palestinian Autonomy in an outcrop near the Jerusalem-Dead Sea highway close to Ma'ale Adummim. At this location, the mineral was detected in altered flamite rocks (+fluormayenite-fluorkyuygenite, brownmillerite, fluorapatite, gehlenite) with heterogeneous development of jasmundite. Flamite is an analog of the synthetic phase α-Ca_2SiO_4 stabilized by K, Na and P impurities [23,24]. Flamite usually shows exsolution structures with individual lamellas enriched in P [24]. Ariegilatite replaces flamite lamellas with increased phosphorus content (Figure 5C) and exhibits a mean composition of $(Ba_{0.97}K_{0.02}Na_{0.01})_{\Sigma 1}(Ca_{11.87}Na_{0.07}Mg_{0.03}Fe^{2+}_{0.03})_{\Sigma 12}(Si_{3.35}P_{0.50}Al_{0.13}Ti^{4+}_{0.02})_{\Sigma 4}(P_{1.72}S^{6+}_{0.21}C_{0.05}V^{5+}_{0.02})_{\Sigma 2}F_{1.48}O_{1.52}$, and it is characterized by increased $(SO_4)^{2-}$ content (Table 1, **MA5b**); accordingly, its Raman spectrum shows a strong $\upsilon_1(SO_4)^{2-}$ band near 995 cm^{-1} (Figure 2, **MA5b**).

Ariegilatite samples showing the highest $(CO_3)^{2-}$ contents were detected in fluorapatite-calcite-spurrite rocks in the Negev Desert near the junction of Hatrurim (Gurim Anticline, Hatrurim Basin), Israel (Figure 5D,E). In these rocks, also cuspidine, fluorite, brownmillerite, chlormayenite, periclase, sphalerite, millerite, U-bearing lakargiite, vorlanite and oldhamite are identified. The empirical formula of ariegilatite from this locality is $(Ba_{0.97}K_{0.01}Na_{0.02})_{\Sigma 1}(Ca_{11.76}Na_{0.24})_{\Sigma 12}(Si_{3.97}Al_{0.03})_{\Sigma 4}(P_{1.53}C_{0.46}Si_{0.01})_{\Sigma 2}F_{1.85}O_{1.15}$ (Table 1, **IS129**). Its Raman spectrum reveals a characteristic band of $(CO_3)^{2-}$ groups (Figure 2, **IS129**). The $(CO_3)^{2-}$ groups can be incorporated according to two schemes: $(PO_4)^{3-} + Ca^{2+} \leftrightarrow (CO_3)^{2-} + Na^+$ and $(PO_4)^{3-} + F^- \leftrightarrow (CO_3)^{2-} + O^{2-}$. Interestingly, in spurrite rock, an ariegilatite relic net of dendrite channels (with an oval cross-section of 0.2–0.4 mm in diameter) filled with coarse-crystalline calcite was observed. Walls of the channels are incrusted by small oldhamite crystals up to 1–2 μm in size. Thanks to intensive oldhamite cathodoluminescence, these channels are well visible within the background of the weakly-luminescent minerals of spurrite rocks (Figure 5F).

4. Discussion

Ariegilatite exhibits an ordered anion distribution. Fluorine and oxygen were assigned to the anion sites utilizing the results of bond valence sum (BVS, Table 5) calculations. Oxygen resides in the center (Figure 3A, blue) and fluorine on the two outer octahedral sites (Figure 3A, green) of the antiperovskite module. New BVS calculations revealed the same preference in nabimusaite, where one fluorine atom seems to be distributed over the two outer octahedral sites, in contrast to what has been reported before [1,2,25]. Fluorine may enter the central octahedra when the charge is compensated as, for example, in arctite $BaCa_7Na_5(PO_4)_6F_3$ [6,8], where Ca atoms (of the central octahedra) are substituted by sodium. Similar behavior is observed in the isostructural synthetic compound $Ca_{5.45}Li_{3.55}[SiO_4]_3O_{0.45}F_{1.55}$ [26], where the outer octahedral sites are occupied by fluorine, whereas the inner site seems to be shared by oxygen and fluorine. For charge compensation, the Ca-sites coordinating the inner octahedral site are partly substituted by lithium.

The main schemes of isomorphism, unifying minerals of the nabimusaite group and arctite are $Ca^{2+} + SiO_4^{4-} = Na^+ + PO_4^{3-}$ and $Ca^{2+} + O^{2-} = Na^+ + F^-$. Arctite was the first mineral with a hexagonal intercalated antiperovskite structure to be discovered. We propose to distinguish an arctite super group (hexagonal intercalated antiperovskites), which at present combines two silicate groups: nabimusaite (nabimusaite, dargaite, ariegilatite, with triple-layer antiperovskite modules) and zadovite (zadovite, aradite, gazeevite, stracherite, with single-layer antiperovskite modules) and arctite, as an ungrouped phosphate.

According to Krivovichev [19], polyphite, $Na_6(Na_4Ca_2)_2Na_2Ti_2Na_2Ti_2(Si_2O_7)_2(PO_4)_6O_4F_4$ [27], also belongs to the hexagonal intercalated antiperovskites. However, we propose that polyphite, the structurally-related sobolevite, $Na_{12}Ca(NaCaMn)Ti_2(Ti,Mn)[Si_2O_7]_2(PO_4)_4O_3F_3$, and quadruphite, $Na_{14}Ca_2Ti_4[Si_2O_7]_2(PO_4)_4O_4F_2$ [27], should not be considered in the arctite super group. In these structures, columns of anion-centered octahedra form continuous layers parallel to titanium-silicate sheets. The proposed arctite super group exhibits $A(TO_4)$ layers, which are perpendicular to the "antiperovskite columns" (Figure 3A).

So far, ariegilatite shows the highest fluorine content (>3 wt %, Table 1) of all members of the nabimusaite group. It occurs in spurrite, as well as in larnite rocks (Figures 1 and 5). Ariegilatite from spurrite rocks contain significant amounts of $(CO_3)^{2-}$ (Table 1). It forms relatively large poikilitic crystals standing out against a fine-grained matrix and grows after fluorapatite (Figures 1 and 5). Minerals of the dargaite-nabimusaite series (and also gazeevite), as well as minerals of the ternesite-silicocarnotite series and jasmundite are high-temperature alteration products of minerals of an early clinker-like association of larnite rocks. These alterations took place under the influence of pyrometamorphism by-products, such as gases and fluids generated by closely-spaced combustion foci [1,28,29].

In our understanding, ariegilatite, as well as associated stracherite are the products of high-temperature alteration of early pyrometamorphic rocks with increased phosphorus (fluorapatite) and decreased sulfur (absence of fluorellestadite and ye'elimite) contents under the influence of high-fluorine, CO_2-bearing fluids/gases. Transport of these fluids/gases is realized through cracks and micro-channel nets (Figure 5F), which may have existed for a long time in heated pyrometamorphic rocks.

Supplementary Materials: The following are available online at http://www.mdpi.com/2075-163X/8/3/19/s1, Table S1: Calculated powder pattern for ariegilatite. Intensities were calculated using the software JANA2006 [30].

Acknowledgments: The investigations were partially supported by the National Science Centre of Poland, Grant No. 2016/23/B/ST10/00869. We would like to thank Jakub Kaminski for important technical assistance during the diffraction experiments at X06DA.

Author Contributions: Evgeny V. Galuskin, Biljana Krüger, Irina O. Galuskina and Hannes Krüger wrote the paper. Evgeny V. Galuskin, Irina O. Galuskina and Yevgeny Vapnik took part in the field works, by the results of which ariegilatite was discovered. Mikhail Murashko found the first specimen with ariegilatite in Jordan. Evgeny V. Galuskin and Irina O. Galuskina performed petrological investigations, measurement of ariegilatite composition from different localities, Raman and optical studies and also selected grains for structural investigations. Biljana Krüger and Hannes Krüger performed SC XRD investigation and refined ariegilatite structure. Justyna A. Wojdyla helped in single-crystal structural investigations using synchrotron radiation at the super-bending magnet beamline X06DA at the Swiss Light Source.

Conflicts of Interest: The authors declare no conflict of interest.

References

1. Galuskin, E.V.; Gfeller, F.; Armbruster, T.; Galuskina, I.O.; Vapnik, Ye.; Murashko, M.; Wodyka, R.; Dzierżanowski, P. New minerals with modular structure derived from hatrurite from the pyrometamorphic Hatrurim Complex, Part I: Nabimusaite, $KCa_{12}(SiO_4)_4(SO_4)_2O_2F$, from larnite rock of the Jabel Harmun, Palestinian Autonomy, Israel. *Mineral. Mag.* **2015**, *79*, 1061–1072. [CrossRef]

2. Galuskina, I.O.; Gfeller, F.; Galuskin, E.V.; Armbruster, T.; Vapnik, Ye.; Dulski, M.; Gardocki, M.; Jeżak, L.; Murashko, M. New minerals with modular structure derived from hatrurite 1 from the pyrometamorphic rocks, part IV: Dargaite, $CaCa_{12}(SiO_4)_4(SO_4)_2O_3$, from Nahal Darga, Palestinian Autonomy. *Mineral. Mag.* **2018**, *82*. in press.

3. Bentor, Y.K. (Ed.) Israel. In *Lexique Stratigraphique International, Asie*; CNRS: Paris, France, 1960; Volume III, Chapter 10.2.

4. Gross, S. *The Mineralogy of the Hatrurim Formation, Israel*; Geological Survey of Israel: Jerusalem, Israel, 1977.

5. Vapnik, Y.; Sharygin, V.V.; Sokol, E.V.; Shagam, R. Paralavas in a combustion metamorphic complex: Hatrurim Basin, Israel. *Rev. Eng. Geol.* **2007**, *18*, 1–21.

6. Sokolova, E.V.; Yamnova, N.A.; Egorov-Tismenko, Y.K.; Khomyakov, A.P. The crystal structure of a new sodium-calcium-barium phosphate of Na, Ca and Ba ($(Na_5Ca)Ca_6Ba(PO_4)_6F_3$). *Dokl. Akad. Nauk SSSR* **1984**, *274*, 78–83.

7. Galuskin, E.V.; Gfeller, F.; Galuskina, I.O.; Armbruster, T.; Krząłała, A.; Vapnik, Ye.; Kusz, J.; Dulski, M.; Gardocki, M.; Gurbanov, A.G.; et al. New minerals with a modular structure derived from hatrurite from the pyrometamorphic rocks. Part III. Gazeevite, $BaCa_6(SiO_4)_2(SO_4)_2O$, from Israel and the Palestine Autonomy, South Levant, and from South Ossetia, Greater Caucasus. *Mineral. Mag.* **2017**, *81*, 499–513. [CrossRef]

8. Sokolova, E.; Hawthorne, F.C. The crystal chemistry of the $[M_3\varphi_{11-14}]$ trimeric structures: From hyperagpaitic complexes to saline lakes. *Can. Mineral.* **2001**, *39*, 1275–1294. [CrossRef]

9. Galuskin, E.V.; Gfeller, F.; Galuskina, I.O.; Pakhomova, A.; Armbruster, T.; Vapnik, Y.; Włodyka, R.; Dzierżanowski, P.; Murashko, M. New minerals with modular structure derived from hatrurite from the pyrometamorphic Hatrurim Complex, Part II: Zadovite, $BaCa_6[(SiO_4)(PO_4)](PO_4)_2F$, and aradite, $BaCa_6[(SiO_4)(VO_4)](VO_4)_2F$, from paralavas of the Hatrurim Basin, Negev Desert, Israel. *Mineral. Mag.* **2015**, *79*, 1073–1087. [CrossRef]

10. Galuskin, E.V.; Krüger, B.; Galuskina, I.O.; Krüger, H.; Vapnik, Ye.; Pauluhn, A.; Olieric, V. Stracherite, $BaCa_6(SiO_4)_2[(PO_4)(CO_3)]F$, a first CO_3-bearing intercalated hexagonal antiperovskite from Negev Desert, Israel. *Amer. Mineral.* **2018**. under review.

11. Waltersperger, S.; Olieric, V.; Pradervand, C.; Glettig, W.; Salathe, M.; Fuchs, M.R.; Curtin, A.; Wang, X.; Ebner, S.; Panepucci, E.; et al. PRIGo: A new multi-axis goniometer for macromolecular crystallography. *J. Synchrotron Radiat.* **2015**, *22*, 895–900. [CrossRef] [PubMed]

12. Wojdyla, J.A.; Kaminski, J.W.; Panepucci, E.; Ebner, S.; Wang, X.; Gabadinho, J.; Wang, M. DA+ data acquisition and analysis software at the Swiss Light Source macromolecular crystallography beamlines. *J. Synchrotron Radiat.* **2018**, *25*, 293–303. [CrossRef] [PubMed]

13. Kabsch, W. XDS. *Acta Crystallogr. D* **2010**, *166*, 125–132. [CrossRef] [PubMed]

14. Sheldrick, G.M. A short history of SHELX. *Acta Crystallogr. A* **2008**, *64*, 112–122. [CrossRef] [PubMed]

15. Krüger, B.; Galuskin, E.V.; Galuskina, I.O.; Krüger, H.; Vapnik, Y.; Olieric, V.; Pauluhn, A. A potentially new mineral with a modular structure based on antiperovskite layers. *Mitt. Österr. Miner. Ges.* **2017**, *163*, 59.

16. Comodi, P.; Liu, Y. CO_3 substitution in apatite: Further insight from new crystal-chemical data of Kasekere (Uganda) apatite. *Eur. J. Mineral.* **2000**, *12*, 965–974. [CrossRef]

17. Banno, Y.; Miyawaki, R.; Momma, K.; Bunno, M. CO_3-bearing member of the hydroxylapatite-hydroxylellestadite series from Tadano, Fukushima Prefecture, Japan: CO_3-SO_4 substitution in the apatite–ellestadite series. *Mineral. Mag.* **2016**, *80*, 363–370. [CrossRef]

18. Jeffery, J.W. The crystal structure of tricalcium silicate. *Acta Crystallogr.* **1952**, *5*, 26–35. [CrossRef]

19. Krivovichev, S.V. Minerals with antiperovskite structure: A review. *Z. Kristallogr.* **2008**, *223*, 109–113. [CrossRef]

20. Geller, Y.I.; Burg, A.; Halicz, L.; Kolodny, Y. System closure during the combustion metamorphic "Mottled Zone" event, Israel. *Chem. Geol.* **2012**, *334*, 25–36. [CrossRef]

21. Novikov, I.; Vapnik, E.; Safonova, I. Mud volcano origin of the Mottled Zone, South Levant. *Geosci. Front.* **2013**, *4*, 597–619. [CrossRef]

22. Khoury, H.N.; Salameh, E.M.; Clark, I.D. Mineralogy and origin of surficial uranium deposits hosted in travertine and calcrete from central Jordan. *Appl. Geochem.* **2014**, *43*, 49–65. [CrossRef]

23. Sokol, E.V.; Seryotkin, Y.V.; Kokh, S.N.; Vapnik, Y.; Nigmatulina, E.N.; Goryainov, S.V.; Belogub, E.V.; Sharygin, V.V. Flamite, $(Ca,Na,K)_2(Si,P)O_4$, a new mineral from ultra high temperature combustion metamorphic rocks, Hatrurim Basin, Negev Desert, Israel. *Mineral. Mag.* **2015**, *79*, 583–596. [CrossRef]

24. Gfeller, F.; Widmer, R.; Krüger, B.; Galuskin, E.V.; Galuskina, I.O.; Armbruster, T. The crystal structure of flamite and its relation to Ca_2SiO_4 polymorphs and nagelschmidtite. *Eur. J. Mineral.* **2015**, *27*, 755–769. [CrossRef]

25. Fayos, J.; Glasser, F.P.; Howie, R.A.; Lachowski, E.; Perez-Mendez, M. Structure of dodecacalcium potassium fluoride dioxide terasilicate bis (sulphate), $KF.2[Ca_6(SO_4)(SiO_4)_2O]$: A fluorine containing phase encountered in cement clinker production process. *Acta Crystallogr. C* **1985**, *C41*, 814–816. [CrossRef]

26. Krüger, H. $Ca_{5.45}Li_{3.55}[SiO4]_3O_{0.45}F_{1.55}$ and $Ca_7K[SiO_4]_3F_3$: Single-crystal synthesis and structures of two trigonal oxyfluorides. *Z. Kristallogr.* **2010**, *225*, 418–424. [CrossRef]

27. Sokolova, E.; Hawthorne, F.C.; Khomyakov, A.P. Polyphite and sobolevite: Revision of their crystal structures. *Can. Mineral.* **2005**, *43*, 1527–1544. [CrossRef]

28. Galuskin, E.V.; Galuskina, I.O.; Gfeller, F.; Krüger, B.; Kusz, J.; Vapnik, Ye.; Dulski, M.; Dzierżanowski, P. Silicocarnotite, $Ca_5[(SiO_4)(PO_4)](PO_4)$, a new "old" mineral from the Negev Desert, Israel, and the ternesite–silicocarnotite solid solution: Indicators of high-temperature alteration of pyrometamorphic rocks of the Hatrurim Complex, Southern Levant. *Eur. J. Mineral.* **2016**, *28*, 105–123. [CrossRef]

29. Galuskina, I.O.; Galuskin, E.V.; Prusik, K.; Vapnik, Ye.; Juroszek, R.; Jeżak, L.; Murashko, M. Dzierżanowskite, $CaCu_2S_2$—A new natural thiocuprate from Jabel Harmun, Judean Desert, Palestine Autonomy, Israel. *Mineral. Mag.* **2017**, *81*, 1073–1085. [CrossRef]

30. Petříček, V.; Dušek, M.; Palatinus, L. Crystallographic computing system JANA2006: General features. *Z. Kristallogr.* **2014**, *229*, 345–352. [CrossRef]

minerals

MDPI

Article

Dynamic Disorder of Fe^{3+} Ions in the Crystal Structure of Natural Barioferrite

Arkadiusz Krzątała [1,*], Taras L. Panikorovskii [2], Irina O. Galuskina [1] and Evgeny V. Galuskin [1]

[1] Faculty of Earth Sciences, University of Silesia, Będzińska 60, 41-200 Sosnowiec, Poland;
    irina.galuskina@us.edu.pl (I.O.G.); evgeny.galuskin@us.edu.pl (E.V.G.)
[2] Kola Science Centre, Russian Academy of Sciences, 14 Fersman street, Apatity 184200, Russia;
    taras.panikorovsky@spbu.ru
* Correspondence: akrzatala@us.edu.pl; Tel.: +48-32-368-9689

Received: 13 July 2018; Accepted: 6 August 2018; Published: 8 August 2018

Abstract: A natural barioferrite, $BaFe^{3+}_{12}O_{19}$, from a larnite–schorlomite–gehlenite vein of paralava within gehlenite hornfels of the Hatrurim Complex at Har Parsa, Negev Desert, Israel, was investigated by Raman spectroscopy, electron probe microanalysis, and single-crystal X-ray analyses acquired over the temperature range of 100–400 K. The crystals are up to 0.3 mm × 0.1 mm in size and form intergrowths with hematite, magnesioferrite, khesinite, and harmunite. The empirical formula of the barioferrite investigated is as follows: $(Ba_{0.85}Ca_{0.12}Sr_{0.03})_{\Sigma 1}(Fe^{3+}_{10.72}Al_{0.46}Ti^{4+}_{0.41}Mg_{0.15}Cu^{2+}_{0.09}Ca_{0.08}Zn_{0.04}Mn^{2+}_{0.03}Si_{0.01})_{\Sigma 11.99}O_{19}$. The strongest bands in the Raman spectrum are as follows: 712, 682, 617, 515, 406, and 328 cm^{-1}. The structure of natural barioferrite ($P6_3/mmc$, $a = 5.8901(2)$ Å, $c = 23.1235(6)$ Å, $V = 694.75(4)$ Å^3, $Z = 2$) is identical with the structure of synthetic barium ferrite and can be described as an interstratification of two fundamental blocks: spinel-like S-modules with a cubic stacking sequence and R-modules that have hexagonal stacking. The displacement ellipsoids of the trigonal bipyramidal site show elongation along the [001] direction during heating. As a function of temperature, the mean apical Fe–O bond lengths increase, whereas the equatorial bond lengths decrease, which indicates dynamic disorder at the Fe2 site.

Keywords: barioferrite; crystal structure; single-crystal investigation; Raman; Hatrurim Complex

---

1. Introduction

Barioferrite, $BaFe_{12}O_{19}$, $P6_3/mmc$, $a = 5.8921(2)$ Å, $c = 23.1092(8)$ Å was described as a new mineral species from a metamorphosed baryte nodule, which was found on the southern slope of Har Ye'elim, a mountain in the Negev Desert, Israel. In this nodule, barioferrite occurs as small platy crystals up to $3 \times 15 \times 15$ μm^3, as well as their aggregates. Therefore, the unit cell dimensions of the holotype specimen were obtained only from powder X-ray diffraction data [1]. There are several descriptions of synthetic $BaFe_{12}O_{19}$ [2–5], whereas there are no previous single-crystal structure studies of natural $BaFe_{12}O_{19}$.

Barioferrite belongs to the magnetoplumbite group. This group is characterized by the general crystal chemical formula $AM_{12}O_{19}$, where A = Ba, Pb, K, Ca, Fe^{2+}, Mn, Mg, REE and $M = Fe^{2+,3+}$, Mg, Al, Ti^{4+}, Cr^{3+}, $Mn^{2+,4+}$, Zn [6]. Minerals of the magnetoplumbite group are characterized by a modular structure composed of three-layered spinel blocks (S) interstratified with R-blocks, in which the larger A-site cations are located [7].

A natural occurrence of $BaFe_{12}O_{19}$ was first reported from Si-undersaturated Ca-Fe-rich buchite in porcelanite deposits at Želénky in the North Bohemian Brown Coal Basin, Czech Republic [8].

Our investigations have shown that barioferrite, occurring in small segregations intergrown with other ferrites, is a common accessory mineral of rankinite-bearing paralavas that are hosted by gehlenite hornfelses that belong to the pyrometamorphic Hatrurim Complex [9–11]. Pyrometamorphic rocks of the Hatrurim Complex are distributed in the territory of Israel, the Palestinian Autonomy, and Jordan along the rift of the Dead Sea [12–15].

Synthetic $BaFe_{12}O_{19}$ is well known under such names as barium ferrite, hexaferrite, barium hexaferrite, ferroxdure, M ferrite, and BaM [7]. An X-ray single-crystal study of synthetic $BaFe_{12}O_{19}$ was first undertaken by Townes [16], and the structure was reinvestigated in detail by Obradors [2]. They found that it is isostructural with magnetoplumbite, $PbFe^{3+}_{12}O_{19}$, which was described by Aminoff (1925) [17] and structurally characterized by Adelsköld in 1938 [7]. Many authors have tried to refine the crystal structure of synthetic $BaFe_{12}O_{19}$ in different space groups (e.g., *P-62c* and *P63mc*), and all of them considered space group *P63/mmc* as the most probable one [2–5]. In space group *P63/mmc*, the Fe2 site has trigonal bipyramid coordination [18]; this site demonstrates splitting or extreme elongation of the displacement ellipsoid along the [001] direction [2]. The relative displacement of iron from the center of the FeO_5 bipyramid induces a local electric and dipole moment, which affects the paraelectric/ferroelectric properties of hexaferrites [19].

In this paper, we present the results of the single-crystal structural investigation of natural barioferrite for the first time. Also, powder diffraction data, chemical composition, and single-crystal Raman spectroscopic data are provided for the studied mineral. We pay attention to the structural deformations of the trigonal bipyramidal site at different temperatures and consider mechanisms for the incorporation of Al, Ti, and Mg into the structure of natural barioferrite.

2. Methods and Description

2.1. Geological Setting and Mineral Description

The barioferrite investigated was found in the outcrops at Har Parsa, which is located in the eastern part of the Hatrurim Basin, about 10 km southeast of Arad, Negev Desert, Israel. The Hatrurim Basin is the one of the largest units in the Hatrurim pyrometamorphic complex. It is a generally recognized fact that the terrigenous-carbonate protolith of the Hatrurim Complex was driven by combustion processes [20–23]. The geology and origin of the Hatrurim Complex have been discussed previously in some detail [21–26]. Many new minerals including aradite, zadovite [9], gurimite, hexacelsian [10], ariegilatite [27], flamite [28], shulamitite [29], barioferrite [1], silicocarnotite [30], and gazeevite [31] have been described from this locality.

Barioferrite forms platy submetallic crystals in a small paralava vein up to 1.5 cm thick in gehlenite hornfels at Har Parsa. Some barioferrite crystals reach 0.3 mm × 0.1 mm in size (Figure 1b,c). Minerals of the larnite-flamite and andradite-schorlomite series, Fe^{3+}-bearing gehlenite, magnesioferrite, and khesinite, are rock-forming minerals in this paralava (Figure 1a). Barioferrite forms intergrowths with hematite, magnesioferrite $MgFe_2O_4$, khesinite $Ca_4Mg_2Fe^{3+}_{10}O_4(Fe^{3+}_{10}Si_2)O_{36}$ (often with Ca > Mg at the Mg site [11]), and harmunite $CaFe_2O_4$ (Figure 1d) and is also present as small inclusions in gehlenite (Figure 1a). Minor and accessory minerals, in addition to the above-listed ferrites, are represented by vorlanite $CaUO_4$, Si–Fe-bearing perovskite, P-bearing fluorellestadite, and baryte.

2.2. Methods of Investigation

The morphology and composition of barioferrite and associated minerals were studied using optical microscopy, scanning electron microscopy (Phenom-World, Eindhoven, The Netherlands)) (Phenom XL, Faculty of Earth Sciences, University of Silesia, Sosonowiec, Poland), and an electron probe microanalyzer (Cameca, Gennevilliers, France) (Cameca SX100, Institute of Geochemistry, Mineralogy and Petrology, University of Warsaw, Warsaw, Poland). Elemental analyses were carried out (WDS-mode, 15 kV, 20 nA, ~1 μm beam diameter) using the following X-ray lines (detection limit, wt %) and standards: Si *Kα* (0.03), Ca *Kα* (0.04), Mg *Kα* (0.03)—diopside; Al *Kα* (0.03)—orthoclase; Ti *Kα* (0.05)—TiO_2; Mn *Kα* (0.10)—rhodonite; Fe *Kα* (0.13)—Fe_2O_3; Ba *Lα* (0.11)—baryte; Sr *Lα*

(0.09)—celestine; Cu *K*α (0.17)—cuprite; and Zn *K*α (0.23)—sphalerite. Corrections were calculated using a PAP procedure provided by CAMECA [32].

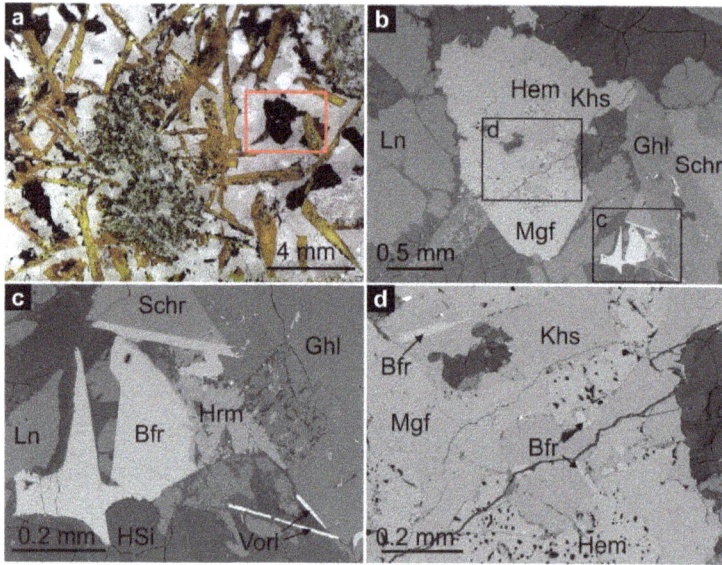

Figure 1. (**a**) Fragments of larnite-gehlenite paralava from Har Parsa with barioferrite aggregates and gehlenite porphyroblasts (megacrysts); fine-grained greenish gehlenite rock occurs in the center. The minerals can be distinguished by color: garnet—dark-brown; gehlenite—orange to yellow-brown; larnite—light-grey to colorless; barioferrite—black; secondary hydrous Ca-silicates—chalk white to light-grey. The area shown in the red frame is magnified in Figure 1b; (**b**) backscattered-electron (BSE) image of larnite-gehlenite paralava with barioferrite crystals. The framed areas are magnified in Figure 1c,d; (**c**) subhedral barioferrite crystals with harmunite; (**d**) aggregate composed of magnesioferrite, hematite, and khesinite with lamellar barioferrite crystals. Labels: Bfr—barioferrite, Ghl—gehlenite, Hem—hematite, Hrm—harmunite, HSi—hydrous silicate, Khs—khesinite, Ln—larnite, Mgf—magnesioferrite, Schr—schorlomite, and Vorl—vorlanite.

The Raman spectra of barioferrite were recorded using a WITec alpha 300R Confocal Raman Microscope (WITec, Ulm, Germany) (Faculty of Earth Sciences, University of Silesia, Sosnowiec, Poland) equipped with an air-cooled solid-state laser (532 nm) and a CCD camera operating at −61 °C. The laser radiation was coupled to a microscope through a single-mode optical fiber with a diameter of 3.5 μm. A Zeiss LD EC Epiplan-Neofluar DIC-100/0.75NA objective was used. Raman scattered light was focused on a broadband single-mode fiber with an effective pinhole size of about 30 μm, and a monochromator with a 600 groove/mm^{-1} grating was used. The power of the laser at the sample position was ca. 25–30 mW. Integration times of 10 s with an accumulation of 20 scans and a resolution 3 cm^{-1} were chosen. The monochromator was calibrated using the Raman scattering line of a silicon plate (520.7 cm^{-1}). Spectra processing, such as baseline correction and smoothing, was performed using the SpectraCalc software package GRAMS (Galactic Industries Corporation, Salem, NH, USA). Band-fitting was performed using a Gauss–Lorentz cross-product function, with a minimum number of component bands used for the fitting process.

Powder X-ray diffraction patterns of barioferrite were obtained using a Rigaku R-AXIS RAPID II diffractometer (Rigaku Co., Tokyo, Japan) (Saint Petersburg State University, Saint Petersburg, Russia) equipped with a cylindrical image plate detector using the Debye-Scherrer geometry (*d* = 127.4 mm;

Co $K\alpha$ radiation) and measured at room temperature. The data were integrated using the software package Osc2Tab/SQRay [33].

The single-crystal X-ray diffraction experiments were carried out at 100, 200, 300, and 400 K using an Agilent Technologies Xcalibur Eos diffractometer (Agilent Technologies, Santa Clara, CA, USA) (Saint Petersburg State University, Saint Petersburg, Russia) operated at 50 kV and 40 mA and equipped with an Oxford Cobra Plus system (Table 1). More than a hemisphere of three-dimensional data was collected using monochromatic Mo $K\alpha$ X-radiation with frame widths of 1° and 40–80 s count time for each frame. The systematic absences observed are consistent with space group $P6_3/mmc$.

Table 1. Crystal data and structure refinement for barioferrite at 100, 200, 300, and 400 K.

Identification Code	100 K	200 K	300 K	400 K
Temperature (K)	100(2)	200(2)	300(2)	400(2)
Crystal system, Z		Hexagonal		
Space group		$P6_3/mmc$		
a (Å)	5.8920(2)	5.8901(2)	5.8901(2)	5.8992(2)
c (Å)	23.1092(7)	23.1171(7)	23.1235(6)	23.1891(9)
Volume (Å³)	694.78(5)	694.57(4)	694.75(4)	698.87(6)
ρ_{calc} (g/cm³)	5.183	5.184	5.183	5.152
μ (mm⁻¹)	14.131	14.136	14.132	14.049
$F(000)$	1017.0	1017.0	1017.0	1017.0
Crystal size (mm³)		$0.20 \times 0.09 \times 0.07$		
Radiation		Mo $K\alpha$ ($\lambda = 0.71073$)		
2θ range (°)	7.054–64.038	7.05–63.804	7.048–61.54	7.028–64.436
hkl range	$-8 \leq h \leq 8, -6 \leq k \leq 6, -33 \leq l \leq 33$	$-8 \leq h \leq 8, -8 \leq k \leq 8, -32 \leq l \leq 32$	$-7 \leq h \leq 7, -8 \leq k \leq 8, -32 \leq l \leq 28$	$-8 \leq h \leq 8, -7 \leq k \leq 8, -32 \leq l \leq 32$
Reflections collected	3907	6363	6065	6804
Independent reflections	501 [$R_{int} = 0.0176$, $R_{sigma} = 0.0106$]	505 [$R_{int} = 0.0300$, $R_{sigma} = 0.0148$]	461 [$R_{int} = 0.0281$, $R_{sigma} = 0.0121$]	519 [$R_{int} = 0.0279$, $R_{sigma} = 0.0118$]
Data/restraints/ parameters	501/0/42	508/0/42	461/0/42	519/0/42
GooF	1.229	1.270	1.288	1.268
Final R indexes [I ≥ 2σ (I)]	$R_1 = 0.0142$, w$R_2 = 0.0364$	$R_1 = 0.0207$, w$R_2 = 0.0493$	$R_1 = 0.0161$, w$R_2 = 0.0425$	$R_1 = 0.0173$, w$R_2 = 0.0433$
Final R indexes [all data]	$R_1 = 0.0164$, w$R_2 = 0.0372$	$R_1 = 0.0234$, w$R_2 = 0.0506$	$R_1 = 0.0179$, w$R_2 = 0.0431$	$R_1 = 0.0193$, w$R_2 = 0.0440$
Largest difference peak/hole (e Å⁻³)	0/53/−0.56	1.13/−0.99	0.43/−0.89	0.79/−1.16

2.3. Chemical Composition and Raman Spectroscopy

To identify barioferrite, we performed electron probe microanalysis and Raman spectroscopy. Barioferrite from the paralava of Har Parsa has a relatively constant composition and contains substantial substitutional elements, including Ti^{4+}, Al, and Ca (Table 2). The presence of Cu and Zn and also the interpreted incorporation of small amounts of Ca in the spinel module is a specific feature not only of barioferrite from Har Parsa paralava, but also barioferrite and magnesioferrite from other localities of pyrometamorphic rocks of the Hatrurim Complex [11,29,34]. The positive charge caused by substituted trivalent iron by tetravalent titanium is compensated by the presence of divalent elements. The empirical crystal chemical formula of barioferrite is as follows: $(Ba_{0.85}Ca_{0.12}Sr_{0.03})_{\Sigma 1}(Fe^{3+}_{10.72}Al_{0.46}Ti^{4+}_{0.41}Mg_{0.15}Cu^{2+}_{0.09}Ca_{0.08}Zn_{0.04}Mn^{2+}_{0.03}Si_{0.01})_{\Sigma 11.99}O_{19}$. This composition of barioferrite is close to the end-member $BaFe_{12}O_{19}$.

In the barioferrite structure, the spinel blocks are composed of $Fe^{(3)}O_4$ tetrahedra and $Fe^{(1,4)}O_6$ octahedra, whereas R-blocks are formed by $Fe^{(2)}O_5$ trigonal bipyramids and $Fe^{(5)}O_6$ octahedra with dodecacoordinated Ba in structural cages. The main bands observed in the Raman spectra of the studied barioferrite are associated with $Fe^{3+}O$ vibrations and shifted towards the higher Raman shift frequency in comparison with analogous bands of synthetic barioferrite (our data, Figure 2/synthetic barioferrite [35], cm^{-1}): 714/713 [A_{1g}, tetrahedron $Fe^{(3)}O_4$]; 689/684 [A_{1g}, bipyramid $Fe^{(2)}O_5$]; 624/614 [A_{1g}, octahedron $Fe^{(5)}O_6$]; 532/527, 520/512 [A_{1g}, octahedron $Fe^{(1)}O_6$ and $Fe^{(4)}O_6$]; 412/409 [A_{1g}, $Fe^{(4)}O_6$ octahedron dominated]; 335/335 + 317 [$E_{1g} + E_{2g}$]. Similar band shifts were observed in Raman spectrum of synthetic barioferrite $Ba_{1-x}Ca_xFe_{12}O_{19}$, with x = 0.1 [36]. In the studied barioferrite, 0.12 Ca pfu is present at the Ba site (Table 2), which most probably determines the band shifts. Also, we cannot exclude an influence of Al and Ti^{4+} impurities substituting for Fe^{3+} (Table 2) on bands position in the Raman spectrum. A similar effect was described before for Ti-bearing synthetic barioferrite [37]. Moreover, wide band centered about 1359 cm^{-1} is overtone (Figure 2). The bands at 164 and 178 cm^{-1} are assigned to the motion of the spinel block as a whole [35]. The band near 673 cm^{-1} is not usually noted in barioferrite spectra [35,37]; it was observed once in the spectrum of pure barioferrite [38], and it is possible, that this band is related to Fe^{3+} in the bipyramid.

Table 2. Chemical composition of barioferrite from Har Parsa.

Component (wt %)	Mean (*n* = 17)	Range	Standard Deviation.	Calculated on 19 O	
TiO_2	3.05	2.90–3.27	0.09	Ba	0.85
SiO_2	0.05	0.00–0.10	0.02	Ca	0.12
Fe_2O_3	80.47	78.89–81.53	0.72	Sr	0.03
Al_2O_3	2.22	2.11–2.38	0.06	*A* site	1
BaO	12.32	11.56–13.41	0.46	Fe^{3+}	10.72
MgO	0.58	0.51–0.68	0.05	Al	0.46
CaO	1.03	0.83–1.32	0.17	Ti	0.41
MnO	0.18	0.11–0.23	0.04	Mg	0.15
SrO	0.32	0.14–0.43	0.07	Cu^{2+}	0.09
CuO	0.65	0.48–0.84	0.11	Ca	0.08
ZnO	0.33	0.19–0.47	0.08	Zn	0.04
	101.2	100.09–101.69		Mn^{2+}	0.03
				Si	0.01
				M sites	11.99

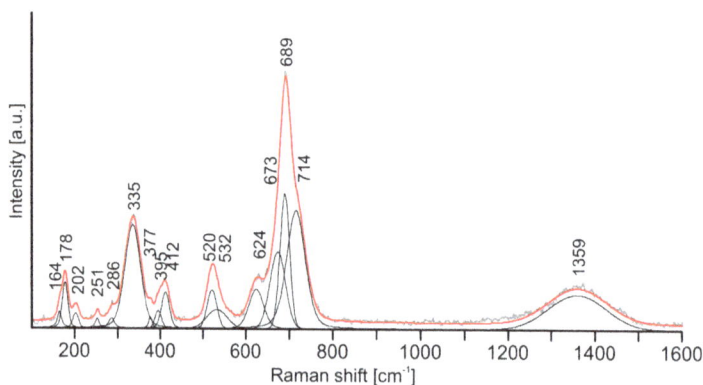

Figure 2. Raman spectrum of barioferrite from Har Parsa; grain is shown in Figure 1c.

3. X-ray Diffraction Studies

3.1. Powder Diffraction Data

All peaks in the powder spectra (Figure 3) of barioferrite correspond to the synthetic $BaFe_{12}O_{0.19}$ (PDF-2, card number 01-084-0757). The unit-cell parameters refined from these data are as follows: $P6_3/mmc$, $a = 5.8908(1)$ Å, $c = 23.1334(6)$ Å, $V = 694.91(3)$ Å^3, and $Z = 2$, which are in good agreement with the single-crystal data $a = 5.8901(6)$ Å, $c = 23.1235(6)$ Å, and $V = 694.75(4)$ Å^3 obtained at 300 K. There are no additional superstructure reflections, as observed in [39].

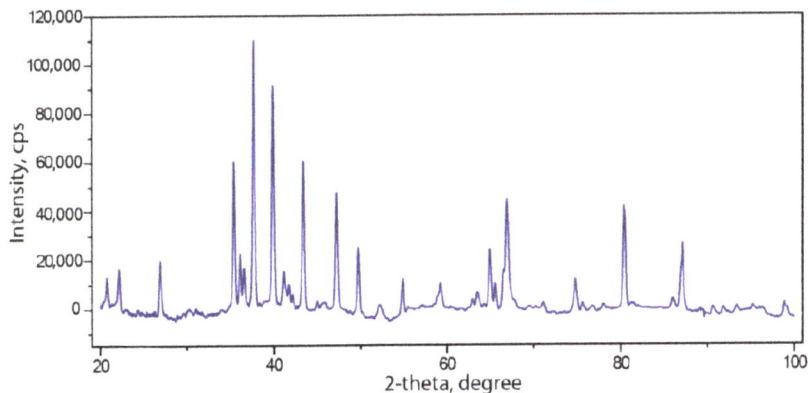

Figure 3. Powder diffraction spectra of barioferrite.

3.2. Single-Crystal X-ray Diffraction Data

A fragment of the barioferrite crystal shown in Figure 1c was used for the single-crystal structural investigations. The experimental details and crystallographic parameters are given in Table 1. The crystal structure of barioferrite was refined with the use of the *SHELX* program [40]. Empirical absorption corrections were applied in the CrysAlisPro [41] suite of programs using spherical harmonics, implemented in the SCALE3 ABSPACK scaling algorithm. The volumes of coordination polyhedra were calculated using the VESTA 3 program [42]. Atomic labels were assigned following the first crystal structure description [3]. The final atomic coordinates and isotropic displacement parameters are reported in Table 3, selected interatomic distances are in Table 4, and anisotropic displacement parameters are given in the Table 5. Table 6 provides information on the polyhedral volumes of cations.

Table 3. Atomic coordinates, displacement parameters (Å^2), and site occupancies for barioferrite at 100, 200, 300, and 400 K.

Temperature: 100 K

Site	population	x/a	y/b	z/c	U_{eq}
Ba1	$Ba_{0.89}Ca_{0.011}$	2/3	1/3	1/4	0.0038(1)
Fe1	$Fe_{0.69}Al_{0.31}$	0	0	0	0.0020(3)
Fe2	$Fe_{0.96}Ti_{0.04}$	0	0	1/4	0.0116(2)
Fe3	$Fe_{0.94}Mg_{0.06}$	1/3	2/3	0.02730(3)	0.0025(1)
Fe4	$Fe_{0.94}Ti_{0.06}$	0.33617(8)	0.16809(4)	0.10792(2)	0.0026(1)
Fe5	Fe	1/3	2/3	0.18977(3)	0.0031(1)
O1	O	0	0	0.1514(1)	0.0043(5)
O2	O	2/3	1/3	0.0555(1)	0.0037(5)
O3	O	0.5022(2)	0.0043(3)	0.1496(1)	0.0035(3)
O4	O	0.1543(2)	0.3086(3)	0.05198(5)	0.0050(3)
O5	O	0.1825(2)	0.3651(5)	1/4	0.0051(4)

Temperature: 200 K

Site	Population	x/a	y/b	z/c	U_{eq}
Ba1	$Ba_{0.89}Ca_{0.011}$	2/3	1/3	1/4	0.0063(1)
Fe1	$Fe_{0.69}Al_{0.31}$	0	0	0	0.0038(2)
Fe2	$Fe_{0.96}Ti_{0.04}$	0	0	1/4	0.0154(2)
Fe3	$Fe_{0.94}Mg_{0.06}$	1/3	2/3	0.02729(3)	0.0043(2)
Fe4	$Fe_{0.94}Ti_{0.06}$	0.33622(7)	0.16811(4)	0.10789(2)	0.0044(1)
Fe5	Fe	1/3	2/3	0.18974(3)	0.0047(2)
O1	O	0	0	0.15126(14)	0.0057(6)
O2	O	2/3	1/3	0.05564(15)	0.0054(6)
O3	O	0.5022(2)	0.0044(4)	0.14967(8)	0.0056(4)
O4	O	0.15399(19)	0.3080(4)	0.05197(9)	0.0062(4)
O5	O	0.1825(3)	0.3649(5)	1/4	0.0066(5)

Temperature: 300 K

Site	population	x/a	y/b	z/c	U_{eq}
Ba1	$Ba_{0.89}Ca_{0.011}$	2/3	1/3	1/4	0.0082(1)
Fe1	$Fe_{0.69}Al_{0.31}$	0	0	0	0.0049(2)
Fe2	$Fe_{0.96}Ti_{0.04}$	0	0	1/4	0.0184(3)
Fe3	$Fe_{0.94}Mg_{0.06}$	1/3	2/3	0.0272(3)	0.0052(2)
Fe4	$Fe_{0.94}Ti_{0.06}$	0.33623(7)	0.16811(4)	0.10789(2)	0.00563(14)
Fe5	Fe	1/3	2/3	0.18975(3)	0.0055(1)
O1	O	0	0	0.15107(13)	0.0063(6)
O2	O	2/3	1/3	0.0557(1)	0.0063(6)
O3	O	0.5022(2)	0.0044(3)	0.14958(7)	0.0064(4)
O4	O	0.1541(2)	0.3082(4)	0.05206(8)	0.0073(4)
O5	O	0.1821(3)	0.3643(5)	1/4	0.0080(5)

Temperature: 400 K

Site	population	x/a	y/b	z/c	U_{eq}
Ba1	$Ba_{0.89}Ca_{0.011}$	2/3	1/3	1/4	0.0104(1)
Fe1	$Fe_{0.69}Al_{0.31}$	0	0	0	0.0063(2)
Fe2	$Fe_{0.96}Ti_{0.04}$	0	0	1/4	0.0222(3)
Fe3	$Fe_{0.94}Mg_{0.06}$	1/3	2/3	0.02729(3)	0.0066(1)
Fe4	$Fe_{0.94}Ti_{0.06}$	0.33624(7)	0.16812(4)	0.10789(2)	0.00720(11)
Fe5	Fe	1/3	2/3	0.18972(3)	0.00744(14)
O1	O	0	0	0.1509(1)	0.0079(6)
O2	O	2/3	1/3	0.0557(1)	0.0076(6)
O3	O	0.5024(2)	0.0048(4)	0.14951(8)	0.0083(3)
O4	O	0.1542(2)	0.3084(4)	0.05187(8)	0.0088(3)
O5	O	0.1820(3)	0.3639(5)	1/4	0.0098(5)

The refined formula is: $(Ba_{0.89}Ca_{0.11})_{1.00}(Fe_{11.17}Ti_{0.40}Al_{0.31}Mg_{0.12})_{12.00}O_{19}$.

Table 4. Selected bond distances (Å) for the crystal structures of barioferrite at 100, 200, 300, and 400 K.

Bond		100 K Mean	200 K Mean	300 K Mean	400 K Mean
Ba1-O3	x6	2.8624(16)	2.8624(19)	2.8646(18)	2.8711(18)
Ba1-O5	x6	2.9505(2)	2.9495(2)	2.9493(16)	2.9537(2)
<Ba1-O>		2.9061	2.906	2.907	2.913
Fe1-O4	x6	1.980(2)	1.9782(19)	1.9796(19)	1.9822(18)
Fe2-O1	x2	2.278(3)	2.283(3)	2.288(3)	2.298(3)
Fe2-O5	x3	1.863(2)	1.862(3)	1.858(3)	1.859(3)
<Fe2-O>		2.028	2.030	2.041	2.034
Fe3-O2		1.914(3)	1.917(4)	1.916(2)	1.922(3)
Fe3-O4	x3	1.914(2)	1.917(2)	1.915(4)	1.9167(19)
<Fe3-O>		1.914	1.917	1.917	1.919
Fe4-O1		1.9886(5)	1.9867(17)	1.9847(16)	1.9866(17)
Fe4-O2		2.0757(17)	2.074(2)	2.0750(19)	2.0783(19)
Fe4-O3	x2	1.9399(10)	1.9394(12)	1.9385(12)	1.9406(12)
Fe4-O4	x2	2.0956(12)	2.0943(15)	2.0949(13)	2.1013(14)
<Fe4-O>		2.022	2.021	2.020	2.025
Fe5-O3	x3	1.9562(16)	1.9560(19)	1.9574(19)	1.9629(18)
Fe5-O5	x3	2.0749(18)	2.076(2)	2.078(12)	2.0847(19)
<Fe5-O>		2.016	2.016	2.018	2.024

Table 5. Anisotropic displacement parameters (Å²) for the structure of barioferrite.

Atom	100 K						200 K					
	U^{11}	U^{22}	U^{33}	U^{23}	U^{13}	U^{12}	U^{11}	U^{22}	U^{33}	U^{23}	U^{13}	U^{12}
Ba1	0.00371(12)	0.00371(12)	0.00389(17)	0.000	0.000	0.00186(6)	0.00695(15)	0.00695(15)	0.0050(2)	0.000	0.000	0.00348(8)
Fe1	0.0024(2)	0.0024(2)	0.00104(4)	0.000	0.000	0.00121(12)	0.0051(3)	0.0051(3)	0.0014(4)	0.000	0.000	0.00253(15)
Fe2	0.0012(2)	0.0012(2)	0.0323(6)	0.000	0.000	0.00062(11)	0.0032(3)	0.0032(3)	0.0400(7)	0.000	0.000	0.00159(14)
Fe3	0.00195(17)	0.00195(17)	0.0037(3)	0.000	0.000	0.00097(8)	0.0042(2)	0.0042(2)	0.0043(3)	0.000	0.000	0.00212(11)
Fe4	0.00195(16)	0.00214(13)	0.00356(16)	−0.00001(6)	−0.00003(11)	0.00098(8)	0.0042(2)	0.00473(17)	0.00411(2)	−0.00010(6)	−0.00021(12)	0.00211(10)
Fe5	0.00272(17)	0.00272(17)	0.0038(3)	0.000	0.000	0.00136(8)	0.0052(2)	0.0052(2)	0.0039(3)	0.000	0.000	0.00258(10)
O1	0.0032(7)	0.0032(7)	0.0064(12)	0.000	0.000	0.0016(4)	0.0054(9)	0.0054(9)	0.0064(14)	0.000	0.000	0.0027(4)
O2	0.0030(7)	0.0030(7)	0.0051(12)	0.000	0.000	0.0015(4)	0.0061(9)	0.0061(9)	0.0042(14)	0.000	0.000	0.0030(5)
O3	0.0022(5)	0.0034(7)	0.0054(7)	0.0009(6)	0.0004(3)	0.0017(3)	0.0051(6)	0.0059(8)	0.0059(9)	0.0011(6)	0.0005(3)	0.0030(4)
O4	0.0049(5)	0.0053(7)	0.0050(7)	0.0010(6)	0.0005(3)	0.0026(4)	0.0068(6)	0.0073(9)	0.0047(9)	0.0008(7)	0.0004(3)	0.0036(5)
O5	0.0052(8)	0.0033(10)	0.0062(10)	0.000	0.000	0.0016(5)	0.0092(10)	0.0055(13)	0.0039(12)	0.000	0.000	0.0028(6)

Atom	300 K						400 K					
	U^{11}	U^{22}	U^{33}	U^{23}	U^{13}	U^{12}	U^{11}	U^{22}	U^{33}	U^{23}	U^{13}	U^{12}
Ba1	0.00855(16)	0.00855(16)	0.0075(2)	0.000	0.000	0.00428(8)	0.01059(14)	0.01059(14)	0.01000(19)	0.000	0.000	0.00529(7)
Fe1	0.0060(3)	0.0060(3)	0.0026(4)	0.000	0.000	0.00300(15)	0.0074(3)	0.0074(3)	0.0041(4)	0.000	0.000	0.00371(14)
Fe2	0.0036(3)	0.0036(3)	0.0480(8)	0.000	0.000	0.00179(14)	0.0044(3)	0.0044(3)	0.0579(8)	0.000	0.000	0.00219(13)
Fe3	0.0050(2)	0.0050(2)	0.0056(3)	0.000	0.000	0.00251(11)	0.00611(19)	0.00611(19)	0.0076(3)	0.000	0.000	0.00305(9)
Fe4	0.0048(2)	0.00538(17)	0.00608(19)	−0.00019(7)	−0.00037(13)	0.00239(10)	0.00589(18)	0.00685(14)	0.00853(18)	−0.00031(6)	−0.00062(12)	0.00295(9)
Fe5	0.0062(2)	0.0062(2)	0.0056(3)	0.000	0.000	0.00312(11)	0.00755(18)	0.00755(18)	0.0072(3)	0.000	0.000	0.00377(9)
O1	0.0054(9)	0.0054(9)	0.0083(13)	0.000	0.000	0.0027(5)	0.0068(8)	0.0068(8)	0.0100(13)	0.000	0.000	0.0034(4)
O2	0.0066(9)	0.0066(9)	0.0057(14)	0.000	0.000	0.0033(5)	0.0075(8)	0.0075(8)	0.0079(13)	0.000	0.000	0.0038(4)
O3	0.0060(6)	0.0060(9)	0.0074(8)	0.0020(7)	0.0010(3)	0.0030(4)	0.0068(5)	0.0078(8)	0.0107(8)	0.0025(6)	0.0012(3)	0.0039(4)
O4	0.0077(6)	0.0078(9)	0.0065(8)	0.0005(7)	0.0003(3)	0.0039(5)	0.0091(6)	0.0093(8)	0.0081(8)	0.0006(6)	0.0003(3)	0.0047(4)
O5	0.0104(10)	0.0054(13)	0.0066(11)	0.000	0.000	0.0027(6)	0.0125(9)	0.0062(12)	0.0085(11)	0.000	0.000	0.0031(6)

Table 6. Data of polyhedral volumes in the crystal structure of barioferrite at 100, 200, 300, and 400 K.

Polyhedron	Vol. (Å^3) 100 K	Vol. (Å^3) 200 K	Vol. (Å^3) 300 K	Vol. (Å^3) 400 K
Ba1	57.060	57.045	57.124	57.463
Fe1	10.318	10.273	10.304	10.340
Fe2	6.847	6.849	6.841	6.877
Fe3	3.591	3.606	3.596	3.615
Fe4	10.908	10.886	10.882	10.941
Fe5	10.680	10.687	10.720	10.815

The site labels and atomic positions were taken from [3], and the crystal structure was refined anisotropically to R_1 = 0.014, 0.021, 0.016, and 0.017 for 501, 508, 461, and 519 unique observed reflections with $|F_o| \geq 4\sigma_F$ at 100, 200, 300, and 400 K, respectively (Supplementary materials). The thermal ellipsoid of the Fe2 site shows elongation along the [001] direction (Table 5) at 100 K, and additional experiments at 200, 300, and 400 K were performed in order to check the thermal behavior of this site. In contrast to hibonite [43], the model where Fe2 split into two $4e$ sites does not improve R_1 values. In the split model, during refinement, both $4e$ sites shift into the central $2b$ position. In this model, the anisotropic displacement parameters will be more reasonable in character, but due to the short Fe2–Fe2 distance of 0.277(2) Å, their thermal ellipsoids overlapped. Because modern refinements of magnetoplumbite-type structures use a split model, we choose the model with the one central Fe2 atom for convenience of calculation of the polyhedral volumes.

The crystal structure of barioferrite belongs to the magnetoplumbite-type structure (Figure 4). In the terms of close packing, it can be described as the stacking of cubic and hexagonal pack layers in the sequence … **BAB'ABCAC'AC** … along the [001] direction [5]. The layers **A**, **B**, and **C** consist exclusively of oxide ions, while the layers B' and C' are represented by the $(\text{BaO}_3)^{4-}$ layers [5,44]. In the view of modular crystallography, the crystal structure of barioferrite consist of two fundamental blocks (Figure 4): spinel-like *S*-modules with a cubic stacking sequence of close packed layers and *R*-modules with a hexagonal stacking sequence [2]. *S*-modules consist of a spinel structure fragment, which is based on the two identical sheets of edge-shared Fe4 octahedra and one octahedral (Fe1)/tetrahedral (Fe3) layer embedded between two octahedral layers. The *S*-module contains three independent iron Fe1 ($2a$), Fe3 ($4f$), and Fe4 ($12k$) sites. The Al atoms are incorporated onto the Fe1 site, for which the refined occupancy at 100 K is $\text{Fe}_{0.69}\text{Al}_{0.31}$. The polyhedral volume of the Fe1 octahedron is significantly smaller than the other octahedral sites (Figure 5), and the Fe1–O bond length is 1.980 Å (100 K), also consistent with the admixture of aluminum. The tetrahedral Fe3 and octahedral Fe4 sites have admixtures of Mg and Ti, respectively. Their mean Fe–O bond lengths are 1.915 and 2.022 Å (at 100 K) and are consistent with the refined occupancies of $\text{Fe}_{0.94}\text{Mg}_{0.06}$ and $\text{Fe}_{0.94}\text{Ti}_{0.06}$. These bond lengths are close to those theoretically expected, 1.87 and 2.05 Å [45], respectively, and close to those observed in synthetic barioferrite: 1.888 and 2.007 Å, respectively [2].

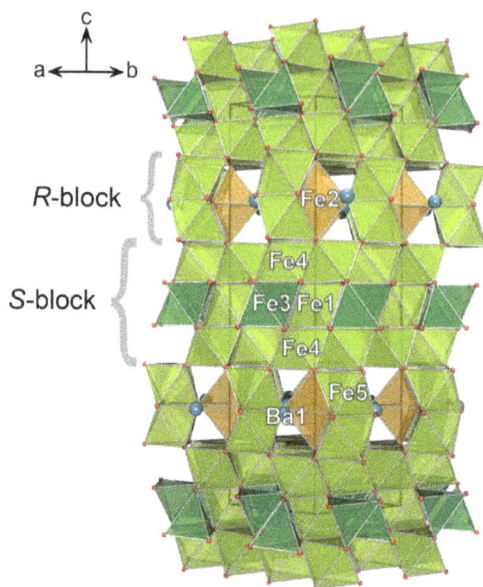

Figure 4. Crystal structure of barioferrite at 100 K. The arrangement of iron-centered polyhedra: octahedral—light green, tetrahedra—dark green, and trigonal bipyramid—orange, and the blue spheres represent Ba.

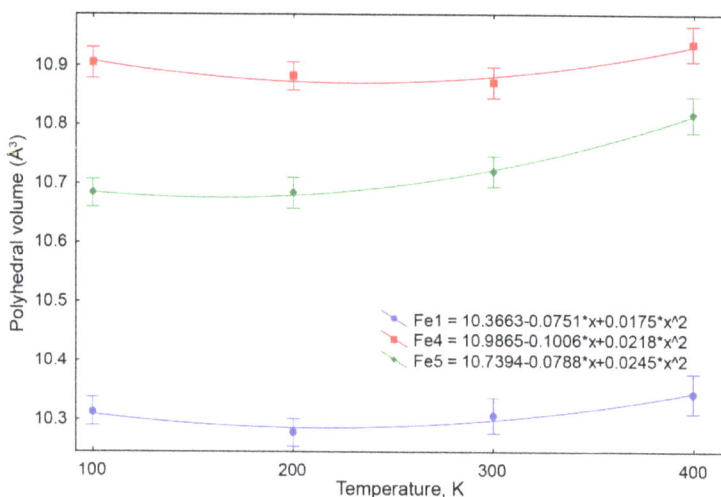

Fe1 = $10.3663-0.0751*x+0.0175*x^2$
Fe4 = $10.9865-0.1006*x+0.0218*x^2$
Fe5 = $10.7394-0.0788*x+0.0245*x^2$

Figure 5. Polyhedral volumes of Fe1, Fe4, and Fe5 octahedra in the crystal structure of barioferrite at 100, 200, 300, and 400 K, respectively.

Each *R*-module contains layers formed by face-shared (Fe_2O_9) octahedra (Fe5 site) that are connected by their apical vertexes with (FeO_5) trigonal bipyramids (Fe2 site). These layers include cavities 5.89×5.89 Å filled by Ba (Ba1) cations. The twelve-coordinated Ba1 site (2*d*) has an admixture of Ca, its refined occupancy is $Ba_{0.904}Ca_{0.096}$ at 100 K. The refined occupancy of the Fe2 site is $Fe_{0.96}Ti_{0.04}$. The Fe5 (4*f*) and Fe2 (2*b*) sites have mean Fe–O bond lengths of 2.016 and 2.028 Å, which are consistent with magnetoplumbite-type compounds [5,18,46]. The admixtures of Sr, Cu, Ca, Zn, Mn, and Si with an amount lower than 0.1 *apfu* were not determined during refinement; their incorporation is discussed below.

The refined formula of barioferrite can be written as $^{Ba1}(Ba_{0.89}Ca_{0.11})^{Fe1}(Fe_{0.69}Al_{0.31})^{Fe2}(Fe_{0.96}Ti_{0.04})$ $^{Fe3}(Fe_{1.88}Mg_{0.12})^{Fe4}(Fe_{5.64}Ti_{0.36})^{Fe5}(Fe_{2.00})O_{19}$ or simplified as $(Ba_{0.89}Ca_{0.11})_{1.00}(Fe_{11.17}Ti_{0.40}Al_{0.31}$ $Mg_{0.12})_{12.00}O_{19}$.

4. Discussion

The natural samples of hexaferrites are of interest to chemists and mineralogists due to the wide range of substitutions, which affect the physical-chemical properties of the titled compounds [43,47]. The present study confirms the incorporation of Mg into the tetrahedral Fe3 site as suggested in [43,48] The mean Fe3–O bond length of 1.914 Å was slightly higher than that in the synthetic barioferrite of 1.888 Å [2]. The incorporation of Ti into magnetoplumbite-like compounds are widely discussed in the literature; according to this data Ti orders into M2 and M4 [49,50] or Fe4 > Fe3 > Fe2 [5]. In our case, Ti is allocated in the Fe2 and Fe4 sites. The polyhedral volume of Fe4 octahedra is significantly larger than other octahedral sites (Figure 5) because of the presence of admixtures of divalent cations, such as Cu, Ca, Zn, and Mn.

With the increase of temperature over the range of 100–300 K, most polyhedra in the structure of natural barioferrite do not significantly change in volume (Tables 5 and 6). That is in agreement with the observations of the magnetoelectric effect in some magnetoplumbite-like compounds at temperatures up to ambient [51]. However, the local negative thermal expansion (NTE) of Fe1 and Fe4 octahedra in the S-block of the structure at the temperatures 100–200 and 100–300 K, respectively—probably, due to local structure deformations—requires further investigation.

The most difficult question is the character of the disorder in the Fe2 site (Figure 6). According to [2,5,52], there are three different models, which can be considered for the structural and dynamic characteristics of the bipyramidal Fe^{3+} ions: (1) a non-disordered configuration, (2) a static disorder between two adjacent pseudotetrahedral sites, and (3) a dynamical disorder between these sites (Figure 7).

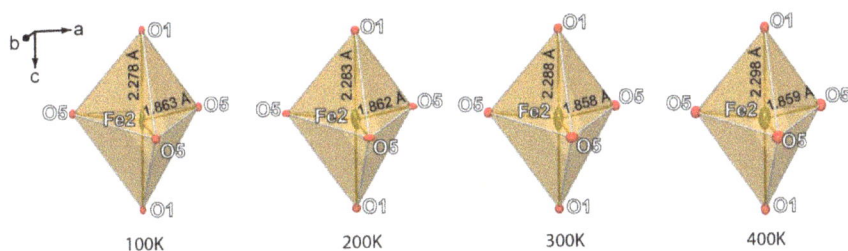

Figure 6. Fe2 trigonal bipyramid and Fe–O bond lengths at 100, 200, 300, and 400 K, respectively. Displacement ellipsoids are drawn at the 50% probability level.

It seems that both refinement models support a dynamic disorder character (Figure 8). In the single-atom model, the elongation of the displacement ellipsoid of the iron atom in the trigonal bipyramid along the [001] direction increased with temperature (Table 5). The decrease of the Fe2–O5 distances from 1.862(3) at 100 K to 1.857(3) Å at 400 K was correlated to the increase of the Fe2–O1 distances from 2.279(3) at 100 K to 2.300(3) Å at 400 K. According to our data, the iron atoms were randomly displaced from the center of the trigonal bipyramid. In this case, both structural models were appropriate (Figure 7). In the split-atom model, the Fe2–Fe2 distance increased from 0.277(2) at 100K to 0.358(7) Å at 400 K. In the static model of disorder, the site on the one side of the triangle (O5–O5–O5) plane is occupied by tetrahedrally coordinated Fe, whereas the site on the other side of the plane is vacant. In our case, the changes of the bond lengths of the FeO$_5$ polyhedra are significantly smaller for the static model of splitting the Fe2 site, and all attempts to refine the model with a split Fe2 site (situated on the center of the tetrahedra) leads to a significant increase of the *R*-values. We consider that the disorder at the trigonal bipyramid in the barioferrite crystal structure has an exclusively dynamic disorder character.

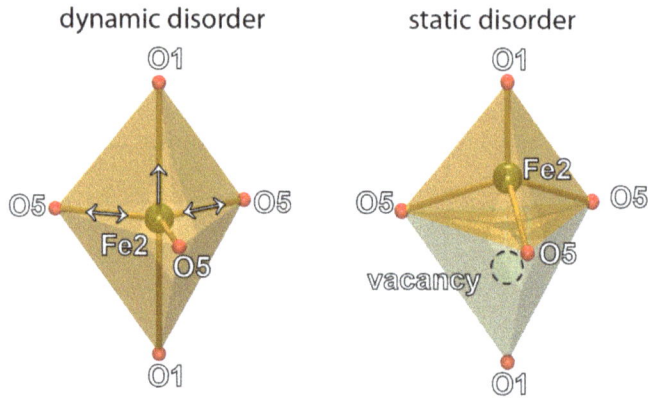

Figure 7. Theoretical models of arrangement of the Fe2 site in the barioferrite crystal structure. Arrows shows direction of the bond elongation.

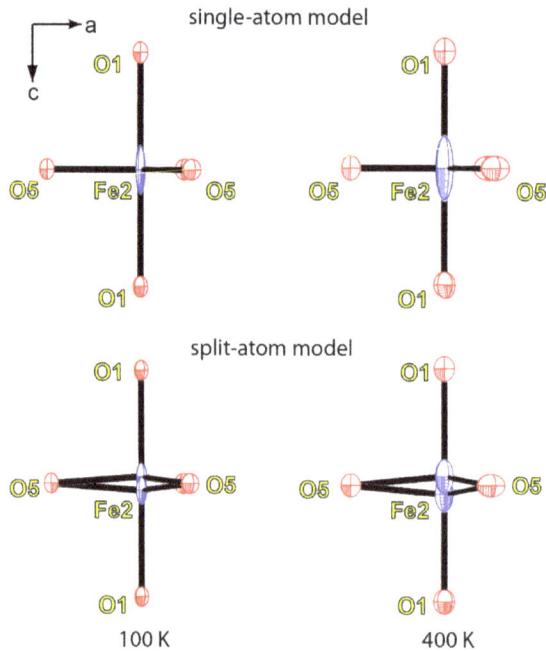

Figure 8. Thermal ellipsoids of Fe2, O1, and O5 atoms at 100 and 400 K. Thermal ellipsoids are drawn at the 90% probability level.

Minerals **2018**, *8*, 340

In conclusion, it should be noted that although barioferrite was first discovered in an altered baryte nodule and considered to be a very rare mineral [1], it turned out to be a common accessory mineral of high-calcium paralavas and gehlenite hornfelses [9,11]. The natural occurrence of the Ca-analogue of barioferrite (or Fe-analogue of hibonite) $CaFe_{12}O_{19}$ has been recently reported [11], and in addition to barioferrite with a $S/R = 3:1$ module ratio, a barium ferrite with the $S/R = 5:1$ ratio $Ba(Fe^{2+},Mg)_2Fe_{16}O_{27}$ also exists [53].

Supplementary Materials: The following are available online at http://www.mdpi.com/2075-163X/8/8/340/s1, barioferrite cif files for 100 K, 200 K, 300 K, and 400 K.

Author Contributions: A.K. and T.L.P. wrote the paper. A.K. performed the Raman study and also selected the grains for the structural investigations. A.K., I.O.G., and E.V.G. collected samples for investigation during the field work, performed petrological investigations, and measured the composition of the barioferrite from Har Parsa. T.L.P. performed the SC-XRD investigation, refined the barioferrite structure, and interpreted the structural data.

Funding: Grant Preludium of the National Science Centre of Poland No. 2016/21/N/ST10/00463.

Acknowledgments: The investigations were partially supported by the National Science Centre of Poland, Grant Preludium No. 2016/21/N/ST10/00463. X-ray diffraction studies were performed at the X-ray Diffraction Centre of Saint Petersburg State University. The authors thank the anonymous reviewers for their careful reviews, which helped to improve a previous version of the manuscript.

Conflicts of Interest: The authors declare no conflict of interest.

References

1. Murashko, M.N.; Chukanov, N.V.; Mukhanova, A.A.; Vapnik, E.; Britvin, S.N.; Polekhovsky, Y.S.; Ivakin, Y.D. Barioferrite $BaFe_{12}O_{19}$: A new mineral species of the magnetoplumbite group from the Haturim Formation in Israel. *Geol. Ore Depos.* **2011**, *53*, 558–563. [CrossRef]

2. Obradors, X.; Collomb, A.; Pernet, M.; Samaras, D.; Joubert, J.C. X-ray analysis of the structural and dynamic properties of $BaFe_{12}O_{19}$ hexagonal ferrite at room temperature. *J. Solid State Chem.* **1985**, *56*, 171–181. [CrossRef]

3. Bertaut, E.F.; Deschamps, A.; Pauthenet, R. Etude de la substitution de Fe par Al, Ga et Cr dans l'hexaferrite de baryum, BaO (Fe_2O_3)$_6$. *J. Phys. Radium* **1959**, 404–408. [CrossRef]

4. Aleshko-Ozhevskii, O.P.; Faek, M.K.; Yamzin, I.I. A neutron diffraction study of the structure of magnetoplumbite. *Sov. Phys. Crystallogr.* **1969**, *14*, 367–369.

5. Vinnik, D.A.; Zherebtsov, D.A.; Mashkovtseva, L.S.; Nemrava, S.; Perov, N.S.; Semisalova, A.S.; Krivtsov, I.V.; Isaenko, L.I.; Mikhailov, G.G.; Niewa, R. Ti-Substituted $BaFe_{12}O_{19}$ single crystal growth and characterization. *Cryst. Growth Des.* **2014**, *14*, 5834–5839. [CrossRef]

6. Lengauer, C.L.; Tillmanns, E.; Hentschel, G. Batiferrite, $Ba[Ti_2Fe_{10}]O_{19}$, a new ferrimagnetic magnetoplumbite-type mineral from the Quaternary volcanic rocks of the western Eifel area, Germany. *Miner. Petrol.* **2001**, *71*, 1–19. [CrossRef]

7. Pullar, R.C. Hexagonal ferrites: A review of the synthesis, properties and applications of hexaferrite ceramics. *Prog. Mater. Sci.* **2012**, *57*, 1191–1334. [CrossRef]

8. Žáček, V.; Skála, R.; Chlupáčová, M.; Dvořák, Z. Ca–Fe^{3+}-rich, Si-undersaturated buchite from Zelenky, North-Bohemian Brown Coal Basin, Czech Republic. *Eur. J. Mineral.* **2005**, *17*, 623–634. [CrossRef]

9. Galuskin, E.V.; Gfeller, F.; Galuskina, I.O.; Pakhomova, A.; Armbruster, T.; Vapnik, Y.; Włodyka, R.; Dzierżanowski, P.; Murashko, M. New minerals with a modular structure derived from hatrurite from the pyrometamorphic Hatrurim Complex. Part II. Zadovite, $BaCa_6[(SiO_4)(PO_4)](PO_4)_2F$ and aradite, $BaCa_6[(SiO_4)(VO_4)](VO_4)_2F$, from paralavas of the Hatrurim Basin, Negev Desert, Israel. *Mineral. Mag.* **2015**, *79*, 1073–1087. [CrossRef]

10. Galuskina, I.O.; Galuskin, E.V.; Vapnik, Y.; Prusik, K.; Stasiak, M.; Dzierżanowski, P.; Murashko, M.; Krivovichev, S. Gurimite, $Ba_3(VO_4)_2$ and hexacelsian, $BaAl_2Si_2O_8$—Two new minerals from schorlomite-rich paralava of the Hatrurim Complex, Negev Desert, Israel. *Mineral. Mag.* **2017**, *81*, 1009–1019. [CrossRef]

11. Galuskina, I.O.; Galuskin, E.V.; Pakhomova, A.S.; Widmer, R.; Armbruster, T.; Krüger, B.; Grew, E.S.; Vapnik, Y.; Dzierażanowski, P.; Murashko, M. Khesinite, $Ca_4Mg_2Fe^{3+}_{10}O_4[(Fe^{3+}_{10}Si_2)O_{36}]$, a new rhönite-group (sapphirine supergroup) mineral from the Negev Desert, Israel—Natural analogue of the SFCA phase. *Eur. J. Mineral.* **2017**, 101–116. [CrossRef]

12. Picard, L. *Geological Research in the Judean Desert*; Goldberg's Press: Jerusalem, Israel, 1931; p. 108.
13. Kolodny, Y.; Gross, S. Thermal metamorphism by combustion of organic matter: Isotopic and petrological evidence. *J. Geol.* **1974**, *82*, 489–506. [CrossRef]
14. Burg, A.; Starinsky, A.; Bartov, Y.; Kolodny, Y. Geology of the Hatrurim Formation ("Mottled Zone") in the Hatruruim basin. *Isr. J. Earth Sci.* **1991**, *40*, 107–124.
15. Burg, A.; Kolodny, Y.; Lyakhovsky, V. Hatrurim-2000: The "Mottled Zone" revisited, forty years later. *Isr. J. Earth Sci.* **2000**, *48*, 209–223.
16. Townes, W.D.; Fang, J.H.; Perrotta, A.J. The crystal structure and refinement of ferrimagnetic barium ferrite, $BaFe_{12}O_{19}$. *Z. Kristallogr. Cryst. Mater.* **1967**, *125*, 437–449. [CrossRef]
17. Aminoff, G. Über ein neues oxydisches mineral aus långban. (Magnetoplumbit.). *Geologiska Föreningen i Stockholm Förhandlingar* **1925**, *47*, 283–289. [CrossRef]
18. Moore, P.B.; Gupta, P.K.S.; Le Page, Y. Magnetoplumbite, $Pb^{2+}Fe^{3+}_{12}O_{19}$: Refinement and lone-pair splitting. *Am. Mineral.* **1989**, *74*, 1186–1194.
19. Shen, S.-P.; Chai, Y.-S.; Cong, J.-Z.; Sun, P.-J.; Lu, J.; Yan, L.-Q.; Wang, S.-G.; Sun, Y. Magnetic-ion-induced displacive electric polarization in FeO_5 bipyramidal units of $(Ba,Sr)Fe_{12}O_{19}$ hexaferrites. *Phys. Rev. B* **2014**, *90*, 180404. [CrossRef]
20. Bentor, Y.K. *Lexique Stratigraphique International: Asie fascicule 10 c 2 Israel*; Centre National de la Recherche Scientifique: Paris, France, 1960; Volume 3.
21. Gross, S. The mineralogy of the Hatrurim Formation, Israel. *Geol. Surv. Isr. Bull.* **1977**, *70*, 1–80.
22. Vapnik, Y.; Sharygin, V.V.; Sokol, E.V.; Shagam, R. Paralavas in a combustion metamorphic complex: Hatrurim Basin, Israel. In *Geology of Coal Fires: Case Studies from Around the World*; Geological Society of America: Boulder, CO, USA, 2007; Volume 18, pp. 133–153, ISBN 978-0-8137-4118-5.
23. Novikov, I.; Vapnik, Y.; Safonova, I. Mud volcano origin of the Mottled Zone, South Levant. *Geosci. Front.* **2013**, *4*, 597–619. [CrossRef]
24. Minster, T.; Yoffe, O.; Nathan, Y.; Flexer, A. Geochemistry, mineralogy, and paleoenvironments of deposition of the Oil Shale Member in the Negev. *Isr. J. Earth Sci.* **1997**, *46*, 41–59.
25. Sokol, E.; Novikov, I.; Zateeva, S.; Vapnik, Y.; Shagam, R.; Kozmenko, O. Combustion metamorphism in the Nabi Musa dome: New implications for a mud volcanic origin of the Mottled Zone, Dead Sea area. *Basin Res.* **2010**, *22*, 414–438. [CrossRef]
26. Sokol, E.V.; Kozmenko, O.A.; Kokh, S.N.; Vapnik, Y. Gas reservoirs in the Dead Sea area: Evidence from chemistry of combustion metamorphic rocks in Nabi Musa fossil mud volcano. *Russ. Geol. Geophys.* **2012**, *3*, 745–762. [CrossRef]
27. Galuskin, E.V.; Krüger, B.; Galuskina, I.O.; Krüger, H.; Vapnik, Y.; Wojdyla, J.A.; Murashko, M. New mineral with modular structure derived from hatruritefrom the pyrometamorphic rocks of the hatrurim complex: Ariegilatite, $BaCa_{12}(SiO_4)_4(PO_4)_2F_2O$, from Negev desert, Israel. *Minerals* **2018**, *8*, 109. [CrossRef]
28. Sokol, E.V.; Seryotkin, Y.V.; Kokh, S.N.; Vapnik, Y.; Nigmatulina, E.N.; Goryainov, S.V.; Belogub, E.V.; Sharygin, V.V. Flamite, $(Ca,Na,K)_2(Si,P)O_4$, a new mineral from ultrahightemperature combustion metamorphic rocks, Hatrurim Basin, Negev Desert, Israel. *Mineral. Mag.* **2015**, *79*, 583–596. [CrossRef]
29. Sharygin, V.V.; Lazic, B.; Armbruster, T.M.; Murashko, M.N.; Wirth, R.; Galuskina, I.O.; Galuskin, E.V.; Vapnik, Y.; Britvin, S.N.; Logvinova, A.M. Shulamitite $Ca_3TiFe^{3+}AlO_8$—A new perovskite-related mineral from Hatrurim Basin, Israel. *Eur. J. Mineral.* **2013**, 97–111. [CrossRef]
30. Galuskin, E.V.; Galuskina, I.O.; Gfeller, F.; Krüger, B.; Kusz, J.; Vapnik, Y.; Dulski, M.; Dzierżanowski, P. Silicocarnotite, $Ca_5[(SiO_4)(PO_4)](PO_4)$, a new "old" mineral from the Negev Desert, Israel, and the ternesite–silicocarnotite solid solution: Indicators of high-temperature alteration of pyrometamorphic rocks of the Hatrurim Complex, Southern Levant. *Eur. J. Mineral.* **2016**, *28*, 105–123. [CrossRef]
31. Galuskin, E.V.; Gfeller, F.; Galuskina, I.O.; Armbruster, T.; Krzątała, A.; Vapnik, Y.; Kusz, J.; Dulski, M.; Gardocki, M.; Gurbanov, A.G.; et al. New minerals with a modular structure derived from hatrurite from the pyrometamorphic rocks. Part III. Gazeevite, $BaCa_6(SiO_4)_2(SO_4)_2O$, from Israel and the Palestine Autonomy, South Levant, and from South Ossetia, Greater Caucasus. *Mineral. Mag.* **2017**, *81*, 499–513. [CrossRef]
32. Pouchou, J.L.; Pichoir, F. Quantitative Analysis of Homogeneous or Stratified Microvolumes Applying the Model "PAP". In *Electron Probe Quantitation*; Springer: Boston, MA, USA, 1991; pp. 31–75, ISBN 978-1-4899-2619-7.

33. Britvin, S.N.; Dolivo-Dobrovolsky, D.V.; Krzhizhanovskaya, M.G. Software for processing the X-ray powder diffraction data obtained from the curved image plate detector of Rigaku RAXIS Rapid II diffractometer. *Proc. Russ. Mineral. Soc.* **2017**, *146*, 104–107.

34. Galuskina, I.O.; Vapnik, Y.; Lazic, B.; Armbruster, T.; Murashko, M.; Galuskin, E.V. Harmunite CaFe$_2$O$_4$: A new mineral from the Jabel Harmun, West Bank, Palestinian Autonomy, Israel. *Am. Mineral.* **2014**, *99*, 965–975. [CrossRef]

35. Kreisel, J.; Lucazeau, G.; Vincent, H. Raman Spectra and Vibrational Analysis of BaFe$_{12}$O$_{19}$ Hexagonal Ferrite. *J. Solid State Chem.* **1998**, *137*, 127–137. [CrossRef]

36. Kumar, S.; Supriya, S.; Pandey, R.; Pradhan, L.K.; Singh, R.K.; Kar, M. Effect of lattice strain on structural and magnetic properties of Ca substituted barium hexaferrite. *J. Magn. Magn. Mater.* **2018**, *458*, 30–38. [CrossRef]

37. Kreisel, J.; Lucazeau, G.; Vincent, H. Raman study of substituted barium ferrite single crystals, BaFe$_{12-2x}$Me$_x$Co$_x$O$_{19}$ (Me = Ir, Ti). *J. Raman Spectrosc.* **1999**, *30*, 115–120. [CrossRef]

38. Silva Júnior, F.M.; Paschoal, C.W.A. Spin-phonon coupling in BaFe$_{12}$O$_{19}$ M-type hexaferrite. *J. Appl. Phys.* **2015**, *116*, 244110. [CrossRef]

39. Ganapathi, L.; Gopalakrishnan, J.; Rao, C.N.R. Barium hexaferrite (M-phase) exhibiting superstructure. *Mater. Res. Bull.* **1984**, *19*. [CrossRef]

40. Sheldrick, G.M. A short history of SHELX. *Acta Crystallogr. Sect. Found. Crystallogr.* **2008**, *64*, 112–122. [CrossRef] [PubMed]

41. Agilent Technologies. *CrysAlisCCD and CrysAlis Red*; Oxford Diffraction Ltd.: Oxford, UK, 2014.

42. Momma, K.; Izumi, F. VESTA 3 for three-dimensional visualization of crystal, volumetric and morphology data. *J. Appl. Crystallogr.* **2011**, *44*, 1272–1276. [CrossRef]

43. Nagashima, M.; Armbruster, T.; Hainschwang, T. A temperature-dependent structure study of gem-quality hibonite from Myanmar. *Mineral. Mag.* **2010**, *74*, 871–885. [CrossRef]

44. Siegrist, T.; Vanderah, T.A. Combining magnets and dielectrics: Crystal chemistry in the BaO-Fe$_2$O$_3$-TiO$_2$ System. *Eur. J. Inorg. Chem.* **2003**, 1483–1501. [CrossRef]

45. Shannon, R.D. Revised effective ionic radii and systematic studies of interatomic distances in halides and chalcogenides. *Acta Crystallogr. Sect. Cryst. Phys. Diffr. Theor. Gen. Crystallogr.* **1976**, *32*, 751–767. [CrossRef]

46. An, S.Y.; Shim, I.-B.; Kim, C.S. Mössbauer and magnetic properties of Co-Ti substituted barium hexaferrite nanoparticles. *J. Appl. Phys.* **2002**, *91*, 8465–8467. [CrossRef]

47. Hofmeister, A.M.; Wopenka, B.; Locock, A.J. Spectroscopy and structure of hibonite, grossite, and CaAl$_2$O$_4$: Implications for astronomical environments. *Geochim. Cosmochim. Acta* **2004**, *68*, 4485–4503. [CrossRef]

48. Bermanec, V.; Holtstam, D.; Sturman, D.; Criddle, A.J.; Back, M.E.; Scavnicar, S. Nezilovite, a new member of the magnetoplumbite group, and the crystal chemistry of magnetoplumbite and hibonite. *Can. Mineral.* **1996**, *34*, 1287–1297.

49. Ardit, M.; Borcănescu, S.; Cruciani, G.; Dondi, M.; Lazău, I.; Păcurariu, C.; Zanelli, C. Ni–Ti codoped hibonite ceramic pigments by combustion synthesis: Crystal structure and optical properties. *J. Am. Ceram. Soc.* **2016**, *99*, 1749–1760. [CrossRef]

50. Giannini, M.; Ballaran, T.B.; Langenhorst, F. Crystal chemistry of synthetic Ti–Mg-bearing hibonites: A single-crystal X-ray study. *Am. Mineral.* **2014**, *99*, 2060–2067. [CrossRef]

51. Kitagawa, Y.; Hiraoka, Y.; Honda, T.; Ishikura, T.; Nakamura, H.; Kimura, T. Low-field magnetoelectric effect at room temperature. *Nat. Mater.* **2010**, *9*, 797–802. [CrossRef] [PubMed]

52. Graetsch, H.; Gebert, W. Positional and thermal disorder in the trigonal bipyramid of magnetoplumbite structure type SrGa$_{12}$O$_{19}$. *Z. Kristallogr.* **1994**, *209*, 338–342. [CrossRef]

53. Galuskin, E.V.; Galuskina, I.O.; Widmer, R.; Armbruster, T. First natural hexaferrite with mixed β$'''$-ferrite (β-alumina) and magnetoplumbite structure from Jabel Harmun, Palestinian Autonomy. *Eur J. Mineral.* **2018**. [CrossRef]

Article

Parafiniukite, $Ca_2Mn_3(PO_4)_3Cl$, a New Member of the Apatite Supergroup from the Szklary Pegmatite, Lower Silesia, Poland: Description and Crystal Structure

Adam Pieczka [1,*], **Cristian Biagioni** [2], **Bożena Gołębiowska** [1], **Piotr Jeleń** [3], **Marco Pasero** [2] and **Maciej Sitarz** [3]

[1]   Department of Mineralogy, Petrography and Geochemistry, AGH University of Science and Technology, 30-059 Kraków, Poland; goleb@agh.edu.pl
[2]   Dipartimento di Scienze Della Terra, Università di Pisa, Via Santa Maria 53, I-56126 Pisa, Italy; cristian.biagioni@unipi.it (C.B.); marco.pasero@unipi.it (M.P.)
[3]   Faculty of Materials Science and Ceramics, AGH University of Science and Technology, 30-059 Kraków, Poland; pjelen@agh.edu.pl (P.J.); msitarz@agh.edu.pl (M.S.)
*   Correspondence: pieczka@agh.edu.pl

Received: 29 September 2018; Accepted: 22 October 2018; Published: 26 October 2018

Abstract: Parafiniukite, ideally $Ca_2Mn_3(PO_4)_3Cl$, is a new apatite-supergroup mineral from the Szklary pegmatite, Lower Silesia, Poland. It occurs as anhedral grains, up to 250 μm in size, dark olive green in colour, embedded in a mixture of Mn-oxides and smectites around beusite. It has a vitreous luster, and it is brittle with irregular, uneven fracture. The calculated density is 3.614 g·cm^{-3}. Parafiniukite is hexagonal, space group $P6_3/m$, with unit-cell parameters $a = 9.4900(6)$, $c = 6.4777(5)$ Å, $V = 505.22(5)$ Å^3, $Z = 2$. The eight strongest reflections in the calculated X-ray powder diffraction pattern of parafiniukite are [d in Å (I) hkl]: 3.239 (39) 002; 2.801 (55) 211; 2.801 (76) 121; 2.740 (100) 300; 2.675 (50) 112; 2.544 (69) 202; 1.914 (31) 222; and 1.864 (22) 132. Chemical analysis by an electron microprobe gave (in wt%) P_2O_5 39.20, MgO 0.19, CaO 24.14, MnO 31.19, FeO 2.95, Na_2O 0.05, F 0.39, Cl 3.13, $H_2O_{(calc)}$ 0.68, O=(Cl,F) -0.87, sum 101.05. The resulting empirical formula on the basis of 13 anions per formula unit is $(Mn_{2.39}Ca_{2.34}Fe_{0.22}Mg_{0.03}Na_{0.01})_{\Sigma4.99}P_{3.00}O_{12}[Cl_{0.48}(OH)_{0.41}F_{0.11}]$. The crystal structure of parafiniukite was refined to an $R_1 = 0.0463$ for 320 independent reflections with $F_o > 4\sigma(F_o)$ and 41 refined parameters. Parafiniukite is isotypic with apatites. Manganese is the dominant cation at the $M(2)$ site, and Ca is the dominant cation at the $M(1)$ site.

Keywords: parafiniukite; apatite supergroup; hedyphane group; manganese; calcium; phosphorus; Szklary pegmatite; Lower Silesia; Poland

---

1. Introduction

Parafiniukite, ideally $Ca_2Mn_3(PO_4)_3Cl$, was discovered in an LCT (Li–Cs–Ta) granitic pegmatite hosted by serpentinites of the Szklary massif, Lower Silesia, SW Poland. The mineral occurs as small relict grains, not exceeding 250 μm in size, embedded in a Mn-oxide, smectite mixture overgrowing partly altered beusite, the most frequent mineral component of phosphate aggregates occurring in graphic zone of the pegmatite. The new mineral and its name have been approved by the Commission on New Minerals, Nomenclature and Classification (CNMNC) of the International Mineralogical Association (IMA 2018-047). The name of the mineral is after Jan Parafiniuk (b. 1954), Professor of mineralogy at the Institute of Geochemistry, Mineralogy and Petrology of the University of Warsaw, Poland. The parafiniukite holotype (specimen Sz 31) is deposited in the collection of the Mineralogical Museum of the University of Wrocław, Faculty of Earth Science and Environmental Management,

Institute of Geological Sciences, 50-205 Wrocław, Cybulskiego 30, Poland, with the catalogue number MMWr IV8024. The crystal used for the collection of single-crystal data is deposited in the mineralogical collection of the Museo di Storia Naturale, Università di Pisa, Via Roma 79, Calci (Pisa), Italy, catalogue number 19902. Parafiniukite was previously reported as "Mn-rich apatite" [1], and codified by the IMA Subcommittee on Unnamed Minerals as UM2007-18:PO:CaClMn mineral, $Mn_3Ca_2(PO_4)_3(Cl,F,OH)$, corresponding to the Mn-dominant analogue of apatite. The present study indicates that the correct end-member formula of this mineral is $Ca_2Mn_3(PO_4)_3Cl$. The aim of this paper is to describe this member of the apatite supergroup.

2. Occurrence

Parafiniukite, was discovered in the Szklary LCT (Li–Cs–Ta) pegmatite (50°39.068' N, 16°49.932' E), ~6 km N of the Ząbkowice Śląskie town, ~60 km south of Wrocław, in Lower Silesia, SW Poland, outcropped in the northern part of the Szklary serpentinite massif. The massif is considered as a part of the tectonically dismembered Central-Sudetic Ophiolite. It adjoins the Góry Sowie Block (GSB) unit on the east and is enclosed as a mega-boudin in the mylonitized Góry Sowie gneisses of the Early Carboniferous Niemcza Shear Zone. The Szklary pegmatite, now completely excavated by mineral collectors, formed a NNE-SSW elongated lens or a boudin ~4 × 1 m large in planar section, which was in primary intrusive contact with an altered aplitic gneiss, up to 2 m thick, to the southwest, and both rocks were surrounded by tectonized serpentinite. A vermiculite-chlorite-talc zone was locally present along the pegmatite-serpentinite contact (for more information about the geological setting of the pegmatite see [2]). The pegmatite represents the beryl–columbite–phosphate subtype of the rare element (*REL*)–Li pegmatite class sensu Černý and Ercit [3]. The age of the pegmatite, estimated at 383 ± 2 Ma (CHIME dating on monazite-(Ce) [4]), is significantly older than the age ~335–340 Ma of the neighbouring small late-syntectonic dioritic, syenitic and granodioritic intrusions occurring in the Niemcza Shear Zone (e.g., [5]). It is rather contemporaneous with the anatectic event in the adjacent GSB [6–8]. The Szklary pegmatite is built mostly of sodic plagioclase (oligoclase-albite), microcline perthite, quartz and biotite, with minor Fe^{3+}-bearing schorl-dravite tourmaline, spessartine garnet and muscovite. It is relatively poorly zoned with: (1) a marginal graphic zone of albite + quartz ± minor-to-accessory biotite commonly altered to clinochlore + black tourmaline; (2) a coarser-grained intermediate graphic zone of microcline perthite + quartz + small quartz-tourmaline nests, with smaller amounts of albite and biotite and, simultaneously, increased abundance of muscovite and spessartine; accessory chrysoberyl is locally present in muscovite aggregates; (3) a central zone of graphic microcline + quartz, in places developed as blocky microcline with interstitial albite, rare muscovite, and no black tourmaline or biotite [4,9]. The zones (2) and (3) contain numerous accessory minerals disseminated as inclusions mainly in quartz, microcline, albite, and muscovite. They include native elements and alloys (arsenic, antimony, bismuth, gold, stibarsen, and paradocrasite), Nb-Ta oxides (columbite-group minerals, stibiocolumbite and stibiotantalite, fersmite, pyrochlore-, microlite- and betafite-group minerals), Mn-oxides (ernienickelite, jianshuiite, cesàrolite, cryptomelane, and others), As and Sb oxides, lepageite (*type locality*, TL) and other arsenites and antimonites, beryllium minerals (chrysoberyl, beryl, and bertrandite), phosphate and arsenate minerals (monazite-(Ce), cheralite, xenotime-(Y), Mn-bearing fluor-, hydroxyl- and chlorapatite, pieczkaite, parafiniukite, beusite and beusite-(Ca), bobfergusonite, fillowite, lithiophilite grading to sicklerite, simferite and purpurite, natrophilite, Pb- and Ba-bearing dickinsonite, triploidite, fairfieldite, phosphohedyphane, plumbogummite, mitridatite, pararobertsite, gorceixite, arsenogorceixite), and a number of silicates, such as e.g., (Be,Mn,Na,Cs)-bearing cordierite, tourmaline-supergroup minerals (dravite/schorl and foitite/oxy-schorl), dumortierite-supergroup minerals (dumortierite, holtite, niboholtite (TL), titanoholtite (TL) and szklaryite (TL)), spessartine, zircon, harmotome and pollucite [4,10–12].

Parafiniukite is closely associated with small aggregates of beusite, up to ~1 cm in size, which underwent intense alteration into a secondary assemblage of Mn-oxides and smectites.

Pieczkaite and rarer parafiniukite usually survived only as small relicts, not exceeding 250 μm in size, surrounded by the mixture (Figure 1).

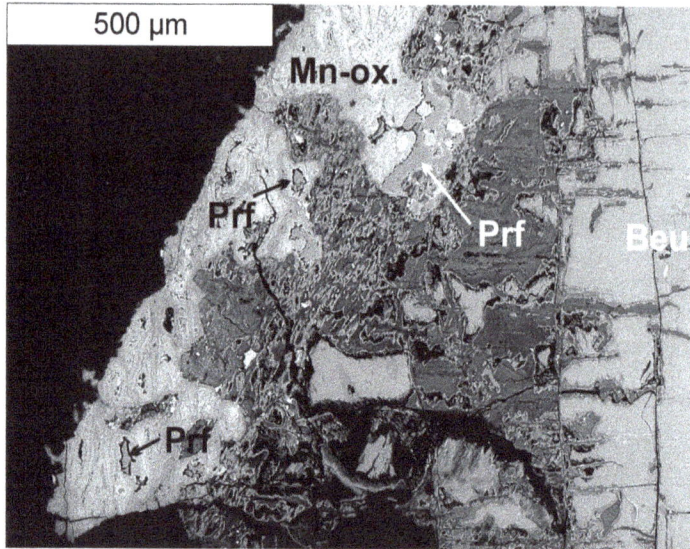

Figure 1. Parafiniukite as relict grains within the Mn-oxide; smectite matrix, around beusite. Holotype specimen MMWrIV8024. Labels: Beu = beusite; Mn-ox. = Mn-oxides; Prf = parafiniukite.

3. Experimental Data

3.1. Mineral Description and Physical Properties

Parafiniukite occurs only as small anhedral grains, up to 250 μm in size (Figure 1). The mineral is transparent with dark olive-green colour sometimes masked by the Mn-oxides. It has a vitreous luster, and it is brittle with irregular, uneven fracture. No forms and twinning were observed. Due to the very small amount of available material and its possible contamination by the surrounding Mn-oxides, streak, hardness, density as well as main optical properties were not determined. Cleavage, parting, and fluorescence were not observed. By analogy with pieczkaite [13], the Mohs hardness of parafiniukite could be estimated at 4–5. The density calculated on the basis of the empirical formula and the refined unit-cell volume of type parafiniukite is 3.614 g·cm^{-3}. The mean refractive index, obtained from the Gladstone–Dale relation [14,15], using the empirical formula and calculated density, is 1.731.

3.2. Chemical Data

Preliminary EDS chemical analysis showed Ca, Mn, P, and Cl as the only elements with Z > 8. Quantitative chemical analyses were carried out using a Cameca SX 100 electron microprobe (WDS mode, 15 kV, 20 nA, 3 μm beam diameter) at the Inter-Institute Analytical Complex for Minerals and Synthetic Substances at the University of Warsaw. Standard materials (element, emission line) were: fluorophlogopite (F $K\alpha$), YbPO$_4$ (P $K\alpha$), hematite (Fe $K\alpha$), rhodonite (Mn $K\alpha$), diopside (Mg $K\alpha$, Si $K\alpha$, Ca $K\alpha$), albite (Na $K\alpha$), tugtupite (Cl $K\alpha$), orthoclase (Al $K\alpha$, K $K\alpha$), celestine (Sr $L\alpha$), baryte (S $K\alpha$, Ba $L\alpha$), crocoite (Pb $M\alpha$) and sphalerite (Zn $K\alpha$) and GaAs (As $L\alpha$). The following diffracting crystals were used: PC0 for F; TAP for Na, Mg, Al, Si and As; LPET for P, S, Cl, K, Ca, Sr and Pb; LLIF for Mn, Fe, Zn and Ba. Aluminium, Si, S, K, Zn, Sr, As, Ba and Pb were sought but found below the detection limits. The raw data were reduced with the PAP routine [16]. Direct H_2O determination was not performed owing to scarcity of material. However, the occurrence of H_2O was confirmed by micro-Raman spectroscopy (see below). The analytical data on the holotype material are given in Table 1.

Table 1. Chemical composition of parafiunikite (in wt%).

Constituent	Mean (*n* = 10)	Range	e.s.d.
P_2O_5	39.20	38.98–39.44	0.14
MgO	0.19	0.12–0.27	0.05
CaO	24.14	23.66–24.64	0.39
MnO	31.19	30.04–31.78	0.62
FeO	2.95	2.72–3.15	0.16
Na_2O	0.05	0.01–0.07	0.02
F	0.39	0.29–0.46	0.05
Cl	3.13	3.00–3.29	0.09
$H_2O_{(calc)}$	0.68	0.61–0.71	0.03
O=(F + Cl)	−0.87		
Total	101.05		

Note: H_2O was calculated according to stoichiometry, in order to have 1 (OH + F + Cl) per formula unit (pfu); e.s.d., estimated standard deviation.

The empirical formula of parafiniukite (with rounding errors) calculated in relation to 12 O^{2-} and 1 (F,Cl,OH)$^-$ anion per formula unit (apfu) is $(Mn_{2.39}Ca_{2.34}Fe_{0.22}Mg_{0.03}Na_{0.01})_{\Sigma 4.99}P_{3.00}O_{12}[Cl_{0.48}(OH)_{0.41}F_{0.11}]$. Taking into account the results of the crystal-structure investigation (see below), the ideal formula of parafiniukite should be written as $Ca_2Mn_3(PO_4)_3Cl$, corresponding to (in wt%): P_2O_5 37.14, CaO 19.56, MnO 37.12, Cl 6.18, sum 100.00.

3.3. Micro-Raman Spectroscopy

The unpolarized micro-Raman spectra in the range 100–4000 cm^{-1} (Figure 2) were collected on a polished sample of parafiniukite with randomly oriented crystals in nearly back-scattered geometry at the Faculty of Materials Science and Ceramics, AGH UST, Cracow, Poland, using a Horiba Labram integrated with an Olympus BX40 confocal microscope with a 100× objective. The 532 nm line of a solid-state Nd-YAG laser (10 mW) was used. The minimum lateral and depth resolution was set to ~1 μm. The system was calibrated using the 520.7 cm^{-1} Raman band of silicon. The spectra were collected through two acquisitions with single counting times of 600 s. The backscattered radiation was analyzed with an 1800 mm^{-1} grating monochromator.

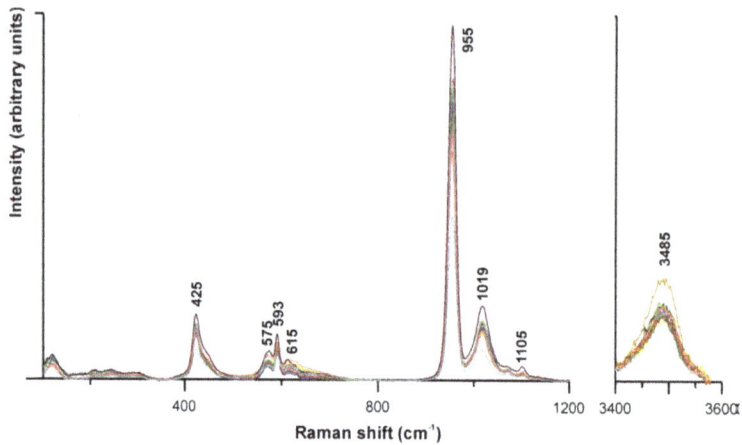

Figure 2. Raman spectra of parafiniukite in the spectral ranges 100–1200 cm^{-1} and 3400–3600 cm^{-1}.

In Figure 2, the Raman spectra are shown, as recorded in various relicts of the holotype parafiniukite. The bands at ~955 (strong), 1019 (medium) and 1105 (weak) cm^{-1} may be assigned to stretching vibrations of the PO$_4$ groups, in agreement with [13]. The peaks at ~615 (weak), 593 (medium), 575 (medium) and 425 cm^{-1} (medium) are due to the bending vibrations of PO$_4$ groups. The weaker peaks below ~300 cm^{-1} are due to deformations of the Ca and Mn polyhedra, whereas the peak at ~3485 cm^{-1} is from stretching vibrations of O–H bonds.

3.4. Crystallography

A crystal fragment (70 × 40 × 30 μm) was extracted from polished section Sz31 and mounted on a carbon fibre. Intensity data were collected using a Bruker Smart Breeze diffractometer (50 kV, 30 mA) equipped with an air-cooled CCD detector, and graphite-monochromatized Mo Kα radiation. The detector-to-crystal working distance was 50 mm. A total of 1418 frames was collected in ω and φ scan modes. The exposure time was 120 s per frame. The data were integrated and corrected for Lorentz and polarization, background effects, and absorption using the package of software Apex3 [17], resulting in a set of 422 independent reflections (see Supplementary Materials). The refinement of unit-cell parameters constrained to hexagonal symmetry gave a = 9.4900(6), c = 6.4777(5) Å, V = 505.22(5) Å^3. The c:a ratio calculated from the unit-cell parameters is 0.6826. The statistical tests on the distribution of $|E|$ values ($|E^2 - 1|$ = 0.784) suggested an acentric structure; taking into account also the systematic absences, the space group $P6_3$ could be proposed. However, any trial to refine the structure in that space group symmetry was not successful and the space group $P6_3/m$ was eventually chosen. The crystal structure was refined starting from the atomic coordinates of turneaureite [18] using Shelxl-2014 [19]. Scattering curves for neutral atoms were taken from the International Tables for Crystallography [20]. The site occupancy factors (s.o.f.) of the three cation sites were initially refined using the following scattering curves: Ca vs. Mn at the M(1) and M(2) sites, and P vs. □ (vacancy) at the T site. The three O sites (O(1), O(2), and O(3)) were refined assuming the full occupancy of O. Owing to the complex chemistry of parafiniukite, the position of the X anions was found through successive difference-Fourier maps, showing the occurrence of an electron density maximum at (0, 0, 1/4). The site scattering of this position was refined using the curves of Cl vs. □. After several cycles of isotropic refinement, the R1 index converged to 0.092. The s.o.f. at M(1) and M(2) sites showed mixed (Ca,Mn) occupancies, whereas the T site was found to be fully occupied by P only. Its s.o.f. was then fixed to 1.

After the introduction of the anisotropic displacement parameters for cations, the R_1 value converged to 0.067. Assuming an anisotropic model also for the O(1), O(2), and O(3) positions, the R_1 value converged to 0.056. However, the anisotropic displacement parameter of O(2) was negatively defined. The displacement parameter of the X anion was refined isotropically, as discussed in [18]. A residual of ~2.5 $e/Å^3$ at (0, 0, 0.1667) was found. By adding this additional position (named Xb), constraining the isotropic U value of Xa and Xb to be equal and the sum of their s.o.f. to be 1, the R_1 value converged to 0.058. Finally, we tried to refine the anisotropic displacement parameters of O(2), resulting in positive defined values. The Xb position was assumed to be occupied by Cl only, owing to the longer $M(2)$–Xb distance, whereas Xa was assumed to have a mixed (OH, F, Cl) occupancy. The site occupancies at the X sub-positions were fixed, taking into account the observed site scattering and the electron microprobe data, to $(OH_{0.40}Cl_{0.30}F_{0.10})$ and $Cl_{0.20}$ at Xa and Xb, respectively.

The refinement converged to $R_1 = 0.0463$ for 320 unique reflections with $F_o > 4\sigma(F_o)$ (0.0676 for all 422 reflections) and 41 refined parameters. The chemical formula derived from the structure refinement (SREF) is $(Ca_{2.56}Mn_{2.44})(PO_4)_3[Cl_{0.50}(OH)_{0.40}F_{0.10}]$. The details of data collection and refinement are given in Table 2.

Table 2. Crystal and experimental details for parafiniukite.

Crystal Data	
Crystal size (mm)	$0.07 \times 0.04 \times 0.03$
Cell setting, space group	Hexagonal, $P6_3/m$
a (Å)	9.4900(6)
c (Å)	6.4777(5)
V (Å^3)	505.22(5)
Z	2
Data Collection and Refinement	
Radiation, wavelength (Å)	Mo $K\alpha$, $\lambda = 0.71073$
Temperature (K)	293
$2\theta_{max}$ (°)	54.89
Measured reflections	6047
Unique reflections	422
Reflections with $F_o > 4\sigma(F_o)$	320
R_{int}	0.1008
$R\sigma$	0.0422
Range of h, k, l	$-12 \leq h \leq 12, -12 \leq k \leq 12, -8 \leq l \leq 8$
$R [F_o > 4\sigma(F_o)]$	0.0463
R (all data)	0.0676
wR (on F_o^2)	0.0933
Goof	1.163
Number of least-squares parameters	41
Maximum and minimum residual peak (e Å^{-3})	0.77 (at 0.98 Å from Xb) −0.80 (at 0.76 Å from T)

The fractional atom coordinates and isotropic or equivalent isotropic displacement parameters are reported in Table 3. Table 4 reports the selected bond distances, whereas Tables 5 and 6 show the proposed site populations and the bond-valence calculations obtained using the bond-valence parameters of Brese and O'Keeffe [21].

Table 3. Site labels, Wyckoff positions, atom coordinates, and equivalent isotropic or isotropic (*) displacement parameters (Å^2) for parafiniukite.

Site	Wyckoff	x/a	y/b	z/c	U_{eq}
M(1)	4f	2/3	1/3	0.0054(3)	0.0082(5)
M(2)	6h	0.0177(2)	0.2645(2)	1/4	0.0206(5)
T	6h	0.3752(2)	0.4034(2)	1/4	0.0103(6)
O(1)	6h	0.5008(7)	0.3465(6)	1/4	0.0119(13)
O(2)	6h	0.4621(7)	0.5920(7)	1/4	0.0247(18)
O(3)	12i	0.2629(5)	0.3424(5)	0.4403(7)	0.0199(12)
Xa	2a	0	0	1/4	0.0132(17) *
Xb	4e	0	0	0.186(4)	0.0132(17) *

Table 4. Selected bond lengths (in Å) for parafiniukite.

M(1)			M(2)			T		
	–O(1)	$2.280(4) \times 3$		–O(3)	$2.204(5) \times 2$		–O(1)	1.535(6)
	–O(2)	$2.366(4) \times 3$		–O(2)	2.259(6)		–O(3)	$1.540(5) \times 2$
	–O(3)	$2.835(4) \times 3$		–O(3)	$2.401(4) \times 2$		–O(2)	1.552(6)
				–Xa	2.431(2)			
				–Xb	2.466(4)			
				–O(1)	3.070(6)			

Table 5. Refined site scattering vs. calculated site scattering (in electrons) and proposed site occupancy for *M* sites in parafiniukite.

Site	Refined Site Scattering	Proposed Site Population	Calculated Site Scattering
M(1)	43.4	$Ca_{1.25}Mn^a_{0.74}Na_{0.01}$	43.6
M(2)	69.0	$Mn^a_{1.90}Ca_{1.06}Mg_{0.04}$	69.2

Note: [a] Includes 0.22 Fe.

Table 6. Bond-valence sums (in valence unit) in parafiniukite.

Site	M(1)	M(2)	T	ΣAnions
O(1)	$^{2\times}\rightarrow 0.37^{\downarrow \times 3}$	0.04	1.21	1.99
O(2)	$^{2\times}\rightarrow 0.29^{\downarrow \times 3}$	0.34	1.15	2.07
O(3)	$0.08^{\downarrow \times 3}$	$0.40^{\downarrow \times 2}$ $0.23^{\downarrow \times 2}$	$1.19^{\downarrow \times 2}$	1.90
Xa		$^{3\times}\rightarrow 0.26$		0.78
Xb		$^{3\times}\rightarrow 0.08$		0.24
ΣCations	2.22	1.98	4.74	

Note: In mixed or partially occupied sites, the BVS has been weighted taking into account the site occupancy.

Owing to the small crystal size, powder X-ray diffraction data were not collected. Table 7 reports the X-ray powder diffraction data (for Cu $K\alpha$) calculated on the basis of the refined structural model.

Table 7. Calculated X-ray powder diffraction data (*d* in Å) for parafiniukite. Intensity and d_{hkl} were calculated using the software PowderCell 2.3 [22] on the basis of the structural model given in Table 3. Only reflections with $I_{calc} > 5$ are listed, if not observed. The eight strongest reflections are given in bold.

I_{calc}	d_{calc}	*h k l*	I_{calc}	d_{calc}	*h k l*
13	8.22	1 0 0	10	2.150	3 1 1
7	5.09	1 0 1	6	1.965	1 1 3
6	4.109	2 0 0	**31**	**1.914**	**2 2 2**
16	3.470	2 0 1	8	1.885	3 2 0
39	**3.239**	**0 0 2**	22	**1.864**	**1 3 2**
14	3.106	1 2 0	6	1.864	3 1 2
16	3.013	1 0 2	15	1.810	3 2 1
55	**2.801**	**2 1 1**	13	1.793	1 4 0
76	**2.801**	**1 2 1**	17	1.773	1 2 3
100	**2.740**	**3 0 0**	16	1.773	2 1 3
50	**2.675**	**1 1 2**	17	1.735	4 0 2
69	**2.544**	**2 0 2**	14	1.629	2 3 2
7	2.523	3 0 1	17	1.619	0 0 4
13	2.279	3 1 0	12	1.466	5 0 2
10	2.242	1 2 2	7	1.439	1 5 1

4. Discussion

4.1. Crystal Structure Description

The crystal structure of parafiniukite is topologically similar to those of the other members of the apatite supergroup (e.g., [23]). The refined site scattering values, as given in Table 5, agree with the chemical formula calculated from the electron microprobe data. The *M*(1) and *M*(2) sites are Ca- and Mn-dominant, respectively, whereas Cl is the dominant *X* anion. Hence, the end-member composition of parafiniukite is $Ca_2Mn_3(PO_4)_3Cl$. A strong partitioning of Mn^{2+} at the *M*(2) site was previously reported in pieczkaite, ideally $Mn_5(PO_4)_3Cl$ [13].

The *T* site is occupied by tetrahedrally-coordinated P^{5+} with distances ranging between 1.535 and 1.556 Å.

The nine-fold coordinated *M*(1) site has a mixed (Ca,Mn) occupancy. This occupancy can be idealized as $Ca_{0.625}Mn_{0.375}$. Taking into account the ionic radii of $^{[9]}Ca^{2+}$ (1.18 Å), $^{[9]}Mn^{2+}$ (1.04 Å), and $^{[3]}O^{2-} = 1.36$ Å (after [13,24]), a <*M*(1)–O> distance of 2.49 Å can be predicted, in agreement with the observed <*M*(1)–O> distance, i.e., 2.494 Å. This average value is slightly larger than that observed in pieczkaite (2.473 Å), where the *M*(1) site is Mn-dominant, with a site occupancy $Mn_{0.575}Ca_{0.425}$ [13], and smaller <*M*(1)–O> distance than in Ca-apatites, e.g., hydroxylapatite (2.55 Å [25]).

The *M*(2) site has a seven-fold coordination and a mixed (Mn,Ca) occupancy. The site occupancy at *M*(2) may be idealized as $Mn_{0.63}Ca_{0.37}$, to be compared with that reported in pieczkaite, $Mn_{0.82}Ca_{0.18}$ [13]. The lower Mn/(Mn + Ca) atomic ratio in parafiniukite results in slightly longer *M*(2)–O bond distances. *M*(2) is coordinated also by the monovalent anions at the *X* site. In parafiniukite, two sub-sites, labelled *Xa* and *Xb*, were located; the former on the mirror plane, at $z = 1/4$, and the latter slightly displaced from the mirror plane. In pieczkaite, the *X* anion ($X = Cl_{0.62}OH_{0.38}$) is only slightly displaced from the mirror plane at $z = 1/4$. Since parafiniukite is a ternary apatite, containing also minor F, the situation is more complicated. Indeed, in order to avoid short anion–anion contacts, only some configurations are possible, confirming that local order within individual [001] anion columns should be present (e.g., [18,26]). In parafiniukite, the anion columns are occupied by ($Cl_{0.50}OH_{0.40}F_{0.10}$).

According to previous studies [27,28], Mn tends to order at the *M*(1) site. However, as discussed above and observed in pieczkaite [13], Mn is more strongly ordered at the *M*(2) site than at *M*(1), with Mn/(Mn + Ca) atomic ratios of 0.375 and 0.63 at *M*(1) and *M*(2), respectively, to be compared with

0.575 and 0.82 in pieczkaite [13]. This preferential distribution seems to be favoured by the occurrence of Cl anion at the *X* site, whereas Mn tends to be ordered at *M*(1) when *X* = F, indicating that the nature of the monovalent anion strongly affects the ordering of Mn and Ca in mixed Ca-Mn apatites [13].

Following [13], the hypothetical short-range ordered distributions could be proposed. The *M*(2) site population can be simplified as $Mn_{1.9}Ca_{1.1}$. All Ca must be locally associated with an OH group and F at the neighbouring *X* sites. In the crystal structure of parafiniukite, there is 0.50 (OH + F) at *X*, and this is associated with $0.50 \times 3 = 1.50$ *M*(2) cations. This means that OH and F are bonded to 1.10 Ca + 0.40 Mn, and all Cl is bonded to Mn.

Parafiniukite corresponds to the end-member composition $Ca_2Mn_3(PO_4)_3Cl$ hypothesized by Tait et al. [13]. It belongs to class "08.BN Phosphates with only large cations, (OH, etc.): $RO_4 = 0.33:1$" in the Strunz-Nickel classification [29] and to the class "41.08 Anhydrous Phosphates, etc. containing hydroxyl or halogen where $(A)_5(XO_4)_3$ Zq, 41.08.01 Apatite group" of the Dana classification [30]. Following [31], parafiniukite is a new member of the Hedyphane Group in the Apatite Supergroup.

4.2. Crystallization of Parafiniukite

Parafiniukite is a primary phosphate mineral that crystallized from a highly fractionated melt. The Mn-Fe fractionation index, expressed as Mn/(Mn + Fe) atomic ratio, indicates beusite + beusite-(Ca) as the earliest crystallized phosphates (Mn/(Mn + Fe) = 0.71–0.92), followed by pieczkaite (0.89–0.96), parafiniukite (0.91–0.95), Mn-bearing chlorapatite (0.95–0.99), and Cl-free, (Mn,OH)-bearing fluorapatite (0.97–0.99), determining the position of parafiniukite in the crystallization sequence of apatite-supergroup minerals in the Szklary pegmatite. Szuszkiewicz et al. [2] noticed that this crystallization sequence is reversed in relation to that observed in the Cross-Lake pegmatite, the type locality for pieczkaite [13], where early crystallized manganoan fluorapatite is succeeded by late Mn-rich chlorapatite and pieczkaite. The latter sequence reflects progressive enrichment in Mn and F, and next also in Cl with progress in crystallization, corresponding to the general observation that the Cl-enrichment in the apatite structure is typically characteristic of late to secondary stages of evolution (e.g., [32]). Studies of the (Mn,Cl)-rich apatites from the Szklary pegmatite [2] showed that, contrary to the Cross Lake pieczkaite, they are primary magmatic phases that formed together or slightly after beusites, and their crystallization predates manganoan fluor- to hydroxylapatites. Therefore, the early crystallization of pieczkaite and parafiniukite in the Szklary pegmatite was probably constrained by another process, which may be related to the interaction of the evolved pegmatite-forming melt with the ultramafic wall-rock (serpentinite), and its contamination by serpentinite-related fluid-mobile elements, particularly Cl [33]. This enabled precipitation of the small-volume droplets of a hydrous melt extremely enriched in Mn, Cl and P, as a precursor for beusites, pieczkaite and parafiniukite, immixing with normal aluminosilicate melt from which Mn-rich fluorapatite to hydroxylapatite devoid of Cl crystallized in the final stages of the pegmatite formation. Metasomatic interaction between the Szklary pegmatite and the host serpentinite is also marked by intensive Mg contamination, appeared by the occurrence of dravite, and the enrichment of the pegmatite in some other rare components like As and Sb (native elements and alloys, oxides, arsenites and antimonites, arsenates, As- and Sb-bearing dumortierite-supergroup minerals) [11,12], Pb (pyrochlore-, microlite- and betafite group minerals), Bi (native element, pyrochlore-, microlite and betafite group minerals), Ba (K-feldspars, harmotome, baryte), Li and Cs (some phosphates, (Mn,Be,Na,Cs)-bearing cordierite, (Cs,Mg)-bearing beryl, (Cs,Mg)-bearing muscovite, Cs-bearing phlogophite and Cs-bearing annite) identified as serpentinite-related fluid-mobile elements [33], playing important role in zones (1) and (2).

Supplementary Materials: The following are available online at http://www.mdpi.com/2075-163X/8/11/485/s1, CIF: parafiniukite.

Author Contributions: A.P. found parafiniukite and made its microprobe studies; B.G., P.J. and M.S. performed micro-Raman spectroscopic studies of the mineral, and C.B. and M.P. collected single-crystal X-ray diffraction data and refined the crystal structure; A.P. and C.B. prepared the manuscript.

Funding: This study was supported by the National Science Centre (Poland) grant 2015/17/B/ST10/03231 to A.P.

Conflicts of Interest: The authors declare no conflict of interest.

References

1. Pieczka, A. Beusite and unusual Mn-rich apatite from the Szklary granitic pegmatite, Lower Silesia, Poland. *Can. Mineral.* **2007**, *45*, 901–914. [CrossRef]
2. Szuszkiewicz, A.; Pieczka, A.; Gołębiowska, B.; Dumańska-Słowik, M.; Marszałek, M.; Szełęg, E. Chemical composition of Mn- and Cl-rich apatites from the Szklary pegmatite, Central Sudetes, SW Poland: Taxonomic and genetic implications. *Minerals* **2018**, *8*, 350. [CrossRef]
3. Černý, P.; Ercit, T.S. The classification of granitic pegmatites revisited. *Can. Mineral.* **2005**, *43*, 2005–2026. [CrossRef]
4. Pieczka, A.; Szuszkiewicz, A.; Szełęg, E.; Janeczek, J.; Nejbert, K. Granitic pegmatites of the Polish part of the Sudetes (NE Bohemian massif, SW Poland). In *Fieldtrip Guidebook, Proceedings of the 7th International Symposium on Granitic Pegmatites, Książ, Poland, 17–19 June 2015*; PEG2015: Książ, Poland, 2015; pp. 73–103.
5. Oliver, G.J.H.; Corfu, F.; Krogh, T.E. U–Pb ages from SW Poland: Evidence for a Caledonian suture zone between Baltica and Gondwana. *J. Geol. Soc. Lond.* **1993**, *150*, 355–369. [CrossRef]
6. Van Breemen, O.; Bowes, D.R.; Aftalion, M.; Żelaźniewicz, A. Devonian tectonothermal activity in the Sowie Góry gneissic block, Sudetes, southwestern Poland: Evidence from Rb-Sr and U-Pb isotopic studies. *J. Pol. Geol. Soc.* **1988**, *58*, 3–10.
7. Timmermann, H.; Parrish, R.R.; Noble, S.R.; Kryza, R. New U–Pb monazite and zircon data from the Sudetes Mountains in SW Poland: Evidence for a single-cycle Variscan orogeny. *J. Geol. Soc. Lond.* **2000**, *157*, 265–268. [CrossRef]
8. Turniak, K.; Pieczka, A.; Kennedy, A.K.; Szełęg, E.; Ilnicki, S.; Nejbert, K.; Szuszkiewicz, A. Crystallisation age of the Julianna pegmatite system (Góry Sowie Block, NE margin of the Bohemian massif): Evidence from U-Th-Pb SHRIMP monazite and CHIME uraninite studies. In *Book of Abstracts, Proceedings of the 7th International Symposium on Granitic Pegmatites, Książ, Poland, 17–19 June 2015*; PEG2015: Książ, Poland, 2015; pp. 111–112.
9. Pieczka, A. A rare mineral-bearing pegmatite from the Szklary serpentynite massif, the Fore-Sudetic Block, SW Poland. *Geol. Stud.* **2000**, *33*, 23–31.
10. Pieczka, A. Primary Nb-Ta minerals in the Szklary pegmatite, Poland: New insights into controls of crystal chemistry and crystallization sequences. *Am. Mineral.* **2010**, *95*, 1478–1492. [CrossRef]
11. Pieczka, A.; Grew, E.S.; Groat, L.A.; Evans, R.J. Holtite and dumortierite from the Szklary pegmatite, Lower Silesia, Poland. *Mineral. Mag.* **2011**, *75*, 303–315. [CrossRef]
12. Pieczka, A.; Evans, R.J.; Grew, E.S.; Groat, L.A.; Ma, C.; Rossman, G.R. The dumortierite supergroup. II. Three new minerals from the Szklary pegmatite, SW Poland: Nioboholtite, $(Nb_{0.6}\square_{0.4})Al_6BSi_3O_{18}$, titanoholtite, $(Ti_{0.75}\square_{0.25})Al_6BSi_3O_{18}$, and szklaryite, $\square Al_6BAs^{3+}_3O_{15}$. *Mineral. Mag.* **2013**, *77*, 2841–2856. [CrossRef]
13. Tait, K.; Ball, N.A.; Hawthorne, F.C. Pieczkaite, ideally $Mn_5(PO_4)_3Cl$, a new apatite-supergroup mineral from Cross Lake, Manitoba, Canada: Description and crystal structure. *Am. Mineral.* **2015**, *100*, 1047–1052. [CrossRef]
14. Mandarino, J.A. The Gladstone-Dale relationship. Part III. Some general applications. *Can. Mineral.* **1979**, *17*, 71–76.
15. Mandarino, J.A. The Gladstone-Dale relationship. Part IV. The compatibility concept and its application. *Can. Mineral.* **1981**, *19*, 441–450.
16. Pouchou, J.-L.; Pichoir, F. Quantitative analysis of homogeneous or stratified microvolumes applying the model "PAP". In *Electron Probe Quantitation*; Heinrich, K.F.J., Newbury, D.E., Eds.; Plenum Press: New York, NY, USA, 1991; pp. 31–75.
17. Bruker AXS Inc. APEX 3. In *Bruker Advanced X-ray Solution*; Bruker AXS Inc.: Madison, WI, USA, 2004.
18. Biagioni, C.; Bosi, F.; Hålenius, U.; Pasero, M. The crystal structure of turneaureite, $Ca_5(AsO_4)_3Cl$, the arsenate analog of chlorapatite, and its relationships with the arsenate apatites johnbaumite and svabite. *Am. Mineral.* **2017**, *102*, 1981–1986. [CrossRef]
19. Sheldrick, G.M. Crystal structure refinement with SHELXL. *Acta Crystallogr.* **2015**, *C71*, 3–8.
20. Wilson, A.J.C. Volume C: Mathematical, Physical and Chemical Tables. In *International Tables for Crystallography*; Kluwer Academic: Dordrecth, The Netherlands, 1992.

21. Brese, N.E.; O'Keeffe, M. Bond-valence parameters for solids. *Acta Crystallogr.* **1991**, *B47*, 192–197. [CrossRef]
22. Kraus, W.; Nolze, G. POWDER CELL—A program for the representation and manipulation of crystal structures and calculation of the resulting X-ray powder patterns. *J. Appl. Crystallogr.* **1996**, *29*, 301–303. [CrossRef]
23. Hughes, J.M.; Rakovan, J. The crystal structure of apatite, $Ca_5(PO_4)_3(F,OH,Cl)$. In *Phosphates: Geochemical, Geobiological and Materials Importance*; Kohn, M.L., Rakovan, J., Hughes, J.M., Eds.; Mineralogical Society of America: Chantilly, VA, USA, 2002.
24. Shannon, R.D. Revised effective ionic radii and systematic studies of interatomic distances in halides and chalcogenides. *Acta Crystallogr.* **1976**, *A32*, 751–767. [CrossRef]
25. Hughes, J.M.; Cameron, M.; Crowley, K.D. Structural variations in natural F, OH, and Cl apatites. *Am. Mineral.* **1989**, *74*, 870–876.
26. Hughes, J.M.; Cameron, M.; Crowley, K.D. Crystal structures of natural ternary apatites: Solid solution in the $Ca_5(PO_4)_3X$ (X = F, OH, Cl) system. *Am. Mineral.* **1990**, *75*, 295–304.
27. Suitch, P.R.; Lacout, J.L.; Hewat, A.W.; Young, R.A. The structural location and role of Mn^{2+} partially substituted for Ca^{2+} in fluorapatite. *Acta Crystallogr.* **1985**, *B41*, 173–179. [CrossRef]
28. Hughes, J.M.; Ertl, A.; Bernhardt, H.J.; Rossman, G.R.; Rakovan, J. Mn-rich fluorapatite from Austria: Crystal structure, chemical analysis and spectroscopic investigations. *Am. Mineral.* **2004**, *89*, 629–632. [CrossRef]
29. Strunz, H.; Nickel, E.H. *Strunz Mineralogical Tables*, 9th ed.; Schweizerbart'sche Verlagsbuchhandlung: Stuttgart, Germany, 2001.
30. Gaines, R.V.; Skinner, H.C.; Foord, E.E.; Mason, B.; Rosenzweig, A. *Dana's New Mineralogy*, 9th ed.; John Wiley & Sons, Inc.: Hoboken, NJ, USA, 1997.
31. Pasero, M.; Kampf, A.R.; Ferraris, C.; Pekov, I.V.; Rakovan, J.; White, T.J. Nomenclature of the apatite supergroup minerals. *Eur. J. Mineral.* **2010**, *22*, 163–179. [CrossRef]
32. Černý, P.; Fryer, B.J.; Chapman, R. Apatite from granitic pegmatite exocontact in Moldanubian serpentinites. *J. Czech Geol. Soc.* **2001**, *46*, 15–20.
33. Deschamps, F.; Godard, M.; Guillot, S.; Hattori, K. Geochemistry of subduction zone serpentynites: A review. *Lithos* **2013**, *178*, 96–127. [CrossRef]

minerals

MDPI

Article

Aurihydrargyrumite, a Natural Au_6Hg_5 Phase from Japan

Daisuke Nishio-Hamane [1],*, Takahiro Tanaka [2] and Tetsuo Minakawa [3]

[1]   Institute for Solid State Physics, the University of Tokyo, Kashiwa, Chiba 277-8581, Japan
[2]   Sunagawa cho, Tachikawa, Tokyo 190-0031, Japan; penta@kuh.biglobe.ne.jp
[3]   Ehime University Museum, Ehime University, Matsuyama, Ehime 790-8577, Japan;
      minagawa@sci.ehime-u.ac.jp
*   Correspondence: hamane@issp.u-tokyo.ac.jp; Tel.: +81-4-7136-3462

Received: 28 August 2018; Accepted: 17 September 2018; Published: 19 September 2018

Abstract: Aurihydrargyrumite, a natural Au_6Hg_5 phase, was found in Iyoki, Uchiko, Ehime Prefecture, Shikoku Island, Japan. Aurihydrargyrumite with a metallic silver luster occurs as a submicron- to 2 μm-thick layer on the outermost surface of the placer gold. A prismatic face may be formed by {001} and {100} or {110}. The streak is also silver white and its Mohs hardness value is ca. 2.5. Its tenacity is ductile and malleable, and its density, as calculated based on the empirical formula and powder unit-cell data, is 16.86 g·cm^{-3}. The empirical formula of aurihydrargyrumite, on the basis of 11 Au + Hg, is $Au_{5.95}Hg_{5.05}$. Aurihydrargyrumite is hexagonal, $P6_3/mcm$, with the lattice parameters $a = 6.9960(10)$ Å, $c = 10.154(2)$ Å and $V = 430.40(15)$ Å^3, which is identical with the synthetic Au_6Hg_5 phase. The seven strongest lines in the powder X-ray diffraction (XRD) pattern [d in Å(I/I_0)(hkl)] were 2.877(29)(112), 2.434(42)(113), 2.337(100)(104), 2.234(87)(211), 1.401(39)(314), 1.301(41)(404), and 1.225(65)(217). Aurihydrargyrumite forms through the weathering of mercury-bearing placer gold by involvement of self-electrorefining. This new mineral has been approved by the IMA-CNMNC (2017-003) and it is named for its composition, being a natural amalgam of gold (Latin: aurum) and mercury (Latin: hydrargyrum).

Keywords: aurihydrargyrumite; Au_6Hg_5 phase; gold; placer; self-electrorefining; Ehime; Japan

1. Introduction

The investigation of the phase relations in the Au–Hg system has a long history. In the latest phase diagram [1], three amalgam solid phases appear in the system: Au_4Hg, Au_3Hg, and Au_2Hg. Additionally, Lindahl [2] reported a metastable Au_6Hg_5 phase synthesized by reducing gold and mercury ions. Au_6Hg_5 phase is stable around room temperature, while it alters by leaching the mercury component at temperatures of above 70 °C [2].

In nature, several amalgam phases without structural identification have been reported from the placer deposit from the Palakharya River, Bulgaria [3,4]. They are polymineral aggregates consisting of stoichiometric compounds among gold, silver, and mercury: (Au,Ag,Hg) phases with mercury contents of up to 3 wt.%, $(Au,Ag)_3Hg$, $(Au,Ag)_2Hg$, $(Au,Ag)_3Hg_2$, and AuAgHg phases. A mineral with $(Au,Ag)_{1.2}Hg_{0.8}$ (=Au_3Hg_2) composition was described as the new mineral weishanite from the Weishancheng ore field, China [5]. Although the stoichiometric compound $(Au,Ag)_3Hg_2$ does not exist on the phase diagram, the symmetry and unit-cell parameters for weishanite correspond to those of a synthetic Au_3Hg phase [5]. Amalgam grains with $Au_{94-88}Hg_{6-12}$ composition were also found in Pleistocene alluvial deposits along the Snake River near Blackfoot, Idaho [6]. The X-ray diffraction (XRD) pattern of this amalgam was indexed as monoclinic symmetry, however this phase also does not appear in the phase diagram.

The anthropogenic mercury that was used to collect gold in the past affects the composition and texture of present gold grains through weathering processes. Barkov et al. [7] reported zoned amalgam of composition $(Au_{1.5-1.9}Ag_{1.1-1.4})_{\Sigma 2.8-3.0}Hg_{1.0-1.2}$ from a placer deposit in the Tulameen–Similkameen river system, British Columbia, Canada. They discussed the concentration of the mercury component at the grain rim by some electrochemical factors that are related to the process of self-electrorefining, although they considered the amalgam grain to be anthropogenic in origin [7]. Recently, Svetlitskaya et al. [8] described native gold from Inagli Pt–Au placer deposit, in the Aldan Shield, Russia. Most gold placer was determined to be natural in origin, while some gold grains with a brain-like appearance were rich in a mercury component brought by humans to recover gold in this region [8].

Many gold deposits, including placer deposits, have been reported in the Japanese Islands [9,10]. For example, the gold deposits of Shikoku Island, Ehime Prefecture, were summarized by Miyahisa and Higaki [11]. Although gold was recovered as a by-product of the Kieslarger-type copper deposit, there are no deposits that are mined primarily for gold in Ehime [11]. Additionally, a small amount of placer gold was reported in the basin of the Dozan River in eastern Ehime. In the present study, we conducted a survey in an area where gold has not been known to occur to date, and discovered a new placer locality in southwest Ehime. There, the surface of some gold particles is partly or completely coated by metal with a silver luster. We subsequently studied this unidentified silver metal and found that it corresponds to a new mineral, a natural Au_6Hg_5 phase.

The new mineral, which we name aurihydrargyrumite, is named after its chemical composition, being a natural amalgam of gold (Latin: aurum) and mercury (Latin: hydrargyrum). Both the new mineral and its name have been approved by the International Mineralogical Association, Commission on New Minerals, Nomenclature and Classification (no. 2017-003). The type specimen has been deposited in the collections of the National Museum of Nature and Science, Japan (specimen number NSM-M45047). We herein describe this new mineral.

2. Occurrence

Aurihydrargyrumite, which is a natural Au_6Hg_5 phase, was found in a placer from Iyoki, Uchiko, Ehime Prefecture, Shikoku Island, Japan. The discovery site is located in the middle of the Oda river; Figure 1 shows the geological map for the investigated area. The placer was carried by the Oda and Tamatani rivers, which flow through areas where Sanbagawa metamorphic rocks and Mikabu greenstone are distributed. Sanbagawa metamorphic rocks mainly consist of mafic, pelitic, and psammitic schists. Mikabu greenstone consists of metamorphosed basaltic tuff and lava, metagabbro, and metadolerite.

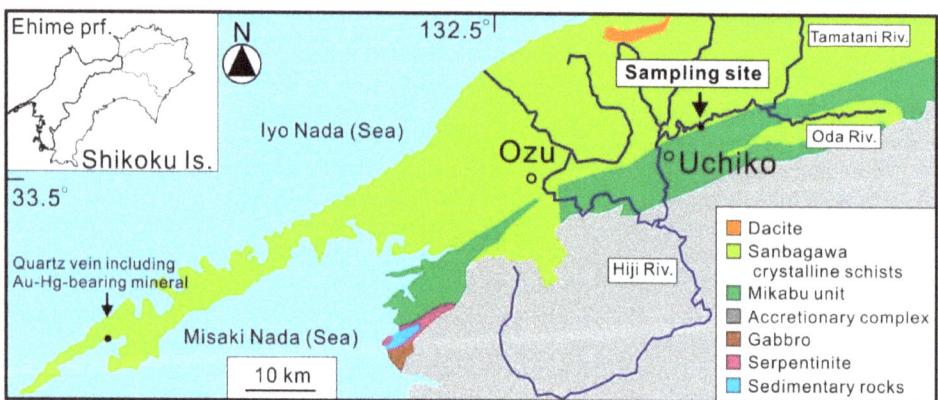

Figure 1. Geological map and water system of the investigated area, southwest Ehime Prefecture, Shikoku Island, Japan.

These units belong to the Sanbagawa metamorphic rocks. Mikabu greenstone contains significant amounts of Kieslarger-type copper deposits and manganese deposits. Although no gold deposit has yet been reported in the investigated area [11], we found a small quartz vein that included the gold- and mercury-bearing mineral from the Sanbagawa metamorphic rocks (Figure 1). Subsequently, we investigated the placer of rivers and found a small amount of gold placer from Iyoki.

The placer collected from Iyoki consists of ilmenite, magnetite, chromite, zircon, scheelite, gold, and the platinum-group minerals (PGMs) iridium, osmium, and irarsite. Aurihydrargyrumite is closely associated with gold and is formed through the weathering of mercury-bearing gold, as discussed later.

3. Appearance and Physical Properties

The surface of some gold particles is partly or completely coated by aurihydrargyrumite with a metallic silver luster (Figure 2a). Such particles show a zoned internal texture (Figure 2b). Aurihydrargyrumite occurs as a thin layer on the outermost surface of the gold particles. Beneath the aurihydrargyrumite layer, a gold-rich zone containing mercury is distributed. The core of the particle consists of silver- and mercury-bearing gold.

Figure 2. (**a**) Photographic image of aurihydrargyrumite (silver material); (**b**) backscattered electron image of a grain cross section; and (**c**) backscattered electron image of the surface of the grain.

Aurihydrargyrumite with a metallic silver luster occurs as a thin, submicron- to 2 μm-thick layer on the outermost surface of the gold particles. The mineral is commonly anhedral, and occasionally shows subhedral hexagonal-like crystals up to 2 μm in width (Figure 2c). Prismatic crystal may be formed by {001} and {100} or {110}. Its streak is silver white, its Mohs hardness value is ca. 2.5, and its tenacity is ductile and malleable. The density, as calculated based on the empirical formula and powder unit-cell data, is 16.86 g·cm^{-3}. Other physical and optical data were not determined due to the small size of the crystals.

4. Chemical Composition

Table 1 summarizes the chemical analysis of aurihydrargyrumite. The data were obtained from the natural surface, because the aurihydrargyrumite layer on the thin section is too small to analyze. Attempts to use WDS could not be used reliably for analysis due to the non-horizontal surface. Therefore, chemical analyses were carried out while using a JEOL JSM-5600 scanning electron microprobe (EDS mode, 15 kV, 0.4 nA, 1 μm beam diameter), and the ZAF method was used for data correction. The slightly high analytical total obtained is probably due to the irregular surface topography. The empirical formula of aurihydrargyrumite, on the basis of 11 Au + Hg, is Au$_{5.95}$Hg$_{5.05}$. The ideal formula is Au$_6$Hg$_5$, which requires Au 54.09 and Hg 45.91 for a total of 100 wt.%.

Table 1 also shows the chemical compositions inside the particles on which the aurihydrargyrumite layer was observed. The gold-rich zone consists of a gold–mercury alloy, while the silver components are generally absent. The core of the particle consists mainly of a gold–silver–mercury alloy.

Table 1. Analytical data for aurihydrargyrumite and the inside of the particles.

	Aurihydrargyrumite		Gold-Rich Zone	Core
	Mean 5 (Range) wt.%	Ideal	Mean 5 (Range) wt.%	Mean 5 (Range) wt.%
Au	54.92 (54.26–55.76)	54.09	96.82 (95.47–98.73)	88.20 (88.15–88.87)
Ag	-	-	-	9.90 (9.83–10.04)
Hg	47.50 (46.54–48.91)	45.91	2.96 (1.41–4.60)	1.69 (1.28–2.17)
Total wt.%	102.42	100	99.78	99.79
	pfu		pfu	pfu
Au	5.95	6	97.09	81.72
Ag	-	-	-	16.75
Hg	5.05	5	2.91	1.53
Σ	11	11	100	100

5. X-ray Crystallography

The micro-XRD measurement technique was applied for the powder XRD, which was carried out using a Rigaku ultrax18 (CrKα radiation, 40 kV, 200 mA, 100 μm collimator) instrument equipped with a curved position sensitive proportional counter and an oscillation sample stage. The powder XRD spectrum of aurihydrargyrumite was successfully obtained from the surface of the particle while using this method. The XRD spectrum is shown in Figure 3, and the resulting data are summarized in Table 2. The seven strongest lines in the powder XRD spectrum [d in Å(I/I_0)(hkl)] are 2.877(29)(112), 2.434(42)(113), 2.337(100)(104), 2.234(87)(211), 1.401(39)(314), 1.301(41)(404) and 1.225(65)(217). Based on these data, aurihydrargyrumite can be indexed to the hexagonal $P6_3/mcm$ space group. The unit-cell parameters as refined from the powder data are a = 6.9960(10) Å, c = 10.154(2) Å and V = 430.40(15) Å^3. The c/a ratio calculated from the unit-cell parameters is 1.451.

Aurihydrargyrumite contains one Au site and two Hg sites. Each distinct site forms a sheet in the ab plane (Figure 4). Gold atoms form triangular trimers, which are arranged in a triangular net in the Au sheets. The Hg atoms form a ditrigonally distorted Kagome net in the Hg1 sheets, but a honeycomb net in the Hg2 sheets. Two Au sheets and one Hg1 sheet form a compound Au–Hg1–Au layer and the next such layer is rotated 60° around the c-axis. The Hg2 sheets occur between the layers.

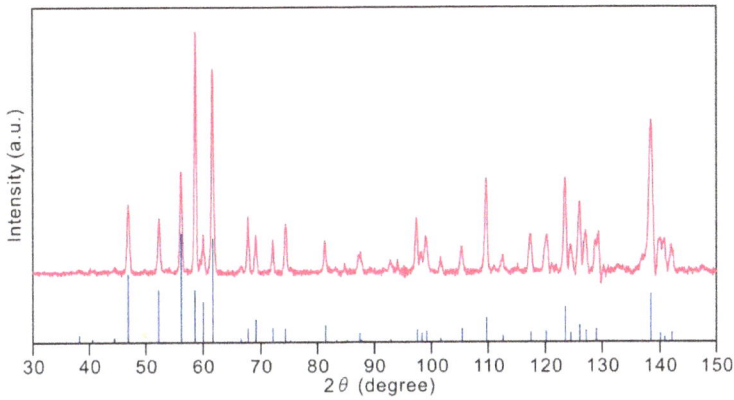

Figure 3. X-ray diffraction spectrum (Cr*Kα* radiation) of aurihydrargyrumite. The bars below the diffraction profile are the simulation calculated using the atomic position by Lindahl [2]. All of the diffraction peaks can be indexed as aurihydrargyrumite.

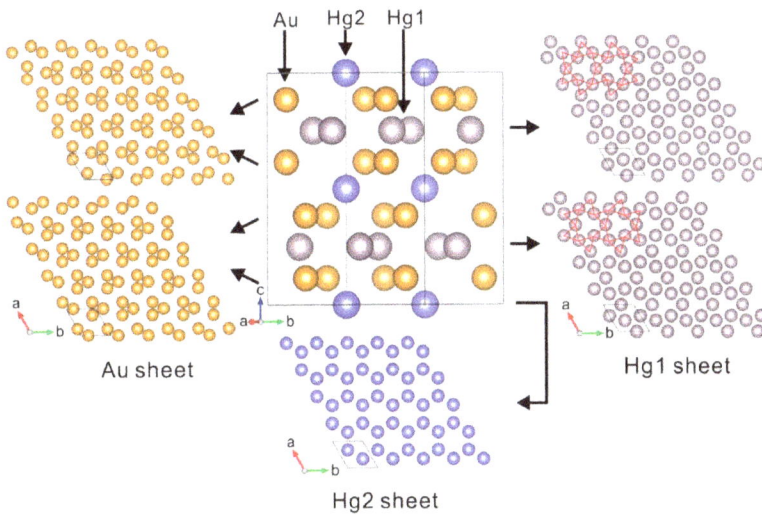

Figure 4. Crystal structure of aurihydrargyrumite.

Table 2. Powder X-ray diffraction data (Cr*Kα* radiation) for aurihydrargyrumite.

	Aurihydrargyrumite			Synthetic Au_6Hg_5 [2]	
hkl	$d_{obs.}$ (Å)	$d_{calc.}$ (Å)	I/I_0	$d_{calc.}$ (Å)	$I_{calc.}$
100		6.059		6.057	4
002		5.077		5.074	<1
102		3.891		3.889	<1
110	3.502	3.498	1	3.497	5
111	3.309	3.307	1	3.306	2
200	3.025	3.029	2	3.028	4
112	2.877	2.881	29	2.879	52

Table 2. *Cont.*

	Aurihydrargyrumite			Synthetic Au$_6$Hg$_5$ [2]	
hkl	$d_{obs.}$ (Å)	$d_{calc.}$ (Å)	I/I_0	$d_{calc.}$ (Å)	$I_{calc.}$
202	2.597	2.602	23	2.600	49
004		2.539		2.537	<1
113	2.434	2.432	42	2.431	100
104	2.337	2.341	100	2.340	46
210	2.290	2.290	13	2.289	38
211	2.234	2.234	87	2.233	91
212	2.087	2.088	3	2.087	5
114	2.053	2.055	22	2.053	12
300	2.019	2.020	15	2.019	21
204	1.947	1.946	13	1.945	14
213	1.895	1.897	21	1.896	18
302	1.876	1.877	2	1.876	2
115	1.758	1.756	13	1.755	13
220		1.749		1.748	<1
221		1.724		1.723	3
214		1.700		1.700	1
006		1.692		1.691	4
310		1.680		1.680	<1
311	1.658	1.658	8	1.657	10
222		1.654		1.653	2
106		1.630		1.629	3
312		1.595		1.595	1
304	1.581	1.580	5	1.580	4
223		1.554		1.553	<1
116	1.525	1.523	23	1.523	18
215		1.519		1.519	<1
400	1.516	1.515	8	1.514	12
313	1.506	1.505	15	1.505	15
206	1.478	1.477	5	1.477	3
402		1.452		1.451	<1
224	1.441	1.440	11	1.440	18
314	1.401	1.401	39	1.401	31
320		1.390		1.390	2
321	1.378	1.377	7	1.377	8
216		1.361		1.360	1
322	1.341	1.341	16	1.340	16
117		1.340		1.339	4
225		1.325		1.325	1
410	1.321	1.322	16	1.322	18
411		0.131		1.311	1
404	1.301	1.301	41	1.300	49
306		1.297		1.296	<1
315	1.294	1.295	11	1.294	14
323	1.286	1.286	29	1.285	29
412	1.280	1.280	18	1.279	21
008	1.269	1.269	15	1.269	18
108		1.242		1.242	<1
413		1.232		1.231	3
217	1.225	1.225	65	1.225	89
324	1.219	1.219	15	1.219	13
226	1.216	1.216	14	1.216	11
500	1.211	1.212	11	1.211	19
118		1.193		1.192	1

Figure 4 shows a crystal structure of aurihydrargyrumite based on the result of Lindahl [2]. Aurihydrargyrumite is identical to the synthetic Au$_6$Hg$_5$ phase.

6. Relation to Other Species

Minerals and synthetic amalgam phases are summarized in Table 3. In the Au–Hg system [1], three amalgam phases were observed: Au_4Hg, Au_3Hg, and Au_2Hg. Above the amalgam phases, the structures have been solved for Au_3Hg [12]. Hexagonal symmetry was reported in Au_4Hg and Au_2Hg phases, while their structures are unsolved [12]. Additionally, Lindahl [2] reported a metastable phase with an Au_6Hg_5 composition that was synthesized by reducing ions in a solution using hydrazine sulfate and electrolytic refining while using a gold anode and a mercury cathode. The Au_6Hg_5 phase is stable at around room temperature, while it alters by leaching the mercury component at above 70 °C [2].

Among natural amalgams, the unit-cell parameters are known for aurihydrargyrumite, weishanite, and UM192-08-EAuHg. In this study, Aurihydrargyrumite is equivalent to the synthetic Au_6Hg_5 phase. Although the composition of weishanite is $(Au,Ag)_3Hg_2$, the symmetry and unit-cell parameters can be indexed as the synthetic Au_3Hg phase [5]. There are no synthetic phases corresponding to UM192-08-EAuHg [6].

Table 3. Comparable data for minerals and synthetic phases in Au–Hg system.

Minerals										
Name	Composition	Symmetry	*a* (Å)	*b* (Å)	*c* (Å)	α (°)	β (°)	γ (°)	*V* (Å³)	Reference
Aurihydrargyrumite	Au_6Hg_5	$P6_3/mcm$	6.996	= a	10.154	90	90	120	430.4	This study
Weishanite	$(Au,Ag)_3Hg_2$	$P6_3/mmc$	2.9265	= a	4.8176	90	90	120	35.7	[5]
UM1992-08-E:AuHg	$Au_{94-88}Hg_{6-12}$	Monoclinic	4.729	5.243	4.546	90	90.9	90	112.7	[6]
Synthetic Phases										
Name	Composition	Symmetry	*a* (Å)	*b* (Å)	*c* (Å)	α (°)	β (°)	γ (°)	*V* (Å³)	Reference
-	Au_6Hg_5	$P6_3/mcm$	6.9937	= a	10.148	90	90	120	430.4	[2]
-	Au_4Hg	Hexagonal	8.736	= a	9.577	90	90	120	633	[12]
-	Au_3Hg	$P6_3/mmc$	2.918	= a	4.8113	90	90	120	35.7	[12]
-	Au_2Hg	Hexagonal	13.98	= a	17.2	90	90	120	2911.2	[12]

7. Discussion

The Au_6Hg_5 phase (=aurihydrargyrumite) could not be formed by the direct reaction between gold and mercury [1], and the special synthetic methods [2] cannot be applied directly to the natural environment. However, aurihydrargyrumite certainly occurs in nature, as demonstrated in this study. The formation of this mineral is discussed in the following.

Aurihydrargyrumite occurs as a thin layer on the outermost surface of the placer gold. This texture is clearly of secondary origin, since the thin surface layer would probably peel off if the placer gold were transported. Therefore, aurihydrargyrumite likely occurred after the placer gold was stably deposited. The accretion of a mercury component from the river water is unlikely, since no mercury was detected in the Oda River in the official water quality survey that was conducted by the Ehime prefectural office. Aurihydrargyrumite is probably formed from an Hg component already present in the placer gold; indeed, the gold grain includes a mercury component in the core. In the synthetic experiment, the Au_6Hg_5 phase (aurihydrargyrumite) was formed by reduction from ions and by electrolytic synthesis. Therefore, in nature, ionization and precipitation may interact in a complex manner on the surface of the placer gold to form aurihydrargyrumite. When considering the situation, self-electrorefining is probably involved in the migration of the gold and mercury components to the surface layer, which is analogous to the process of forming a gold-rich rim in the placer gold [13]. We suggest that the placer gold dissolves at the interface, and that subsequently the gold and mercury components immediately precipitate back onto the surface. The gold and mercury component that was concentrated through this process first formed the gold-rich zone, and aurihydrargyrumite was subsequently crystallized by further self-electrorefining based on the interface of the gold-rich zone. In this way, aurihydrargyrumite probably occurs in the process of weathering of the mercury-bearing placer gold.

These formation processes were also suggested to explain the zoned texture of anthropogenic amalgam grains from the Tulameen–Similkameen river system, Canada [7]. The zoning texture in amalgam phases was also reported in placer gold from the Palakharya River, Bulgaria [3,4], and Inagli Pt–Au placer deposit, Russia [8]. In natural weathering environments, the gold and mercury components in placer gold seem to concentrate on the surface by repeated leaching and deposition. However, the mechanism of leaching and deposition is still controversial, and various mechanisms might be closely involved. Although self-electrorefining is one of the most probable mechanisms, there is no evidence to discount other mechanisms, such as the involvement of microbially assisted reactions in nugget and layer alloy formations. Ionization and re-metallization of gold and mercury components must be a contributory factor in the formation of aurihydrargyrumite.

8. Conclusions

Aurihydrargyrumite is found in placer gold from Iyoki, Uchiko, Ehime Prefecture, Japan. The chemical composition and X-ray diffraction profile confirmed that aurihydrargyrumite and the Au_6Hg_5 phase were identical. Aurihydrargyrumite occurs as thin layers on the outermost surface of the placer gold and its formation relates to the natural weathering environment, which involves ionization and re-metallization of gold and mercury components in the placer gold.

Author Contributions: D.N.-H. wrote the paper. D.N.-H., T.T. and T.M. took part in the field works. T.M. found the first specimen, and subsequently T.T. and D.N.-H. collected the placer samples. D.N.-H. performed the observation, chemical analysis, and XRD study.

Acknowledgments: We would like to thank Hirotada Gotou for technical assistance during the X-ray diffraction experiments at ISSP. The investigations were partly supported by a Grant-in-Aid for Young Scientists B (Grant No. 15K17785).

Conflicts of Interest: The authors declare no conflict of interest.

References

1. Okamoto, H.; Massalski, T.B. Au-Hg (Gold-Mercury). In *Binary Alloy Phase Diagrams*, 2nd ed.; Massalski, T.B., Ed.; ASM International: Materials Park, OH, USA, 1990; Volume 1, pp. 376–379.
2. Lindahl, T. Crystal structure of Au_6Hg_5. *Acta Chem. Scand.* **1970**, *24*, 946–952. [CrossRef]
3. Atanasov, V.; Iordanov, I. Amalgams of gold from the Palakharya River alluvial sands, Distinct of Sofia. *Doklady Bolgarskoj Akademii Nauk.* **1983**, *36*, 465–468.
4. Atanasov, V.; Jordanov, J.A.; Vitov, O.C.; Atanasov, A.V. Part I: Geology. In *Amalgams of Gold in Some of the Bulgarian Alluvial Sands*; Higher Institute of Mining and Geology: Sofia, Bulgaria, 1988; Volume 34, pp. 227–238.
5. Li, Y.; Ouyang, S.; Tian, P. Weishanite—A new gold-bearing mineral. *Acta Mineral. Sin.* **1984**, *4*, 102–105.
6. Desborough, G.A.; Foord, E.E. A monoclinic, pseudo-orthorhombic Au-Hg mineral of potential economic significance in Pleistocene Snake River alluvial deposits of southeastern Idaho. *Can. Mineral.* **1992**, *30*, 1033–1038.
7. Barkov, A.Y.; Nixon, G.T.; Levson, V.M.; Martin, R.F. A cryptically zoned amalgam $(Au1.5-1.9Ag1.1-1.4)\Sigma2.8-3.0Hg1.0-1.2$ from a placer deposit in the Tulameen-Similkameen river system, British Columbia, Canada: Natural or Man-made? *Can. Mineral.* **2009**, *47*, 433–440. [CrossRef]
8. Svetlitskaya, T.V.; Nevolko, P.A.; Kolpakov, V.V.; Tolstykh, N.D. Native gold from the Inagli Pt-Au placer deposit (the Aldan Shield, Russia): Geochemical characteristics and implications for possible bedrock sources. *Miner. Depoz.* **2018**, *53*, 323–338. [CrossRef]
9. Shikazono, N.; Shimizu, M. *Electrum: Chemical Composition, Mode of Occurrence, and Depositional Environment*; The University Museum, University of Tokyo: Tokyo, Japan, 1988; Volume 32, pp. 1–81.
10. Yokoyama, K.; Takeuchi, S.; Nakai, I.; Tsutsumi, Y.; Sano, T.; Shigeoka, M.; Miyawaki, M.; Matsubara, M. *Chemical Compositions of Electrum Grains in Ore and Placer Deposits in the Japanese Islands*; National Museum of Nature and Science Monographs: Tokyo, Japan, 2011; Volume 42, pp. 1–80.
11. Miyahisa, M.; Higaki, J. Gold-Silver Mineral Resource of Ehime Prefecture. *Ehime-Ken. Chika. Shigen. Shiryo.* **1981**, *10*, 11–24.

12. Rolfe, C.; Hume-Rothery, W. The constitution of alloys of gold and mercury. *J. Less-Common Met.* **1967**, *13*, 1–10. [CrossRef]

13. Groen, J.C.; Craig, J.R.; Rimstidt, J.D. Gold-rich rim formation on electrum grains in placers. *Can. Mineral.* **1990**, *28*, 207–228.

minerals

MDPI

Article

Oyonite, Ag$_3$Mn$_2$Pb$_4$Sb$_7$As$_4$S$_{24}$, a New Member of the Lillianite Homologous Series from the Uchucchacua Base-Metal Deposit, Oyon District, Peru

Luca Bindi [1],*, Cristian Biagioni [2] and Frank N. Keutsch [3]

[1] Dipartimento di Scienze della Terra, Università degli Studi di Firenze, Via G. La Pira 4, I-50121 Firenze, Italy
[2] Dipartimento di Scienze della Terra, Università di Pisa, Via Santa Maria 53, I-56126 Pisa, Italy; cristian.biagioni@unipi.it
[3] Paulson School of Engineering and Applied Sciences and Department of Chemistry & Chemical Biology, Harvard University, 12 Oxford Street, Cambridge, MA 02138, USA; keutsch@seas.harvard.edu
* Correspondence: luca.bindi@unifi.it; Tel.: +39-055-275-7532

Received: 18 April 2018; Accepted: 1 May 2018; Published: 2 May 2018

Abstract: The new mineral species oyonite, ideally Ag$_3$Mn$_2$Pb$_4$Sb$_7$As$_4$S$_{24}$, has been discovered in the Uchucchacua base-metal deposit, Oyon district, Catajambo, Lima Department, Peru, as very rare black metallic subhedral to anhedral crystals, up to 100 μm in length, associated with orpiment, tennantite/tetrahedrite, menchettiite, and other unnamed minerals of the system Pb-Ag-Sb-Mn-As-S, in calcite matrix. Its Vickers hardness (VHN$_{100}$) is 137 kg/mm^2 (range 132–147). In reflected light, oyonite is weakly to moderately bireflectant and weakly pleochroic from dark grey to a dark green. Internal reflections are absent. Reflectance values for the four COM wavelengths [R_{min}, R_{max} (%) (λ in nm)] are: 33.9, 40.2 (471.1); 32.5, 38.9 (548.3); 31.6, 38.0 (586.6); and 29.8, 36.5 (652.3). Electron microprobe analysis gave (in wt %, average of 5 spot analyses): Cu 0.76 (2), Ag 8.39 (10), Mn 3.02 (7), Pb 24.70 (25), As 9.54 (12), Sb 28.87 (21), S 24.30 (18), total 99.58 (23). Based on 20 cations per formula unit, the chemical formula of oyonite is Cu$_{0.38}$Ag$_{2.48}$Mn$_{1.75}$Pb$_{3.79}$Sb$_{7.55}$As$_{4.05}$S$_{24.12}$. The main diffraction lines are (*d* in Å, *hkl* and relative intensity): 3.34 (−*312*; 40), 3.29 (−*520*; 100), 2.920 (−*132*; 40), 2.821 (−*232*; 70), 2.045 (*004*; 50). The crystal structure study revealed oyonite to be monoclinic, space group *P*2$_1$/*n*, with unit-cell parameters *a* = 19.1806 (18), *b* = 12.7755 (14), *c* = 8.1789 (10) Å, β = 90.471 (11)°, *V* = 2004.1 (4) Å^3, *Z* = 2. The crystal structure was refined to a final R_1 = 0.032 for 6272 independent reflections. Oyonite belongs to the Sb-rich members of the andorite homeotypic sub-series within the lillianite homologous series. The name oyonite is after the Oyon district, Lima Department, Peru, the district where the type locality (Uchucchacua mine) is located.

Keywords: oyonite; lillianite homologous series; sulfosalt; copper; antimony; arsenic; Oyon district; Lima department; Peru

1. Introduction

Oyonite has been found during a mineralogical investigation of the ore minerals from the Uchucchacua base-metal deposit, Oyon district, Catajambo, Lima Department, Peru. This deposit is the type locality for other four Mn-bearing sulfosalts: uchucchacuaite, AgMnPb$_3$Sb$_5$S$_{12}$ [1], benavidesite, MnPb$_4$Sb$_6$S$_{14}$ [2], manganoquadratite, AgMnAsS$_3$ [3], and menchettiite, AgMn$_{1.60}$Pb$_{2.40}$Sb$_3$As$_2$S$_{12}$ [4].

The sample containing oyonite was not found in situ but originates from a sample given to one of the authors (F.N.K.) by the mineral dealer John Veevaert. The material was found in October 2010 from the Nivel 890, Uchucchacua base-metal deposit, Oyon district, Catajambo, Lima Department, Peru, and it is the same material where menchettiite [4] was found. Geological and metallogenic data concerning this mining district have been reported by [2].

Oyonite is associated with orpiment, tennantite/tetrahedrite, menchettiite, and other unnamed minerals of the system Pb-Ag-Sb-Mn-As-S, in calcite matrix.

The new mineral was named oyonite, after the Oyon district, Lima Department, Peru, where the type locality (Uchucchacua mine) is located. The mineral and its name have been approved by the IMA CNMNC, under the number 2018-002. The holotype specimen of oyonite is deposited in the mineralogical collections of the Museo di Storia Naturale, Università degli Studi di Firenze, Via G. La Pira 4, Florence, Italy, under catalogue number 3283/I.

The mineralogical description of oyonite, as well as its crystal structure, are described in this paper.

2. Mineral Description and Physical Properties

Oyonite (Figure 1) occurs as black subhedral to anhedral crystals, up to 100 μm in length. Streak is black, and the luster is metallic.

Figure 1. Andorite-like minerals occurring with orpiment on calcite. The crystals are not a single phase but a mixture of mainly different andorite-group minerals (i.e., oyonite, menchettiite, ...). The width is 1 mm. Photo by Christian Rewitzer.

In plane-polarized incident light, oyonite is weakly to moderately bireflectant and weakly pleochroic from dark grey to dark green. Internal reflections are absent. Between crossed polars, the mineral is anisotropic, without characteristic rotation tints.

The reflectance was measured in air by means of an MPM-200 Zeiss microphotometer equipped with an MSP-20 system processor on a Zeiss Axioplan ore microscope. Filament temperature was approximately 3350 K. Readings were taken for specimen and standard (SiC) maintained under the same focus conditions. The diameter of the circular measuring area was 0.1 mm. Reflectance percentages in the form [R_{min}, R_{max} (%) (λ in nm)] are: 33.9, 40.2 (471.1); 32.5, 38.9 (548.3); 31.6, 38.0 (586.6); and 29.8, 36.5 (652.3).

Oyonite is brittle. Its Vickers hardness (VHN_{100}) is 137 kg/mm^2 (range 132–147), corresponding to a Mohs hardness of ~3–3.5. Based on the empirical formula, the calculated density is 5.237 g/cm^3. The density, calculated based on the ideal chemical formula (see below), is 5.275 g/cm^3.

3. Chemical Data

A preliminary EDS (Energy Dispersive Spectrometry) analysis performed on the crystal grain used for the structural study did not indicate the presence of elements ($Z > 9$) other than Cu, Ag, Mn, Pb, As, Sb and S.

Quantitative chemical analyses were carried out using a JEOL JXA-8200 electron-microprobe, operating in WDS (Wavelength Dispersive Spectrometry) mode. Major and minor elements were determined at 15 kV accelerating voltage and 30 nA beam current, with 10 s as counting time. For the WDS analyses, the following lines were used: $PbM\alpha$, $AgL\alpha$, $CuK\alpha$, $MnK\alpha$, $SbL\beta$, $AsL\alpha$, $SK\alpha$. The standards employed were: galena (Pb), Ag- pure element (Ag), Cu- pure element (Cu), synthetic MnS (Mn), pyrite (Fe,S), synthetic ZnS (Zn), synthetic Sb_2S_3 (Sb), synthetic As_2S_3 (As), synthetic Bi_2S_3 (Bi), and synthetic $PtSe_2$ (Se). Iron, Zn, Bi, and Se were found below detection limit. The crystal fragment was found to be homogeneous within analytical error. Table 1 gives analytical data (average of 5 spot analyses).

Table 1. Chemical data of oyonite.

Element	wt (%) ($n = 5$)	Range	Estimated Standard Deviation
Cu	0.76	0.50–1.05	0.02
Ag	8.39	8.08–8.91	0.10
Mn	3.02	2.89–3.22	0.07
Pb	24.70	24.55–25.81	0.25
As	9.54	9.11–9.82	0.12
Sb	28.87	28.24–29.61	0.21
S	24.30	23.63–24.71	0.18
Total	99.58	99.18–100.41	0.23

The empirical formula, based on 20 cations per formula unit, is $Cu_{0.38}Ag_{2.48}Mn_{1.75}Pb_{3.79}Sb_{7.55}As_{4.05}S_{24.12}$, with rounding errors. The $Ev(\%)$ value, defined as $[Ev(+) - Ev(-)] \times 100/Ev(-)$, is +1.0. After subtracting minor Cu according to the substitution $Cu^+ \rightarrow Ag^+$, the formula can be written as $Ag_{2.86}Mn_{1.75}Pb_{3.79}Sb_{7.55}As_{4.05}S_{24.12}$. The ideal formula is $Ag_3Mn_2Pb_4Sb_7As_4S_{24}$, which requires (in wt %) Ag 10.17, Mn 3.45, Pb 26.03, As 9.41, Sb 26.77, S 24.17, sum 100.00.

4. Crystallography

For the X-ray single-crystal diffraction study, the intensity data were collected using an Oxford Diffraction Xcalibur 3 diffractometer, equipped with a Sapphire 2 CCD area detector, with $MoK\alpha$ radiation. The detector to crystal working distance was 6 cm. Intensity integration and standard Lorentz-polarization corrections were performed with the *CrysAlis* RED [5] software package. The program ABSPACK in *CrysAlis* RED [5] was used for the absorption correction. Tests on the distribution of $|E|$ values agree with the occurrence of an inversion center ($|E^2 - 1| = 0.895$). This information, together with the systematic absences, suggested the space group $P2_1/n$. We decided to keep this non-standard setting of the space group to make easier the comparison with the other members of the lillianite homologous series [6]. The refined unit-cell parameters are $a = 19.1806$ (18), $b = 12.7755$ (14), $c = 8.1789$ (10) Å, $\beta = 90.471$ (11)°, $V = 2004.1$ (4) Å^3.

The crystal structure was refined with *Shelxl-97* [7] starting from the atomic coordinates of menchettiite [4]. The occurrence of twinning on {100} was considered. The site occupancy factors (s.o.f.) were refined using the scattering curves for neutral atoms given in the *International Tables for Crystallography* [8]. After several cycles of anisotropic refinement, a final $R_1 = 0.0227$ for 2199 reflections with $F_o > 4\sigma(F_o)$ was achieved (0.0317 for all 6272 reflections). Crystal data and details of the intensity data collection and refinement are reported in Table 2.

Table 2. Crystal and experimental details for oyonite.

Crystal Data	
Crystal size (mm^3)	$0.045 \times 0.055 \times 0.070$
Cell setting, space group	Monoclinic, $P2_1/n$
a (Å)	19.1806 (18)
b (Å)	12.7755 (14)
c (Å)	8.1789 (10)
β (°)	90.471 (11)
V (Å^3)	2004.1 (4)
Z	2
Data collection and Refinement	
Radiation, wavelength (Å)	Mo$K\alpha$, $\lambda = 0.71073$
Temperature (K)	293
$2\theta_{max}$ (°)	64.63
Measured reflections	27568
Unique reflections	6272
Reflections with $F_o > 4\sigma(F_o)$	2199
R_{int}	5.01
$R\sigma$	6.16
Range of h, k, l	$-27 \leq h \leq 28$, $0 \leq k \leq 18$, $0 \leq l \leq 12$
$R [F_o > 4\sigma(F_o)]$	0.0227
R (all data)	0.0317
wR (on $F_o{}^2$)	0.0756
Number of least-squares parameters	206
Maximum and minimum residual peak (e Å^{-3})	1.22 (at 1.26 Å from S9) $-$ 2.98 (at 1.00 Å from M8)

Atomic coordinates, site occupancies, and equivalent isotropic displacement parameters are given in Table 3. Selected bond distances are given in Table 4 and bond valence sums are given in Table 5. The crystallographic information file (CIF) is given as Supplementary Material.

Table 3. Atomic coordinates, site occupancies, and equivalent isotropic displacement parameters (in Å^2) for oyonite.

Site	s.o.f.	x	y	z	U_{eq}
M1	Sb$_{0.69(1)}$Mn$_{0.31(1)}$	0.13588 (7)	0.33007 (10)	0.12257 (19)	0.0516 (6)
M2	As$_{0.96(1)}$Sb$_{0.04(1)}$	0.06326 (7)	0.62044 (9)	0.1240 (2)	0.0285 (4)
M3	Pb$_{0.91(1)}$Sb$_{0.09(1)}$	0.24633 (3)	0.58190 (5)	0.38468 (9)	0.0473 (2)
M4	Ag$_{0.80(1)}$Cu$_{0.20(1)}$	0.35946 (6)	0.32999 (10)	0.1290 (2)	0.0526 (6)
M5	Sb$_{1.00}$	0.45771 (5)	0.60464 (6)	0.12818 (14)	0.0335 (2)
M6	Sb$_{1.00}$	0.11431 (5)	0.33212 (7)	0.61891 (14)	0.0379 (3)
M7	Sb$_{0.92(1)}$As$_{0.08(1)}$	0.05163 (5)	0.63732 (8)	0.64699 (15)	0.0378 (4)
M8	Pb$_{1.00}$	0.24437 (3)	0.57781 (4)	0.88178 (8)	0.0376 (2)
M9	Mn$_{0.55(1)}$Ag$_{0.45(1)}$	0.36399 (6)	0.34426 (9)	0.61981 (19)	0.0206 (4)
M10	As$_{1.00}$	0.43438 (6)	0.59947 (9)	0.6332 (2)	0.0254 (3)
S1	S$_{1.00}$	0.10164 (17)	0.5080 (3)	0.3380 (5)	0.0387 (10)
S2	S$_{1.00}$	-0.00527 (16)	0.2329 (2)	0.1480 (5)	0.0307 (8)
S3	S$_{1.00}$	0.25202 (18)	0.4302 (2)	0.1411 (4)	0.0360 (8)
S4	S$_{1.00}$	0.16552 (16)	0.7182 (2)	0.1134 (5)	0.0294 (8)
S5	S$_{1.00}$	0.3869 (2)	0.4998 (3)	0.4219 (6)	0.0470 (12)
S6	S$_{1.00}$	0.3468 (15)	0.7112 (2)	0.1078 (6)	0.0302 (9)
S7	S$_{1.00}$	0.09078 (17)	0.5105 (3)	0.8998 (6)	0.0446 (12)
S8	S$_{1.00}$	0.48785 (19)	0.2893 (3)	0.0896 (4)	0.0410 (10)
S9	S$_{1.00}$	0.23509 (15)	0.4183 (2)	0.6086 (5)	0.0327 (9)
S10	S$_{1.00}$	0.16473 (15)	0.7199 (2)	0.6453 (6)	0.0325 (9)
S11	S$_{1.00}$	0.39479 (17)	0.4896 (2)	0.8338 (5)	0.0397 (11)
S12	S$_{1.00}$	0.34339 (15)	0.7228 (2)	0.6462 (5)	0.0283 (8)

Table 4. Selected bond distances (in Å) in the crystal structure of oyonite.

M1			M2			M3			M4			M5		
	−S3	2.575 (4)		−S4	2.327 (3)		−S3	2.781 (4)		−S3	2.429 (3)		−S8	2.476 (4)
	−S12	2.624 (4)		−S7	2.372 (4)		−S9	2.789 (4)		−S8	2.540 (4)		−S6	2.531 (3)
	−S6	2.698 (4)		−S1	2.376 (4)		−S5	2.907 (4)		−S4	2.594 (4)		−S2	2.615 (3)
	−S1	2.952 (4)		−S2	3.108 (4)		−S1	2.953 (3)		−S10	2.686 (4)		−S11	3.061 (4)
	−S2	2.985 (3)		−S8	3.337 (4)		−S10	3.187 (4)		−S11	3.236 (4)		−S5	3.076 (4)
	−S7	3.056 (4)		−S7	3.400 (3)		−S4	3.208 (4)		−S5	3.271 (4)		−S11	3.088 (3)
							−S12	3.350 (3)						
							−S6	3.411 (4)						

M6			M7			M8			M9			M10		
	−S12	2.505 (4)		−S10	2.412 (3)		−S3	2.840 (3)		−S4	2.556 (4)		−S11	2.293 (4)
	−S6	2.531 (4)		−S2	2.525 (4)		−S4	3.023 (4)		−S10	2.557 (4)		−S5	2.326 (4)
	−S9	2.568 (3)		−S7	2.727 (5)		−S9	3.028 (4)		−S5	2.603 (4)		−S12	2.354 (3)
	−S8	2.887 (4)		−S8	2.840 (4)		−S10	3.053 (4)		−S11	2.616 (4)		−S8	3.054 (4)
	−S1	3.222 (4)		−S1	3.175 (5)		−S7	3.074 (4)		−S9	2.648 (3)		−S2	3.178 (4)
	−S7	3.271 (5)		−S1	3.480 (4)		−S11	3.125 (4)		−S2	2.703 (3)		−S5	3.687 (4)
							−S6	3.181 (4)						
							−S12	3.288 (4)						

Table 5. Bond valence sums (in valence unit) in oyonite.

Site	M1	M2	M3	M4	M5	M6	M7	M8	M9	M10	Σ_{anion}
S1	0.22	0.75	0.33			0.12	0.14 0.06				1.62
S2	0.20	0.10			0.64		0.79		0.24	0.08	2.05
S3	0.61		0.52	0.42				0.46			2.01
S4		0.86	0.17	0.27				0.28	0.36		1.94
S5			0.37	0.04	0.18				0.32	0.84 0.02	1.77
S6	0.43		0.10		0.80	0.80		0.18			2.31
S7	0.16	0.76 0.05				0.11	0.46	0.24			1.78
S8		0.06		0.31	0.93	0.31	0.34			0.12	2.07
S9			0.51			0.73		0.27	0.28		1.79
S10			0.17	0.21			1.07	0.26	0.36		2.07
S11				0.05	0.19 0.18			0.21	0.31	0.92	1.86
S12	0.53		0.11			0.86		0.14		0.78	2.42
Σ_{cation}	2.15	2.58	2.17	1.30	2.92	2.93	2.86	2.04	1.87	2.64	
Theor.	2.69	3.00	2.09	1.00	3.00	3.00	3.00	2.00	1.55	3.00	

X-ray powder diffraction data (Table 6) were obtained on the same fragment used for the single-crystal study with an Oxford Diffraction Excalibur PX Ultra diffractometer fitted with a 165 mm diagonal Onyx CCD detector and using copper radiation (CuKα, λ = 1.54138 Å). The working conditions were 40 kV and 40 nA with 1 hour of exposure; the detector-to-sample distance was 7 cm. The program *Crysalis* RED was used to convert the observed diffraction rings to a conventional powder diffraction pattern. The least squares refinement gave the following unit-cell values: a = 19.175 (1), b = 12.7775 (9), c = 8.1817 (8) Å, β = 90.26 (1)°, V = 2004.6 (2) Å^3.

Table 6. Observed and calculated X-ray powder diffraction data (*d* in Å) for oyonite. The five strongest reflections are given in bold.

I_{obs}	d_{obs}	I_{calc}	d_{calc}	*h*	*k*	*l*
-	-	13	9.5900	2	0	0
15	6.05	21	6.0605	1	2	0
10	5.32	14	5.3164	2	2	0
-	-	12	3.8947	0	1	2
25	3.83	25	3.8348	4	2	0
20	3.62	21	3.6183	−2	1	2
-	-	16	3.5986	2	1	2
40	**3.34**	40	3.3377	−3	1	2
-	-	42	3.3146	3	1	2
100	**3.29**	100	3.2886	5	2	0
-	-	12	3.1967	6	0	0
10	3.20	14	3.1939	0	4	0
30	3.15	30	3.1505	1	4	0
40	**2.920**	25	2.9179	−1	3	2
-	-	28	2.9127	1	3	2
30	2.860	30	2.8587	6	2	0
70	**2.821**	55	2.8240	−2	3	2
-	-	54	2.8146	2	3	2
35	2.678	20	2.6843	−3	3	2
-	-	24	2.6723	3	3	2
20	2.241	11	2.2493	−7	1	2
-	-	12	2.2328	7	1	2
-	-	20	2.0488	−8	1	2
50	**2.045**	35	2.0447	0	0	4
-	-	20	2.0346	8	1	2
20	1.980	16	1.9779	−4	5	2
10	1.971	14	1.9714	4	5	2
25	1.945	25	1.9460	4	6	0
10	1.919	9	1.9180	10	0	0
5	1.863	9	1.8659	−8	3	2
-	-	9	1.8551	8	3	2
15	1.740	14	1.7419	−5	2	4
-	-	14	1.7309	5	2	4

5. Results and Discussion

5.1. Crystal Structure Description

The crystal structure of oyonite (Figure 2) agrees with those of the $^{4,4}L$ homologue in the lillianite homologous series [6]. There are 10 metal sites and 12 S sites in the unit cell.

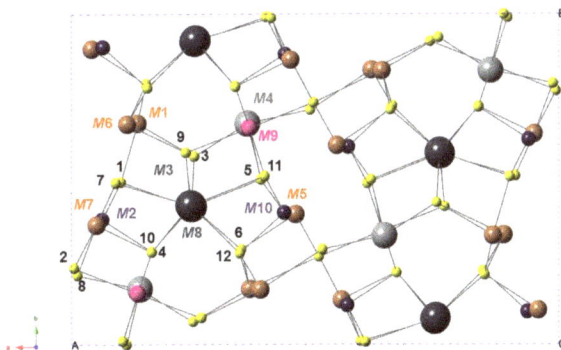

Figure 2. Unit-cell content of oyonite as seen down **c**. Circles: dark grey = Pb sites; light grey = Ag sites; orange = Sb sites; violet = As site; purple = Mn site; yellow = S sites. Atom labels as in Table 3 (for S sites, only the number is reported).

The general organization of oyonite, as seen down **c**, is shown in Figure 3. The crystal structure is formed by the alternation of $(311)_{PbS}$ slabs, four octahedra thick along $(100)_{PbS}$ and unit-cell twinned by reflection on $(311)_{PbS}$ planes. The crystal structure of oyonite shows a two-fold superstructure with respect to the short 4 Å axis of the substructure.

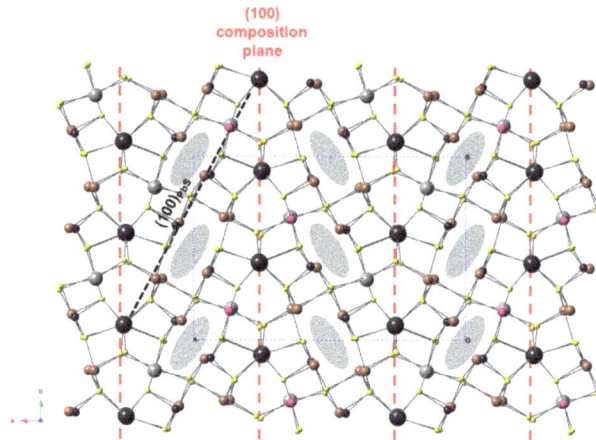

Figure 3. General organization of oyonite, as seen down **c**. Within each slab, separated by the (100) composition plane, $(100)_{PbS}$ planes occur (one of them is shown by the black dashed line). Lone electron-pair micelles are shown as grey ellipses.

Two independent sites are located on the composition planes (100) of the unit-cell twinning, i.e., $M3$ and $M8$. They show a bicapped trigonal prismatic coordination, with a pure Pb occupancy (at $M8$) or a mixed (Pb,Sb) occupancy (at $M3$), with a Pb:Sb atomic ratio of 91:9. The chemical composition of the (100) composition plane is $[(Pb_{1.91}Sb_{0.09})S_2]^{+0.09}$, that can be simplified as (Pb_2S_2).

Only one kind of $(311)_{PbS}$ slab occurs in the crystal structure of oyonite, formed by one kind of diagonal $(100)_{PbS}$ plane (Figure 4).

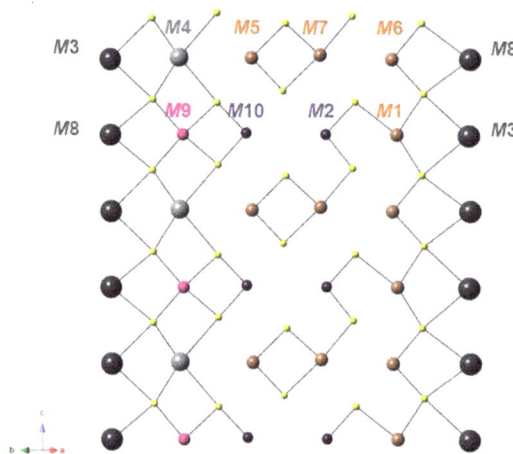

Figure 4. Organization of the $(100)_{PbS}$ layer in oyonite, as seen perpendicular to (110).

On the border of the plane, mixed (Ag,Cu) and (Mn,Ag)-centered octahedra (i.e., *M*4 and *M*9 sites, respectively) on one side, and Sb and (Sb,Mn)-centered sites alternate (*M*6 and *M*1 sites, respectively). In the center of the plane, a pure Sb site (*M*5) alternates with a pure As site (*M*10) along **c** in one column and mixed (Sb,As) and (As,Sb) sites in the other column (*M*7 and *M*2 sites, respectively). The composition of the diagonal $(100)_{PbS}$ plane is $(Cu_{0.20}Ag_{1.25}Mn_{0.86}Sb_{3.65}As_{2.04}S_{10})$, that can be simplified as $(Ag_{1.5}MnSb_{3.5}As_2S_{10})$. Thus, the simplified chemical composition of oyonite is given by $2 \times Pb_2S_2 + 2 \times (Ag_{1.5}MnSb_{3.5}As_2S_{10}) = Ag_3Mn_2Pb_4Sb_7As_4S_{24}$.

The chemical formula, as obtained through the single-crystal X-ray diffraction study, is $Cu_{0.40}Ag_{2.48}Mn_{1.74}Pb_{3.82}Sb_{7.48}As_{4.08}S_{24}$ ($Z = 2$).

5.2. Relation to the Other Species

Oyonite belongs to the andorite sub-series of the Sb-rich homeotypic members within the lillianite homologous series ([9]). Following the calculation procedure for the order N and the $Ag^+ + Me^{3+} = 2Pb^{2+}$ substitution proposed by [6], considering minor cations, the homologue order from chemical analysis is $N = 3.58$, slightly smaller than the crystallographic value $N = 4$ indicated by the crystal structure refinement, whereas the substitution percentage is 83%. The idealized chemical composition corresponds to $N = 4$ and a substitution percentage of 75%.

The simplified chemical formula of the members of the andorite homeotypic series is $Ag_xPb_{3-2x}Me^{3+}_{2+x}S_6$, where $Me =$ (Sb,Bi). Within this series, the classification of its members can be indicated by the percentage of the andorite component And_n, where n (%) = 100 x. In ideal oyonite, $n = 75$. Following [6] and [10], mineral species belonging to the andorite series have compositions close to integer values of m in the general formula $Ag_{16-m}Pb_{16+2m}Sb_{48-m}S_{96}$, where $m = 16 \times [1 - (n/100)]$. The known species have $m = 8$ (uchucchacuaite [11] and menchettiite [4]), 6 (fizélyite [12]), 5 (ramdohrite [13]), 1 (quatrandorite [14]), 0 (senandorite [15]), −1 (andreadiniite [16]), −2 (arsenquatrandorite [17]), −3 (roshchinite [18]), and −5 (jasrouxite [19]). Oyonite, having an andorite component And_{75}, corresponds to a new member in the sequence, having $m = 4$.

Oyonite is the third member of the andorite series where Mn plays a structural role. Based on 12 S atoms per formula unit ($Z = 2$), these three minerals have formulae $AgMnPb_3Sb_5S_{12}$ (uchucchacuaite [1,11]), $AgMn_{1.5}Pb_{2.5}Sb_3As_2S_{12}$ (menchettiite [4]), and $Ag_{1.5}MnPb_2Sb_{3.5}As_2S_{12}$ (oyonite). Oyonite is related to uchucchacuaite through the heterovalent substitutions $0.5^{M1}Sb^{3+} + {}^{M5}Pb^{2+} + 0.5^{M9}Mn^{2+} = 0.5^{M1}Mn^{2+} + {}^{M5}Sb^{3+} + 0.5^{M9}Ag^+$ and by the substitution of Sb-by-As at the *M*2 and *M*10 sites. It differs from menchettiite for a higher And component, related to a higher (Ag+Sb) content in the $(311)_{PbS}$ layers (e.g., Sb at *M*1 and Ag at *M*9). A comparison between site occupancies in the three Mn-rich members of the andorite group is given in Table 7.

Table 7. Summary of site occupancies in Mn-bearing ramdohrite-type minerals.

	Uchucchacuaite [11]	Menchettiite [4]	Oyonite (This Work)
*M*1	Sb	$Mn_{0.60}Pb_{0.40}$	$Sb_{0.69}Mn_{0.31}$
*M*2	Sb	$As_{0.57}Sb_{0.43}$	$As_{0.96}Sb_{0.04}$
*M*3	Pb	$Pb_{0.65}Sb_{0.35}$	$Pb_{0.91}Sb_{0.09}$
*M*4	Ag	Ag	$Ag_{0.80}Cu_{0.20}$
*M*5	Pb	$Sb_{0.77}Pb_{0.23}$	Sb
*M*6	Sb	Sb	Sb
*M*7	Sb	$Sb_{0.68}As_{0.32}$	$Sb_{0.92}As_{0.08}$
*M*8	Pb	Pb	Pb
*M*9	$Mn_{0.9}Ag_{0.05}Sb_{0.05}$	Mn	$Mn_{0.55}Ag_{0.45}$
*M*10	Sb	As	As

Supplementary Materials: The following are available online at http://www.mdpi.com/2075-163X/8/5/192/s1, CIF: oyonite.

Author Contributions: F.N.K. acquired the sample; L.B. found and described the new mineral; L.B. and C.B. conceived and designed the experiments; L.B. performed the experiments; L.B. and C.B. analyzed the data; L.B. wrote the paper with input from C.B. and F.N.K.

Acknowledgments: The research was supported by "progetto d'Ateneo 2015, University of Firenze" to Luca Bindi.

Conflicts of Interest: The authors declare no conflict of interest.

References

1. Moëlo, Y.; Oudin, E.; Picot P Caye, R. L'uchucchacuaite, $AgMnPb_3Sb_5S_{12}$, una nouvelle espèce minérale de la série de l'andorite. *Bull. Mineral.* **1984**, *107*, 597–604.

2. Oudin, E.; Picot, P.; Pillard, F.; Moëlo, Y.; Burke, E.A.J.; Zakrzewski, M.A. La bénavidésite, $Pb_4(Mn,Fe)Sb_6S_{14}$, un noveau minéral de la série de la jamesonite. *Bull. Mineral.* **1982**, *105*, 166–169.

3. Bonazzi, P.; Keutsch, F.N.; Bindi, L. Manganoquadratite, $AgMnAsS_3$, a new manganese-bearing sulfosalt from the Uchucchacua polymetallic deposit, Lima Department, Peru: Description and crystal structure. *Am. Mineral.* **2012**, *97*, 1199–1205. [CrossRef]

4. Bindi, L.; Keutsch, F.N.; Bonazzi, P. Menchettiite, $AgPb_{2.40}Mn_{1.60}Sb_3As_2S_{12}$, a new sulfosalt belonging to the lillianite series from the Uchucchacua polymetallic deposit, Lima Department, Peru. *Am. Mineral.* **2012**, *97*, 440–446. [CrossRef]

5. Oxford Diffraction. *CrysAlis RED (Version 1.171.31.2) and ABSPACK in CrysAlis RED*; Oxford Diffraction Ltd.: Abingdon, UK, 2006.

6. Makovicky, E.; Karup-Møller, S. Chemistry and crystallography of the lillianite homologous series; Part I, General properties and definitions. *Neues Jahrb. Mineral. Abh.* **1977**, *130*, 264–287.

7. Sheldrick, G.M. A short history of SHELX. *Acta Crystallogr.* **2008**, *64*, 112–122. [CrossRef] [PubMed]

8. Wilson, A.J.C. (Ed.) *International Tables for Crystallography Volume C, Mathematical, Physical and Chemical Tables*; Kluwer Academic: Dordrecth, NL, USA, 1992.

9. Moëlo, Y.; Makovicky, E.; Mozgova, N.N.; Jambor, J.L.; Cook, N.; Pring, A.; Paar, W.H.; Nickel, E.H.; Graeser, S.; Karup-Møller, S.; et al. Sulfosalt systematics: A review. Report of the sulfosalt sub-committee of the IMA Commission on Ore Mineralogy. *Eur. J. Mineral.* **2008**, *20*, 7–46. [CrossRef]

10. Moëlo, Y.; Makovicky, E.; Karup-Møller, S. Sulfures complexes plombo-argentifères: Minéralogie et cristallochimie de la série andorite-fizelyite, $(Pb,Mn,Fe,Cd,Sn)_{3-2x}(Ag,Cu)_x(Sb,Bi,As)_{2+x}(S,Se)_6$. *Doc. BRGM* **1989**, *167*, 107.

11. Yang, H.; Downs, R.T.; Evans, S.H.; Feinglos, M.N.; Tait, K.T. Crystal structure of uchucchacuaite, $AgMnPb_3Sb_5S_{12}$, and its relationship with ramdohrite and fizélyite. *Am. Mineral.* **2011**, *96*, 1186–1189. [CrossRef]

12. Yang, H.; Downs, R.T.; Burt, J.B.; Costin, G. Structure refinement of an untwined single crystal of Ag-excess fizélyite, $Ag_{5.94}Pb_{13.74}Sb_{20.84}S_{48}$. *Can. Mineral.* **2009**, *47*, 1257–1264. [CrossRef]

13. Makovicky, E.; Mumme, W.G.; Gable, R.W. The crystal structure of ramdohrite, $Pb_{5.9}Fe_{0.1}Mn_{0.1}In_{0.1}Cd_{0.2}Ag_{2.8}Sb_{10.8}S_{24}$: A new refinement. *Am. Mineral.* **2013**, *98*, 773–779. [CrossRef]

14. Nespolo, M.; Ozawa, T.; Kawasaki, Y.; Sugiyama, K. Structural relations and pseudosymmetries in the andorite homologous series. *J. Mineral. Petrol. Sci.* **2012**, *107*, 226–243. [CrossRef]

15. Sawada, H.; Kawada, I.; Hellner, E.; Tokonami, M. The crystal structure of senandorite (andorite VI): $PbAgSb_3S_6$. *Z. Kristallogr.* **1987**, *180*, 141–150. [CrossRef]

16. Biagioni, C.; Moëlo, Y.; Orlandi, P.; Paar, W.H. Andreadiniite, IMA 2014-049. CNMNC Newsletter No. 22, October 2014, page 1244. *Mineral. Mag.* **2014**, *78*, 1241–1248.

17. Topa, D.; Makovicky, E.; Putz, H.; Zagler, G.; Tajjedin, H. Arsenquatrandorite, IMA 2012-087. CNMNC Newsletter No. 16, August 2013, page 2696. *Mineral. Mag.* **2013**, *77*, 2695–2709.

18. Makovicky, E.; Stöger, B.; Topa, D. The incommensurately modulated crystal structure of roshchinite, $Cu_{0.09}Ag_{1.04}Pb_{0.65}Sb_{2.82}As_{0.37}S_{6.08}$. *Z. Kristallogr.-Cryst. Mater.* **2018**, *233*, 255–267. [CrossRef]

19. Makovicky, E.; Topa, D. The crystal structure of jasrouxite, a Pb-Ag-As-Sb member of the lillianite homologous series. *Eur. J. Mineral.* **2014**, *26*, 145–155. [CrossRef]

![minerals logo] *minerals*

MDPI

Article

Cerromojonite, CuPbBiSe₃, from El Dragón (Bolivia): A New Member of the Bournonite Group

Hans-Jürgen Förster [1],*, Luca Bindi [2], Günter Grundmann [3] and Chris J. Stanley [4]

[1] Helmholtz Centre Potsdam German Research Centre for Geosciences GFZ, DE-14473 Potsdam, Germany

[2] Dipartimento di Scienze della Terra, Università degli Studi di Firenze, Via G. La Pira 4, I-50121 Firenze, Italy; luca.bindi@unifi.it

[3] Chair of Engineering Geology, Technical University Munich, Arcisstr. 23, DE-80333 Munich, Germany; grundmann.g@gmx.de

[4] Department of Earth Sciences, Natural History Museum, Cromwell Road, London SW7 5BD, UK; c.stanley@nhm.ac.uk

* Correspondence: forhj@gfz-potsdam.de; Tel.: +49-0331-288-28843

Received: 25 July 2018; Accepted: 30 August 2018; Published: 21 September 2018

Abstract: Cerromojonite, ideally $CuPbBiSe_3$, represents a new selenide from the El Dragón mine, Department of Potosí, Bolivia. It either occurs as minute grains (up to 30 μm in size) in interstices of hansblockite/quijarroite intergrowths, forming an angular network-like intersertal texture, or as elongated, thin-tabular crystals (up to 200 μm long and 40 μm wide) within lath-shaped or acicular mineral aggregates (interpreted as pseudomorphs) up to 2 mm in length and 200 μm in width. It is non-fluorescent, black, and opaque, with a metallic luster and black streak. It is brittle, with an irregular fracture, and no obvious cleavage and parting. In plane-polarized incident light, cerromojonite is grey to cream-white, and weakly pleochroic, showing no internal reflections. Between crossed polarizers, cerromojonite is weakly anisotropic, with rotation tints in shades of brown and grey. Lamellar twinning on {110} is common. The reflectance values in air for the COM standard wavelengths (R_1 and R_2) are: 48.8 and 50.3 (470 nm), 48.2 and 51.8 (546 nm), 47.8 and 52.0 (589 nm), and 47.2 and 52.0 (650 nm). Electron-microprobe analyses yielded a mean composition of: Cu 7.91, Ag 2.35, Hg 7.42, Pb 16.39, Fe 0.04, Ni 0.02, Bi 32.61, Se 33.37, total 100.14 wt %. The empirical formula (based on 6 atoms *pfu*) is $(Cu_{0.89}Hg_{0.11})_{\Sigma=1.00}(Pb_{0.56}Ag_{0.16}Hg_{0.15}Bi_{0.11}Fe_{0.01})_{\Sigma=0.99}Bi_{1.00}Se_{3.01}$. The ideal formula is $CuPbBiSe_3$. Cerromojonite is orthorhombic (space group $Pn2_1m$), with $a = 8.202(1)$ Å, $b = 8.741(1)$ Å, $c = 8.029(1)$ Å, $V = 575.7(1)$ Å³, $Z = 4$. Calculated density is 7.035 g·cm⁻³. The five strongest measured X-ray powder diffraction lines (*d* in Å (I/I_0) (*hkl*)) are: 3.86 (25) (120), 2.783 (100) (122), 2.727 (55) (212), 2.608 (40) (310), and 1.999 (25) (004). Cerromojonite is a new member of the bournonite group, representing the Se-analogue of součekite, $CuPbBi(S,Se)_3$. It is deposited from strongly oxidizing low-*T* hydrothermal fluids at a f_{Se2}/f_{S2} ratio >1, both as primary and secondary phase. The new species has been approved by the IMA-CNMNC (2018-040) and is named for Cerro Mojon, the highest mountain peak closest to the El Dragón mine.

Keywords: cerromojonite; selenium; copper; lead; mercury; bismuth; součekite; bournonite group; El Dragón; Bolivia

1. Introduction

The Bolivian Andes host two mineralogically important selenide occurrences: Pacajake, district of Hiaco de Charcas; and El Dragón, Province of Antonio Quijarro; both in the Department of Potosí. The El Dragón mineralization represents a multi-phase assemblage of primary and secondary minerals, among which Se-bearing phases are the most prominent. It is the type locality for eldragónite, $Cu_6BiSe_4(Se_2)$ [1]; favreauite, $PbBiCu_6O_4(SeO_3)_4(OH)\cdot H_2O$ [2]; grundmannite,

CuBiSe$_2$ [3]; hansblockite, (Cu,Hg)(Bi,Pb)Se$_2$ [4]; alfredopetrovite, Al$_2$(Se^{4+}O$_3$)$_3$·6H$_2$O [5]; quijarroite, Cu$_6$HgPb$_2$Bi$_4$Se$_{12}$ [6]; and also contains the lately discovered rare orthorhombic dimorph of CuSe$_2$, petříčekite [7]. A comprehensive survey of the geology and origin of the El Dragón Se-mineralization was published by Grundmann and Förster (2017) [8], who also provided a full list of minerals recorded as from this locality.

This paper provides a description of a new species in the Cu–Hg–Pb–Bi–Se system, cerromojonite, ideally CuPbBiSe$_3$, from El Dragón. This new species and its name have been approved by the Commission on New Minerals, Nomenclature and Classification (CNMNC) of the IMA, proposal 2018-040. The X-rayed crystal is preserved by one of the authors (L.B.) at the Dipartimento di Scienze della Terra, Università degli Studi di Firenze. The polished section, from which the crystal was extracted (holotype), is housed in the collections of the Natural History Museum, London, with the catalogue number BM 2018, 11. Cotype material, consisting of another cerromojonite-bearing polished section, is deposited within the Mineralogical State Collection Munich (Mineralogische Staatssammlung München, Museum "Reich der Kristalle"), with the inventory number MSM 73583.

The species is named for Cerro Mojon, the highest mountain peak nearest to El Dragón (4292 m above sea level), located about 800 m northeast of the mine.

Unnamed phase "C", described at El Dragón already in 2016 [3], for which no structural data were obtained, compositionally resembles cerromojonite. A species similar to cerromojonite was speculated to occur in carbonate veins in the Schlema–Alberoda U–Se–polymetallic deposit (Erzgebirge, Germany), forming tiny inclusions which are intimately intergrown with berzelianite [9,10]. However, neither compositional data of pure material nor structural data were provided for this material.

2. Geology

The abandoned El Dragón mine is situated in the Cordillera Oriental (southwestern Bolivia), about 30 km southwest of Cerro Rico de Potosí. It is located at 19°49′23.90″ S (latitude), 65°55′00.60″ W (longitude), at an altitude of 4160 m above sea level. The adit of the mine is on the orographic left side of the Rio Jaya Mayu, cutting through a series of thinly-stratified, pyrite-rich black shales, and reddish-grey, hematite-bearing siltstones, dipping 40° to the north. The almost vertical ore vein is located in the center of a 1.5-m-wide shear zone (average trend 135 degrees). In 1988, the selenium mineralization consisted of a single vein of small longitudinal extension (maximum 15-m-long gallery), ranging mostly from 0.5 to 2 cm in thickness.

3. Physical and Optical Properties

Cerromojonite occurs within two different mineral assemblages, which are interpreted as representing genetically distinct types.

Type-I cerromojonite (maximum grain size ~30 μm) crystallized in the interstices of quijarroite/hansblockite intersertal intergrowths, partly together with penroseite (NiSe$_2$), klockmannite (CuSe), watkinsonite (Cu$_2$PbBi$_4$Se$_8$), clausthalite (PbSe), and, more rarely, petrovicite (Cu$_3$HgPbBiSe$_5$) (Figure 1). These aggregates are cemented by umangite (Cu$_3$Se$_2$) and klockmannite. They were deposited at the surfaces of krut'aite–penroseite (CuSe$_2$–NiSe$_2$) solid solutions.

Cerromojonite type-II occurs within lath-shaped or acicular mineral aggregates (up to 2 mm in length and 200 μm in width), that are interpreted as pseudomorphs after the above described intersertal aggregates (Figure 2). In this case, it forms elongated thin-tabular crystals (up to 200 μm long and 40 μm wide), intimately (subparallel) intergrown with watkinsonite, and less frequently, quijarroite (Figure 3), clausthalite, unnamed CuNi$_2$Se$_4$ [11], and (according to energy-dispersive electron-microprobe analysis) two new Cu–(Ag)–Hg–Pb–Bi selenides (Figure 4), which were all cemented by klockmannite. These pseudomorphs occasionally show parallel intergrowth of grains, which is implied by the serrated prismatic grain surfaces. They are usually deposited in interstices in brecciated krut'aite–penroseite (CuSe$_2$–NiSe$_2$) grains. The appearance of the cerromojonite grains resembles a spinifex texture, indicating fast crystallization. Type-II cerromojonite and associated minerals are themselves altered by late klockmannite, fracture-filling chalcopyrite, covellite, goethite, endmember petříčekite and krut'aite ($CuSe_2$), and native selenium (Figure 5).

Cerromojonite is black in color and possesses a black streak. The mineral is opaque in transmitted light, exhibits a metallic luster, and is non-fluorescent. No cleavage and parting is observed, and the fracture is irregular. Density and Mohs hardness could not be measured owing to the small crystal size. The calculated density is 7.035 g/cm^3 (for Z = 4), based on the empirical formula (see below) and the unit-cell parameters derived from X-ray single-crystal refinement.

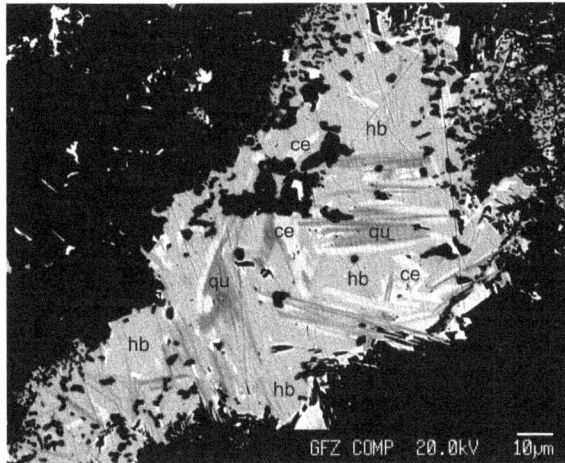

Figure 1. Back-scattered electron (BSE) image of type-I cerromojonite intergrown with hansblockite and quijarroite, forming an angular network-like intersertal texture. Abbreviations: ce = cerromojonite, hb = hansblockite, qu = quijarroite.

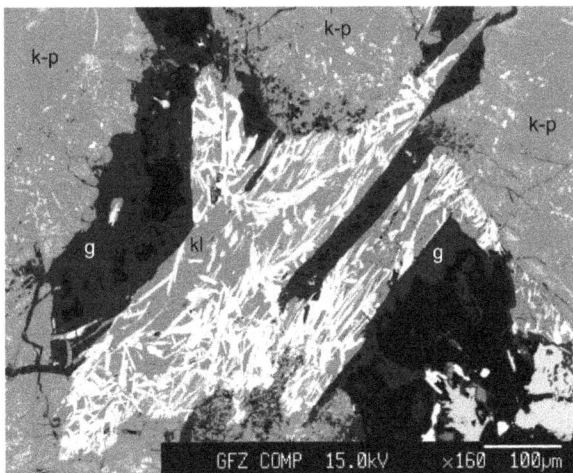

Figure 2. BSE image showing a pseudomorph composed of bright type-II cerromojonite, watkinsonite, and quijarroite, cemented by medium-grey klockmannite. Abbreviations: k–p = krut'aite–penroseite solid solutions, kl = klockmannite, g = goethite.

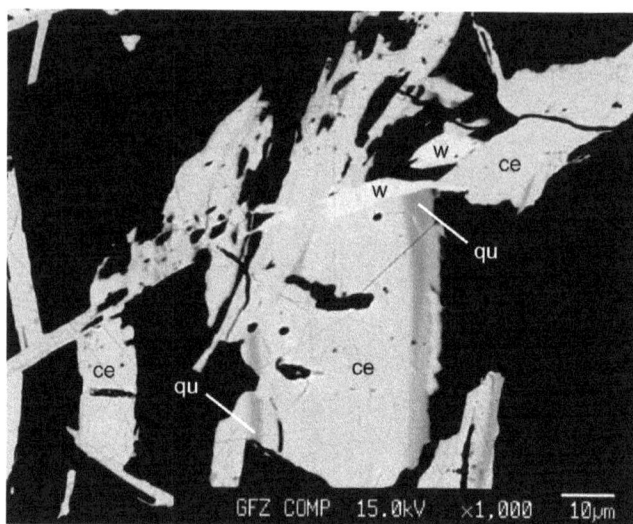

Figure 3. BSE image of type-II cerromojonite (ce) intergrown with quijarroite (qu) and watkinsonite (w).

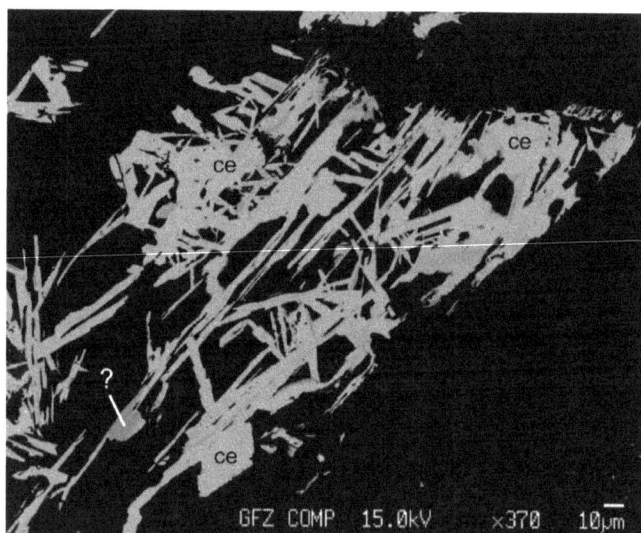

Figure 4. BSE image of parallel-intergrown type-II cerromojonite (ce) grains with darker domains of an unknown Cu–(Ag)–Hg–Pb–Bi selenide. The biggest grain of this potentially new species is marked by a question mark.

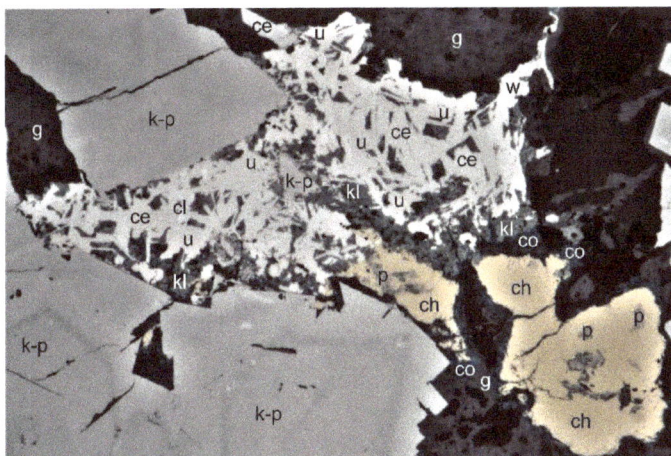

Figure 5. Type-II cerromojonite-bearing pseudomorph progressively altered by goethite and sulfides (reflected light, horizontal field of view is 500 μm). Abbreviations: ce = cerromojonite, ch = chalcopyrite, cl = clausthalite, co = covellite, g = goethite, kl = klockmannite, k–p = krut'aite–penroseite solid solution, p = penroseite, w = watkinsonite, u = unnamed $CuNi_2Se_4$.

In plane-polarized incident light, cerromojonite is grey to cream-white. In the assemblage with klockmannite and watkinsonite, it is weakly pleochroic or bireflectant. The mineral does not show any internal reflections. Between crossed polarizers, cerromojonite is weakly anisotropic, with rotation tints in shades of brown and grey (Figure 6). Twinning of cerromojonite is expressed either by distinct sharp polysynthetic lamellae (Figure 7) or by finely divided twinning in fan-shaped aggregates.

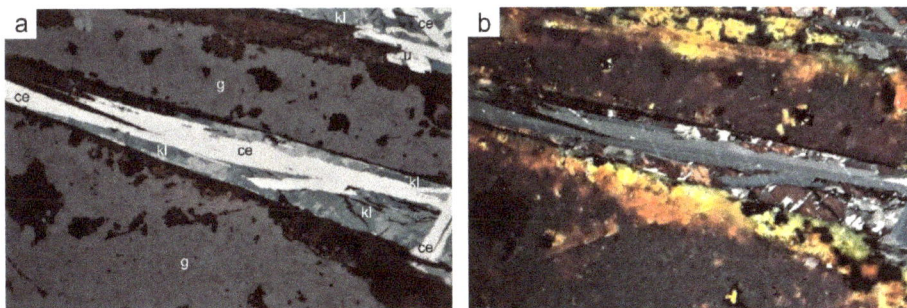

Figure 6. Type-II cerromojonite and associated minerals in plane polarized light (**a**), and at partially crossed polarizers (**b**) (horizontal field of view is 200 μm). Lamellar twinning on {110} of cerromojonite is well displayed in (**b**). Abbreviations: ce = cerromojonite, g = goethite, kl = klockmannite, u = unnamed $CuNi_2Se_4$.

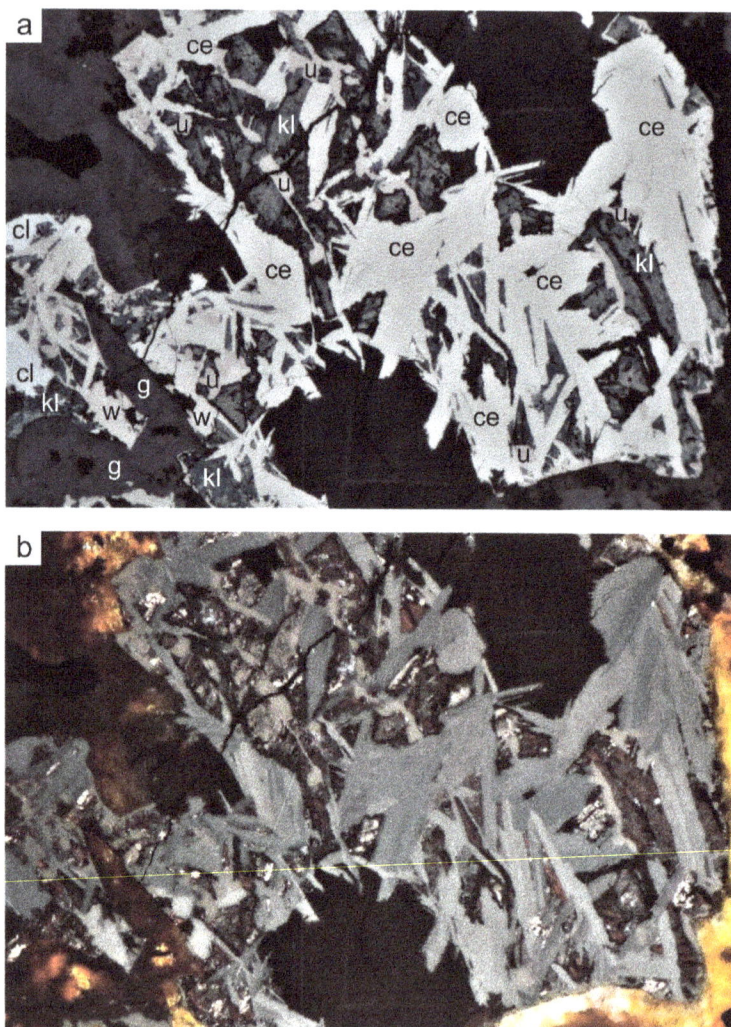

Figure 7. Type-II cerromojonite associated with klockmannite, unnamed $CuNi_2Se_4$, clausthalite, and later-formed goethite in reflected light: (**a**) 1 polarizer, (**b**) partially crossed polarizers. Horizontal field of view is 200 μm. Abbreviations: ce = cerromojonite, kl = klockmannite, w = watkinsonite, cl = clausthalite, u = unnamed $CuNi_2Se_4$, g = goethite.

Quantitative reflectance measurements were performed in air relative to a WTiC standard (Zeiss number 314) by means of a J & M TIDAS diode array spectrometer (J & M Analytik AG, Essingen, Germany), running ONYX software (Version 1.1, Cavendish Instruments Ltd., Sheffield, UK) on a Zeiss Axioplan ore microscope (Carl Zeiss AG, Oberkochen, Germany) (Table 1, Figure 8). Measurements were made on unoriented grains at extinction positions leading to the designation of R_1 (minimum) and R_2 (maximum).

Table 1. Reflectance data.

λ (nm)	R_1 (%)	R_2 (%)	λ (nm)	R_1 (%)	R_2 (%)
400	47.0	48.0	560	48.1	51.9
420	47.2	48.6	580	47.9	52.0
440	47.5	49.3	600	47.7	52.1
460	47.8	50.0	620	47.5	52.1
480	48.1	50.6	640	47.3	52.0
500	48.3	51.1	660	47.1	51.9
520	48.3	51.5	680	46.9	51.7
540	48.3	51.7	700	46.8	51.6

Reflectance percentages (R_1 and R_2) for the four Commission on Minerals (COM) wavelengths are: 48.8, 50.3 (470 nm); 48.2, 51.8 (546 nm); 47.8, 52.0 (589 nm); 47.2, 52.0 (650 nm).

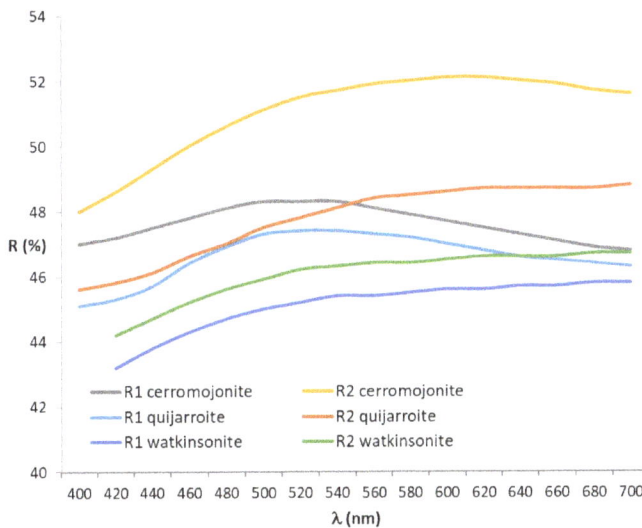

Figure 8. Reflectance spectra of type-II cerromojonite and its most frequently associated Cu–Bi selenides: quijarroite [6] and watkinsonite [12].

4. Chemical Data

Cerromojonite was checked for concentrations of Cu, Ag, Pb, Hg, Fe, Co, Ni, As, Sb, Bi, S, and Se. Twenty-four spot analyses of type-II cerromojonite from the holotype section, were performed using a JEOL JXA-8230 electron microprobe (WDS mode, 20 kV, 20 nA, 1–2 μm beam size) (JEOL Ltd., Akishima, Japan). The composition of the grain used for the structural study corresponds chemically to the other grains analyzed by microprobe, which were proved to be homogeneous within 2σ standard deviations of the analyzed elements. The counting time on the peak was 20 s, with half that time on background on both sites of the peak. The following standards, emission lines, and analyzing crystals (in parentheses) were used: Cu–eskebornite, *K*α (LIFL); Ag–bohdanowiczite, *L*α (PETJ); Pb–galena, *M*α (PETH); Hg–cinnabar, *L*α (LIFL); Fe–chalcopyrite, *K*α (LIFL); Co–cobaltite, *K*α (LIFL); Ni–pentlandite, *K*α (LIFL); As–skutterudite, *L*α (TAP); Sb–skutterudite, *L*α (PETJ); Bi–synthetic Bi_2Se_3, *M*α (PETH); S–chalcopyrite, *K*α (PETJ); Se–synthetic Bi_2Se_3, *K*α (LIFL). The software-implemented PRZ (XPP metal) data-correction routine (JEOL EPMA version 10) (JEOL Ltd., Akishima, Japan), which is based on the $\phi(\rho Z)$ method [13], was used for data processing. Table 2 compiles the analytical data for cerromojonite (means of 24 spot analyses, ranges, and standard deviations). Table 3 provides a selection of results from microprobe spot analyses of cerromojonite, together with the elemental detection limits (d.l.).

Table 2. Chemical data for cerromojonite.

Element	Mean	Range	e.s.d.
Cu (wt %)	7.91	7.40–8.16	0.18
Ag	2.35	2.16–2.54	0.11
Hg	7.42	7.19–7.60	0.10
Pb	16.39	16.15–16.77	0.13
Fe	0.04	0.00–0.18	0.04
Ni	0.02	0.00–0.18	0.04
Bi	32.61	32.19–32.91	0.20
Se	33.37	32.93–33.81	0.24
Total	100.11	99.24–100.79	0.42

$(Cu_{0.89}Hg_{0.11})_{\Sigma = 1.00}(Pb_{0.56}Ag_{0.16}Hg_{0.15}Bi_{0.11}Fe_{0.01})_{\Sigma = 0.99}Bi_{1.00}Se_{3.01}$ is the empirical formula of cerromojonite (based on 6 atoms *pfu*). The ideal formula of the mineral is $CuPbBiSe_3$, corresponding to the ideal contents of the elements (in wt %) Cu 8.87, Pb 28.92, Bi 29.15, Se 33.06, sum 100.00.

Table 3. Representative results of electron-microprobe spot analyses of cerromojonite.

Element	d.l. (ppm)	1	2	3	4	5	6
Cu (wt %)	250	7.40	7.75	8.09	8.03	7.91	7.91
Ag	200	2.45	2.54	2.30	2.33	2.54	2.27
Hg	1100	7.48	7.46	7.37	7.38	7.32	7.33
Pb	400	16.29	16.31	16.50	16.46	16.34	16.36
Fe	150	b.d.l.	0.14	b.d.l.	b.d.l.	b.d.l.	b.d.l.
Ni	200	b.d.l.	0.04	b.d.l.	b.d.l.	0.11	0.03
Bi	300	32.74	32.76	32.78	32.80	32.42	32.51
Se	800	33.48	33.27	33.58	33.31	33.07	33.56
Total		99.86	100.28	100.64	100.22	99.74	99.98
Cu (*a.p.f.u.*)		0.84	0.87	0.90	0.90	0.89	0.89
Ag		0.16	0.17	0.15	0.15	0.17	0.15
Hg		0.27	0.26	0.26	0.26	0.26	0.26
Pb		0.56	0.56	0.56	0.57	0.56	0.56
Fe		-	0.02	-	-	-	-
Ni		-	-	-	-	0.01	-
Bi		1.12	1.12	1.11	1.12	1.11	1.11
Se		3.04	3.00	3.01	3.01	2.99	3.03

a.p.f.u. = atoms per formula unit; b.d.l. = below limit of detection.

5. X-ray Crystallography and Description of the Crystal Structure

X-ray powder diffraction data (Table 4) were obtained from the same fragment used for the single-crystal study (see below), with an Oxford Diffraction Excalibur PX Ultra diffractometer (Oxford Diffraction, Oxford, UK), fitted with a 165 mm diagonal Onyx CCD detector, and using copper radiation (Cu$K\alpha$, λ = 1.54138 Å). The working conditions were 40 kV and 40 mA, with 1 hour of exposure; the detector-to-sample distance was 7 cm.

Table 4. Measured and calculated X-ray powder diffraction data (d in Å) for cerromojonite. The strongest measured diffraction lines are given in bold.

hkl	d_{meas}	I_{meas}	d_{calc}	I_{calc}
020	-	-	4.3705	4
-	4.08	10	4.1010	9
002	4.00	20	4.0145	15
120	**3.86**	**25**	**3.8571**	**20**
210	3.70	10	3.7127	9
112	3.32	10	3.3333	9
220	2.991	10	2.9906	10
022	-	-	2.9565	6
202	-	-	2.8688	7
122	**2.783**	**100**	**2.7814**	**100**
130	2.747	10	2.7456	12
212	**2.727**	**55**	**2.7257**	**50**
310	**2.608**	**40**	**2.6093**	**37**
222	-	-	2.3983	5
320	-	-	2.3178	3
132	-	-	2.2663	9
312	-	-	2.1878	5
040	2.186	10	2.1853	14
004	**1.999**	**25**	**2.0073**	**21**
330	1.992	20	1.9937	21
042	-	-	1.9193	4
142	1.867	10	1.8688	12
412	1.788	20	1.7875	22
332	-	-	1.7856	6
124	-	-	1.7806	4
242	-	-	1.7384	6
422	-	-	1.6849	3
224	-	-	1.6666	3
134	-	-	1.6204	4
314	1.592	20	1.5910	14
252	1.494	10	1.4928	10
044	-	-	1.4783	7
522	-	-	1.4344	4
530	-	-	1.4294	4
334	1.415	10	1.4145	12
260	-	-	1.3728	4
600	-	-	1.3670	4
126	-	-	1.2642	6
452	-	-	1.2628	8
216	-	-	1.2589	3
534	-	-	1.1644	3
264	-	-	1.1331	4
604	-	-	1.1299	4
416	-	-	1.1115	3
722	-	-	1.0893	5
182	-	-	1.0457	3
456	-	-	0.9434	3

Note: calculated diffraction pattern obtained with the atom coordinates reported in Table 6 (only reflections with $I_{rel} \geq 3$ are listed).

The program *Crysalis* RED [14] was used to convert the observed diffraction rings to a conventional powder diffraction pattern. Least squares refinement gave the following orthorhombic unit-cell values: $a = 8.2004(6)$ Å, $b = 8.7461(5)$ Å, $c = 8.0159(5)$ Å, $V = 574.91(5)$ Å^3, and $Z = 4$.

A small crystal ($0.040 \times 0.055 \times 0.060$ mm^3) of type-II cerromojonite was handpicked from the holotype specimen (it is registered under the number #123 in the mineralogical collection of one of the Authors, G.G.). The crystal was preliminarily examined with a Bruker-Enraf MACH3 single-crystal diffractometer (Bruker, Karlsruhe, Germany), using graphite-monochromatized Mo$K\alpha$ radiation. Single-crystal X-ray diffraction intensity data were collected using an Oxford Diffraction Xcalibur diffractometer equipped with an Oxford Diffraction CCD detector, with graphite-monochromatized Mo$K\alpha$ radiation ($\lambda = 0.71073$ Å). The data were integrated, and corrected for standard Lorentz and polarization factors, with the *CrysAlis* RED package [12]. The program ABSPACK in *CrysAlis* RED [12] was used for the absorption correction. Table 5 reports details of the selected crystal, data collection, and refinement.

Table 5. Data and experimental details for the selected cerromojonite crystal.

Crystal Data	
Ideal formula	$CuPbBiSe_3$
Crystal size (mm^3)	$0.040 \times 0.055 \times 0.060$
Form	Block
Color	Black
Crystal system	Orthorhombic
Space group	$Pn2_1m$
a (Å)	8.202(1)
b (Å)	8.741(1)
c (Å)	8.029(1)
V (Å^3)	575.7(1)
Z	4
Data Collection	
Instrument	Oxford Diffraction Xcalibur 3
Radiation type	Mo$K\alpha$ ($\lambda = 0.71073$ Å)
Temperature (K)	293(3)
Detector to sample distance (cm)	6
Number of frames	889
Measuring time (s)	50
Maximum covered 2θ ($^\circ$)	59.30
Absorption correction	multi-scan [12]
Collected reflections	3504
Unique reflections	1359
Reflections with $F_o > 4\sigma(F_o)$	701
R_{int}	0.0356
R_σ	0.0412
Range of h, k, l	$0 \leq h \leq 11, -12 \leq k \leq 8, 0 \leq l \leq 10$
Refinement	Full-matrix least squares on F^2
Final R_1 [$F_o > 4\sigma(F_o)$]	0.0256
Final R_1 (all data)	0.0315
S	1.09
Number refined parameters	68
$\Delta\rho_{max}$ (e Å^{-3})	1.81
$\Delta\rho_{min}$ (e Å^{-3})	−2.06

Statistical tests ($|E^2 - 1| = 0.821$) and systematic absences agreed with the acentric space group $Pn2_1m$. The crystal structure was refined starting from the atomic coordinates of bournonite [15]. Given the observed larger unit-cell volume of cerromojonite (i.e., 575.7 Å^3) compared to bournonite (i.e., 552.3 Å^3; [15]), the site occupancy factor (s.o.f.) at the crystallographic sites was allowed to vary (Pb vs. Ag and Bi vs. Ag for the Pb and Bi sites; Cu vs. Hg for the Cu site; Se vs. S for the anionic site), using scattering curves for neutral atoms taken from the International Tables for Crystallography [16]. After several cycles of anisotropic refinement, a final $R_1 = 0.0256$ for 701 reflections with $F_o > 4$ $\sigma(F_o)$ and 68 refined parameters was achieved (0.0315 for all 1359 reflections). Atomic coordinates, site occupancies, and equivalent isotropic displacement parameters are listed in Table 6, whereas anisotropic displacement parameters are given in Table 7. Selected bond distances and bond-valence sums are provided in Table 8. The Crystallographic Information File (CIF) is available as Supplementary Material.

Table 6. Atoms, site occupancy factors (s.o.f.), fractional atomic coordinates (x, y, z), and equivalent isotropic displacement parameters (U_{eq}, Å^2) for the selected cerromojonite crystal.

Atom	s.o.f.	x	y	z	U_{eq}
Pb1	$Pb_{0.80(2)}Ag_{0.20}$	0.07291(13)	0.9709(3)	0	0.0108(5)
Pb2	$Pb_{0.74(2)}Ag_{0.26}$	0.56972(12)	0.1758(3)	$\frac{1}{2}$	0.0115(5)
Bi1	$Bi_{1.00}$	0.07446(12)	0.9807(2)	$\frac{1}{2}$	0.0118(4)
Bi2	$Bi_{1.00}$	0.55647(12)	0.1819(2)	0	0.0134(4)
Cu	$Cu_{0.870(8)}Hg_{0.130}$	0.27526(17)	0.42151(15)	0.2439(3)	0.0120(6)
Se1	$Se_{1.00}$	0.2450(3)	0.2494(4)	0	0.0103(6)
Se2	$Se_{1.00}$	0.2316(3)	0.2595(4)	$\frac{1}{2}$	0.0105(6)
Se3	$Se_{1.00}$	0.08564(19)	0.65945(19)	0.2374(4)	0.0102(4)
Se4	$Se_{1.00}$	0.57919(18)	0.4826(2)	0.2675(4)	0.0096(4)

Table 7. Anisotropic displacement parameters (U) of the atoms for the selected cerromojonite crystal.

Atom	U^{11}	U^{22}	U^{33}	U^{12}	U^{13}	U^{23}
Pb1	0.0116(7)	0.0118(11)	0.0089(7)	0.0005(5)	0.000	0.000
Pb2	0.0102(8)	0.0135(10)	0.0107(8)	−0.0008(7)	0.000	0.000
Bi1	0.0123(6)	0.0122(10)	0.0110(7)	−0.0001(4)	0.000	0.000
Bi2	0.0149(6)	0.0136(8)	0.0118(6)	−0.0001(6)	0.000	0.000
Cu	0.0131(8)	0.0125(7)	0.0104(9)	0.0007(4)	0.0007(7)	−0.0003(10)
Se1	0.0108(11)	0.0119(16)	0.0081(12)	−0.0007(10)	0.000	0.000
Se2	0.0114(10)	0.0111(15)	0.0089(12)	0.0001(10)	0.000	0.000
Se3	0.0101(7)	0.0108(7)	0.0096(11)	−0.0004(6)	0.0000(7)	0.0036(13)
Se4	0.0111(7)	0.0108(7)	0.0070(12)	−0.0002(7)	−0.0009(6)	−0.0004(12)

Table 8. Bond distances (in Å) and bond valence sums (BVS in valence units) in the structure of cerromojonite.

Pb1-Se1	2.814(4)	Bi2-Se1	2.622(3)
Pb1-Se3 (×2)	2.975(3)	Bi2-Se4 (×2)	2.785(3)
Pb1-Se2	3.107(3)	Bi2-Se4 (×2)	3.395(3)
Pb1-Se3 (×2)	3.334(3)	Bi2-Se3 (×2)	3.620(3)
Pb1-Se4 (×2)	3.408(3)	BVS	3.22
BVS	1.93		
		Cu-Se1	2.482(3)
Pb2-Se2	2.868(3)	Cu-Se2	2.522(3)
Pb2-Se4 (×2)	2.993(3)	Cu-Se4	2.557(2)
Pb2-Se4 (×2)	3.268(3)	Cu-Se3	2.598(2)
Pb2-Se3 (×2)	3.412(3)	BVS	1.29
BVS	1.62		
Bi1-Se2	2.757(4)		
Bi1-Se3 (×2)	2.793(3)		
Bi1-Se1	3.314(3)		
Bi1-Se3 (×2)	3.519(3)		
Bi1-Se4 (×2)	3.563(3)		
BVS	2.96		

v.u. = valence units.

The crystal structure of cerromojonite (Figure 9) is identical to those of the three members of the bournonite group: bournonite ($PbCuSbS_3$), seligmannite ($PbCuAsS_3$), and součekite ($PbCuBi(S,Se)_3$). It consists of [7,9]Pb-polyhedra, [3+2,3+3]Bi-polyhedra, and $CuSe_4$ tetrahedra, which share corners and edges to form a 3-dimensional framework; CuSe$_4$ tetrahedra share corners to form chains parallel to [001] (Figure 10). The two Pb sites were found to exhibit a mean electron number of 75.0 and 72.9 electrons, respectively. According to their structural environments, and taking into account the bond-valence sums calculated using the parameters of Breese and O'Keeffe [17], the following site-populations were determined: Pb$_{0.52}$Ag$_{0.20}$Bi$_{0.16}$Hg$_{0.12}$ and Pb$_{0.60}$Hg$_{0.16}$Ag$_{0.12}$Bi$_{0.05}\square_{0.07}$. Bi-sites were thought to be filled by Bi only (according to the site-occupancy refinement), whereas the Cu site (35.6 electrons) was determined to be Cu$_{0.88}$Hg$_{0.12}$. Such a cation distribution is in agreement with the observed bond distances.

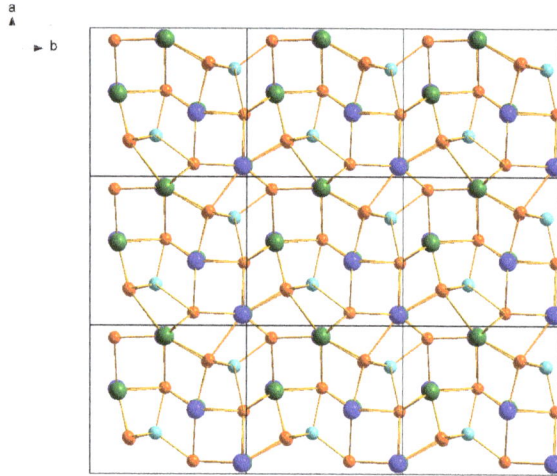

Figure 9. The crystal structure of cerromojonite projected down [001] (six unit-cells). The unit-cell and orientation of the figure are outlined. Symbols: Bi = green dots, Pb = blue dots, Cu = light blue dots, Se = orange dots.

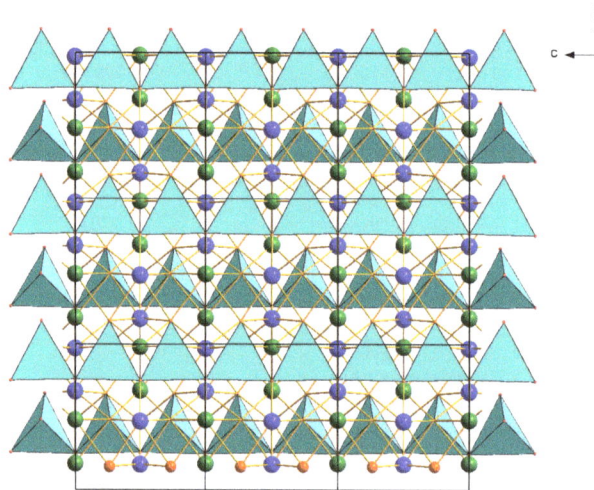

Figure 10. The crystal structure of cerromojonite projected down [100] (six unit-cells). Symbols as in Figure 9. CuSe$_4$ tetrahedra are depicted as light blue polyhedra. The unit-cell and orientation of the figure are outlined.

The overall crystallochemical formula, as obtained through the single-crystal X-ray diffraction study, is $[Cu_{0.880}Hg_{0.120}]Bi[Pb_{0.560}Ag_{0.155}Hg_{0.142}Bi_{0.107}\square_{0.036}]Se_3$ (Z = 4), and is in excellent agreement with that obtained from electron microprobe data.

6. Discussion

Cerromojonite is a new member of the bournonite group and represents the Se-analogue of součekite, $CuPbBi(S,Se)_3$ [12,18]. Interestingly, whereas previously analyzed součekite always contained appreciable amounts of Se (together with minor Te) substituting for S, cerromojonite from El Dragón is practically devoid of S, containing S at concentrations below its detection limit of ~200 ppm. Moreover, součekite is characterized by an almost ideal occupancy of the Cu-, Pb-, and Bi-sites, in contrast to cerromojonite, where significant amounts of other cations, in particular Ag and Hg, entered the structure.

Conclusions on the physico-chemical environment of cerromojonite formation could be drawn from the associated Cu selenides. It is apparent that type-I cerromojonite crystallized together with umangite and klockmannite, implying that selenium fugacities (f_{Se2}) fluctuated around values defined by the umangite–klockmannite univariant reaction. Type-II cerromojonite precipitated in equilibrium with klockmannite, outside the stability fields of umangite and krut'aite. At T = 100 °C, a temperature typical for the formation of vein-type selenide deposits, and an elevated oxygen fugacity defined by the magnetite–hematite buffer, these paragenetic relations are consistent with the range of log f_{Se2} between -14.6 and -11.6 [19]. The absence of krut'aite and sulfides (chalcopyrite, pyrite) define the maximum log of sulfur fugacity (f_{S2}) to be roughly -19.

Supplementary Materials: The following are available online at http://www.mdpi.com/2075-163X/8/10/420/s1, CIF: Cerromojonite.

Author Contributions: G.G. collected the samples and manufactured the polished sections; H.-J.F. and G.G. found the new mineral; H.-J.F. conducted the electron-microprobe analyses; L.B. performed the X-ray structural investigations; C.J.S. and G.G. determined the optical and physical properties; H.-J.F. wrote the paper.

Funding: This research received no external funding.

Acknowledgments: The research was supported by "progetto d'Ateneo 2015, University of Firenze" to L.B. C.J.S. acknowledges Natural Environment Research Council grant NE/M010848/1 Tellurium and Selenium Cycling and Supply. Oona Appelt (GFZ) provided assistance with the electron-microprobe work. Constructive comments of two anonymous reviewers helped to improve the paper.

Conflicts of Interest: The authors declare no conflict of interest.

References

1. Paar, W.H.; Cooper, M.A.; Moëlo, Y.; Stanley, C.J.; Putz, H.; Topa, D.; Roberts, A.C.; Stirling, J.; Raith, J.G.; Rowe, R. Eldragónite, $Cu_6BiSe_4(Se)_2$, a new mineral species from the El Dragón mine, Potosí, Bolivia, and its crystal structure. *Can. Mineral.* **2012**, *50*, 281–294. [CrossRef]
2. Mills, S.J.; Kampf, A.R.; Christy, A.G.; Housley, R.M.; Thorne, B.; Chen, Y.; Steele, I.M. Favreauite, a new selenite mineral from the El Dragón mine, Bolivia. *Eur. J. Mineral.* **2014**, *26*, 771–781. [CrossRef]
3. Förster, H.-J.; Bindi, L.; Stanley, C.J. Grundmannite, $CuBiSe_2$, the Se-analogue of emplectite: A new mineral from the El Dragón mine, Potosí, Bolivia. *Eur. J. Mineral.* **2016**, *28*, 467–477. [CrossRef]
4. Förster, H.-J.; Bindi, L.; Stanley, C.J.; Grundmann, G. Hansblockite, $(Cu,Hg)(Bi,Pb)Se_2$, the monoclinic polymorph of grundmannite, a new mineral from the Se mineralization at El Dragón (Bolivia). *Mineral. Mag.* **2017**, *81*, 629–640. [CrossRef]
5. Kampf, A.R.; Mills, S.J.; Nash, B.P.; Thorne, B.; Favreau, G. Alfredopetrovite: A new selenite mineral from the El Dragón mine. *Eur. J. Mineral.* **2016**, *28*, 479–484. [CrossRef]
6. Förster, H.-J.; Bindi, L.; Grundmann, G.; Stanley, C.J. Quijarroite, $Cu_6HgPb_2Bi_4Se_{12}$, a new selenide from the El Dragón mine, Bolivia. *Minerals* **2016**, *6*, 123. [CrossRef]
7. Bindi, L.; Förster, H.-J.; Grundmann, G.; Keutsch, F.N.; Stanley, C.J. Petříčekite, $CuSe_2$, a new member of the marcasite group from the Předbořice deposit, Central Bohemia Region, Czech Republic. *Minerals* **2016**, *6*, 33. [CrossRef]

8. Grundmann, G.; Förster, H.-J. Origin of the El Dragón selenium mineralization, Quijarro province, Potosí, Bolivia. *Minerals* **2017**, *7*, 68. [CrossRef]

9. Dymkov, Y.M.; Ryzhov, B.I.; Begizov, V.I.; Dubakina, L.S.; Zav'yalov, E.N.; Ryabeva, V.G.; Tsvetkova, M.V. Mgriite, bismuth petrovicite and associated selenides from carbonate veins of the Erzgebirge. *Novye Dannye o Mineralakh* **1991**, *37*, 81–101. (In Russian)

10. Jambor, J.L.; Pertsev, N.N.; Roberts, A.C. New mineral names. *Amer. Mineral.* **1995**, *80*, 845–850.

11. IMA-CNMNC proposal. 2018; submitted.

12. Johan, Z.; Picot, P.; Ruhlmann, F. The ore mineralogy of the Otish Mountains uranium deposit, Quebec: Skippenite, Bi_2Se_2Te, and watkinsonite, $Cu_2PbBi_4(Se,S)_8$, two new mineral species. *Can. Mineral.* **1987**, *25*, 625–638.

13. Heinrich, K.F.J.; Newbury, D.E. *Electron Probe Quantitation*; Plenum Press: New York, NY, USA, 1991.

14. Oxford Diffraction. *CrysAlis RED (Version 1.171.31.2) and ABSPACK in CrysAlis RED*; Oxford Diffraction Ltd.: Oxfordshire, UK, 2006.

15. Edenharter, A.; Nowacki, W.; Takéuchi, Y. Verfeinerung der Kristallstruktur von Bournonit $[(SbS_3)_2 \mid Cu^{IV}_2Pb^{VII}Pb^{VIII}]$ und von Seligmannit $[(AsS_3)_2 \mid Cu^{IV}_2Pb^{VII}Pb^{VIII}]$. *Z. Krist.* **1970**, *131*, 397–417. [CrossRef]

16. Maslon, E.N.; Fox, A.G.; O'Keefe, M.A. Mathematical, physical and chemical tables. In *International Tables for Crystallography*; Wilson, A.J.C., Ed.; Kluwer Academic: Dordrecht, The Netherlands, 1992; Volume C.

17. Breese, N.E.; O'Keeffe, M. Bond-Valence parameters for solids. *Acta Cryst.* **1991**, *B47*, 192–197. [CrossRef]

18. Čech, F.; Vavřín, I. Součekite, $CuPbBi(S,Se)_3$, a new mineral of the bournonite group. *Neues Jahrbuch für Mineralogie, Monatshefte* **1979**, *1979*, 289–295.

19. Simon, G.; Kesler, S.E.; Essene, E.J. Phase relations among selenides, sulphides, tellurides, and oxides: II. Applications to selenide-bearing ore deposits. *Econ. Geol.* **1997**, *92*, 468–484. [CrossRef]

Article

Fiemmeite Cu$_2$(C$_2$O$_4$)(OH)$_2$·2H$_2$O, a New Mineral from Val di Fiemme, Trentino, Italy

Francesco Demartin [1],*, Italo Campostrini [1], Paolo Ferretti [2] and Ivano Rocchetti [2]

[1] Dipartimento di Chimica, Università degli Studi di Milano, Via C. Golgi 19, I-20133 Milano, Italy; italo.campostrini@unimi.it

[2] MUSE, Museo delle Scienze di Trento, Corso del Lavoro e della Scienza 3, I-38122 Trento, Italy; paolo.ferretti@muse.it (P.F.); ivanorocchetti@tiscali.it (I.R.)

* Correspondence: francesco.demartin@unimi.it; Tel.: +39-02-503-14457

Received: 28 May 2018; Accepted: 11 June 2018; Published: 12 June 2018

Abstract: The new mineral species fiemmeite, Cu$_2$(C$_2$O$_4$)(OH)$_2$·2H$_2$O, was found NE of the Passo di San Lugano, Val di Fiemme, Carano, Trento, Italy (latitude 46.312° N, longitude 11.406° E). It occurs in coalified woods at the base of the Val Gardena Sandstone (upper Permian) which were permeated by mineralizing solutions containing Cu, U, As, Pb and Zn. The oxalate anions have originated from diagenesis of the plant remains included in sandstones. The mineral forms aggregate up to 1 mm across of sky blue platelets with single crystals reaching maximum dimensions of about 50 μm. Associated minerals are: baryte, olivenite, middlebackite, moolooite, brochantite, cuprite, devilline, malachite, azurite, zeunerite/metazeunerite, tennantite, chalcocite, galena. Fiemmeite is monoclinic, space group: $P2_1/c$ with a = 3.4245(6), b = 10.141(2), c = 19.397(3) Å, β = 90.71(1)°, V = 673.6(2) Å^3, Z = 4. The calculated density is 2.802 g/cm^3 while the observed density is 2.78(1) g/cm^3. The six strongest reflections in the X-ray powder diffraction pattern are: [d_{obs} in Å (I)(hkl)] 5.079(100)(020), 3.072(58)(112), 9.71(55)(002), 4.501(50)(022), 7.02(28)(012), 2.686(25)(114). The crystal structure was refined from single-crystal data to a final R_1 = 0.0386 for 1942 observed reflections [$I > 2\sigma(I)$] with all the hydrogen atoms located from a Difference–Fourier map. The asymmetric unit contains two independent Cu^{2+} cations that display a distorted square-bipyramidal (4+2) coordination, one oxalate anion, two hydroxyl anions and two water molecules. The coordination polyhedra of the two copper atoms share common edges to form polymeric rows running along [100] with composition [Cu$_2$(C$_2$O$_4$)(OH)$_2$·2H$_2$O]$_n$. These rows are held together by a well-established pattern of hydrogen bonds between the oxalate oxygens not involved in the coordination to copper, the hydrogen atoms of the water molecules and the hydroxyl anions.

Keywords: fiemmeite; new oxalate mineral; Val di Fiemme; Trentino; Italy

---

1. Introduction

The presence of small Cu ore deposits, in the area close to the Passo di San Lugano, Val di Fiemme, Carano, Trento, Italy is well known since the XV and XVI century, as documented by the remains of the old mining sites. From a stratigraphic point of view, the deposits are located within the Val Gardena Sandstone (upper Permian) a few meters above the limit with the ignimbrites of the Athesian Volcanic Group (lower Permian). The sedimentary sequence of the upper Permian consists of the continental deposits of alluvial plain of the Val Gardena Sandstone. The unconformity at the base of the Val Gardena Sandstone indicates a prolonged sub-aerial exposure with erosion of the volcanic substrate and consequent articulated topography. In this context, the deposition of the Val Gardena Sandstone began, in an environment of alluvial plain. The basal portion of the Val Gardena Sandstone corresponds to the first of the five third-order depositional sequences identified by Massari et al. [1], and is represented mainly by deposits of alternating alluvial conoids and reddened mudstones with pedogenized horizons and evaporites. The mineralization is set at the height of a Cu and U rich level, located at the base of the Val Gardena Sandstone (upper Permian). The major ore concentrations are found in deposits of carbon frustules and especially inside coalified trunks of up to metric sizes, impregnated with framboidal pyrite, covellite, tennantite and uraninite and surrounded by evident colored halos of supergenic minerals. This mineralization, referable to "sandstone-uranium type" deposits, can be explained by a genetic model given by a continental source made up of granite or acid volcanites that are eroded in an arid continental climate and then transported with U and other heavy metals, such as Cu, Pb, Zn, in the form of ions dissolved in clastic aquifers, in our case the alluvial conoid deposits of the Val Gardena Sandstone. The deposition of these ions is due to the strong decrease in solubility resulting from the reaction between mineralized groundwater and the strongly reducing environment given by the accumulation of trunks and organic matter in the channels or deposits of overbank. The peculiar greyish color of the sandstones for a range of a few meters around the levels with coalified trunks is a typical example of a reduction front (roll front) [2]. In this environment (latitude 46.312° N, longitude 11.406° E) we have recently identified the second world occurrence of middlebackite $Cu_2C_2O_4(OH)_2$ [3], a new copper oxalate discovered at the Iron Monarch quarry, Middleback Range, Australia and approved by the IMA CNMNC in 2016 (IMA 2015-115) [4]. A systematic investigation by micro-Raman spectroscopy of the mineralogical phases deposited (see Appendix A), allowed us to recognize that, besides middlebackite, other copper oxalates were present such as moolooite $CuC_2O_4 \cdot H_2O$ and another new mineral with chemical formula $Cu_2(C_2O_4)(OH)_2 \cdot 2H_2O$, that was approved as a new species by the IMA Commission on New Minerals, Nomenclature and Classification (No. 2017-115) with the name fiemmeite, after the type locality where it was found. This paper deals with the description of the new mineral fiemmeite, together with its crystal structure determination. Holotype material of fiemmeite is deposited in the Reference Collection of MUSE, Museo delle Scienze di Trento, sample No. 5249.

2. Experimental Data

2.1. Mineral Description and Physical Properties

Fiemmeite occurs in coalified woods as aggregates up to 1 mm across (Figure 1) made of sky blue elongated platelets with maximum dimensions about 50 μm. Associated minerals are: baryte, olivenite, middlebackite, moolooite, brochantite, cuprite, devilline, malachite, azurite, zeunerite/metazeunerite, tennantite, chalcocite and galena. The streak is pale blue and the lustre is from vitreous to waxy. It is brittle with cleavage almost perfect parallel to {010} or {001}, according to the weakest bonds observed in the crystal structure, and fracture uneven in the other directions. Its hardness was not determined due to the minute size of the crystals. Twinning was not observed. The mineral is non-fluorescent both under long- and short-wave ultraviolet radiation. A measurement of the density, obtained by flotation in a diiodomethane-benzene mixture gives the value of 2.78(1) g/cm^3. The density calculated using the empirical formula and single-crystal unit-cell data is 2.802 g/cm^3.

Crystals appear as highly birefringent platelets but we could not perform a precise optical characterization. The maximum and minimum refractive index we could measure are 1.90 and 1.54. A Gladstone-Dale calculation using the data of Mandarino [5,6] gives a mean refractive index of 1.64.

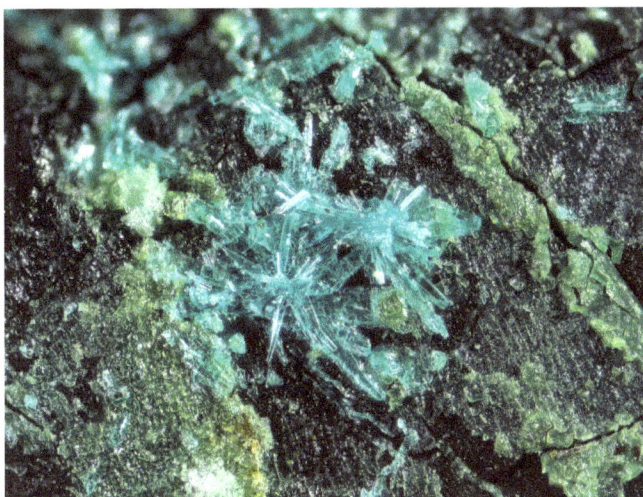

Figure 1. Fiemmeite aggregates with olivenite on coalified wood (base width 4 mm).

2.2. Chemical Data

Insufficient material is available for a direct determination of H_2O or C_2O_3 with a CHN analyzer. The presence of H_2O and C_2O_3 was confirmed by crystal structure analysis and Raman spectroscopy. Crystals rapidly decompose under a microprobe electron beam even using a low voltage current and a wide electron beam, therefore quantitative determination of Cu by microprobe analysis was not possible. However five chemical analyses of the copper and zinc content could be carried out, before damage of the sample, by means of a JEOL JSM 5500LV electron microscope equipped with an IXRF EDS 2000 microprobe (Table 1) with the following conditions: 20 kV, 10^{-11} A, 2 μm beam diameter. No other significant element quantities, besides Cu, Zn, O and C, were detected.

Table 1. Chemical data for fiemmeite.

Constituent	wt %	Range	Stand. Dev.	Probe Standard	wt % **
Cu	44.00	43.79–44.24	0.19	Synth. CuO	44.57
Zn	0.09	0.06–0.12	0.02	Synth. ZnO	0
O	44.40 *				44.89
C	8.34 *				8.42
H	2.10 *				2.12
Total	98.93				100.00

* theoretical for the empirical formula (based on 8 anions) $Cu_{1.996}Zn_{0.004}(C_2O_4)(OH)_2 \cdot 2H_2O$; ** theoretical for the empirical formula $Cu_2(C_2O_4)(OH)_2 \cdot 2H_2O$.

2.3. Micro Raman Spectroscopy

The Raman spectrum (Figure 2a,b) was obtained with an ANDOR 303 spectrometer equipped with a CCD camera iDus DV420A-OE, and with a Nikon CF plan 50×/0.55 objective. The 532 nm line of an OXXIUS solid state laser was used for excitation. The laser power was set to 10 mW in order to prevent damage of the crystal, with an aperture of 75 μm and 8.2 mm working distance. The two bands at 1683 and 1705 cm^{-1} can be assigned to ν_a(C=O), that at 1457 cm^{-1} to ν_s(C-O) + ν_s(C-C), that at 903 cm^{-1} to ν_s(C-O) + δ(O-C=O), that at 853 cm^{-1} to ν_s(C-O)/δ(O-C=O). Those at 466, 517 and 543 cm^{-1} are attributable to ν(Cu-O) + ν(C-C). The remaining below 298 cm^{-1} are assigned to out-of-plane bends and to lattice modes [7]. The observed bands at 3438 and 3471 cm^{-1} (Figure 2b) of the OH/H$_2$O region are consistent with the range of hydrogen bond lengths found (2.655–2.903 Å), according to the Libowitzky correlation [8].

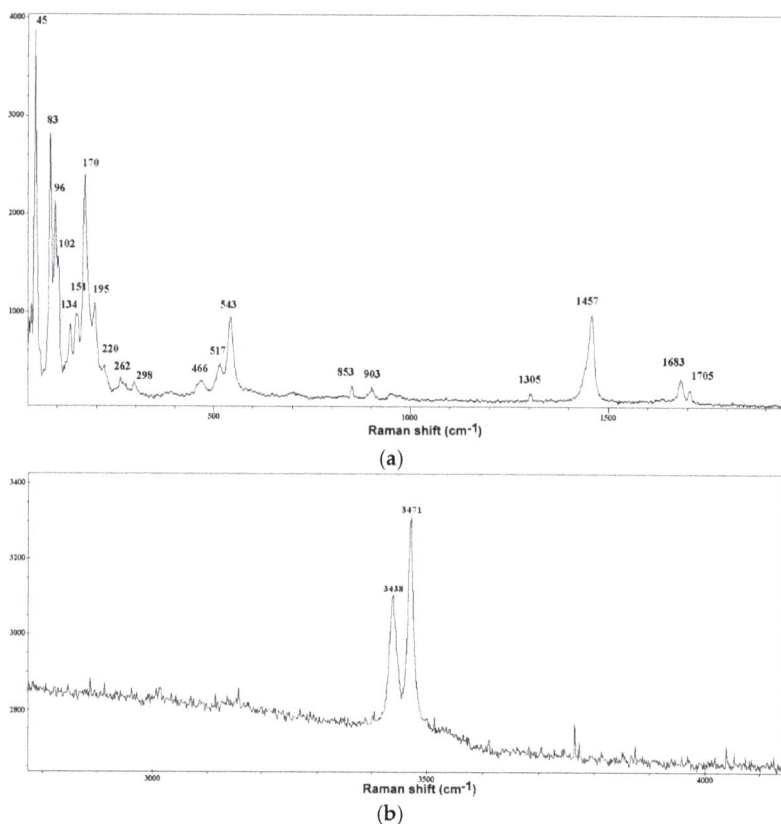

Figure 2. Micro-Raman spectrum of fiemmeite. (**a**) Oxalate and lattice mode bands; (**b**) hydrogen bond bands.

2.4. Crystallography

X-ray powder diffraction data of fiemmeite were obtained using a Bruker D8 diffractometer with graphite monochromatized CuKα radiation (Table 2). The unit-cell parameters, refined from powder data using the UNITCELL software [9], are: a = 3.4345(5), b = 10.159(2), c = 19.412(3) Å, β = 90.83(1)°, V = 677.5(1) Å^3.

Details of the single-crystal X-ray diffraction data collection and refinement are given in Table 3. 7512 intensities were collected at room temperature on a Bruker Apex II diffractometer with MoKα radiation (λ = 0.71073 Å) up to 2θ = 63.16°, of which 2118 unique. SADABS [10] absorption correction was applied. The structure was solved with direct methods [11] and refined with SHELXL-2017 [12] to a final R_1 = 0.0386 for 1942 observed reflections [$I > 2\sigma(I)$]. All the non-hydrogen atoms were refined anisotropically. All the hydrogen atoms were located in a Difference-Fourier map and refined. The coordinates and displacement parameters of the atoms are reported in Table 4; selected interatomic distances and angles and bond-valence values are listed in Table 5. The CIF file of fiemmeite is available as Supplementary Material.

Table 2. X-ray powder diffraction data for fiemmeite.

d_{obs} (Å)	I_{obs}	d_{calc} (Å) $^\$$	I_{calc} $^\$$	h, k, l	d_{obs} (Å)	I_{obs}	d_{calc} (Å) $^\$$	I_{calc} $^\$$	h, k, l
9.71	55	9.698	68	0 0 2	2.251	5	2.247	3	0 4 4
7.02	28	7.009	34	0 1 2	2.190	12	2.187	16	0 2 8
		5.452	3	0 1 3			2.162	6	−1 −3 4
5.079	100	5.071	100	0 2 0	2.151	13	2.147	17	1 3 4
4.914	12	4.906	10	0 2 1			2.122	3	1 2 6
4.855	2	4.849	5	0 0 4			2.052	1	−1 −3 5
4.501	50	4.493	40	0 2 2			2.036	2	1 3 5
3.996	4	3.990	6	0 2 3			2.028	6	−1 −4 1
		3.330	2	0 3 1			2.017	2	0 5 1
		3.241	4	−1 0 2			1.997	1	−1 −4 2
3.237	5	3.233	7	0 0 6	1.998	3	1.995	7	0 4 6
		3.193	1	−1 −1 1			1.972	1	1 2 7
3.198	5	3.192	7	0 3 2			1.953	1	−1 1 8
3.072	58	3.087	65	1 1 2	1.943	8	1.938	13	−1 3 6
3.001	2	2.996	6	0 3 3			1.935	4	0 5 3
2.891	20	2.886	18	1 1 3			1.905	3	0 1 10
		2.813	2	−1 0 4			1.883	2	−1 −4 4
		2.811	5	−1 −2 1			1.873	2	1 4 4
		2.731	2	−1 −2 2	1.873	5	1.870	6	0 4 7
2.730	15	2.726	9	0 2 6			1.824	6	−1 3 7
		2.711	13	−1 −1 4			1.809	2	−1 −4 5
2.686	25	2.682	32	1 1 4			1.797	4	0 5 5
		2.608	2	−1 −2 3	1.755	7	1.752	10	0 4 8
		2.589	1	1 2 3			1.745	1	1 5 0
2.552	3	2.549	6	0 3 5			1.719	1	−1 −5 2
		2.514	1	0 4 1			1.716	3	1 5 2
2.511	2	2.503	6	−1 −1 5			1.697	2	−1 0 10
2.468	2	2.474	7	−1 −2 4	1.695	9	1.689	7	−2 0 2
		2.453	2	0 4 2			1.688	6	2 1 0
2.442	4	2.438	5	1 2 4			1.687	3	−1 −5 3
		2.431	3	0 2 7			1.666	2	−2 −1 2
		2.424	1	0 0 8	1.657	3	1.656	6	1 1 10
		2.390	1	−1 −3 1			1.639	5	−1 −5 4
		2.360	2	0 4 3			1.642	4	0 4 9
		2.340	2	−1 −3 2			1.636	3	1 4 7
		2.336	3	0 3 6			1.609	3	−1 2 10
		2.336	2	1 0 6			1.603	4	−2 −2 2
		2.330	3	1 3 2			1.598	4	1 3 9
2.310	6	2.303	9	−1 1 6			1.596	2	0 6 4
		2.261	2	−1 −3 3			1.578	2	−2 −2 3
		2.248	4	1 3 3			1.544	3	−2 −2 4

Note: $^\$$ Pattern calculated on the basis of the single crystal data and structure refinement.

Table 3. Single-crystal diffraction data and refinement parameters for fiemmeite.

Crystal System	Monoclinic
Space Group	$P2_1/c$ (No. 14)
a (Å)	3.4245(6)
b (Å)	10.141(2)
c (Å)	19.397(3)
β (°)	90.71(1)
V (Å^3)	673.6(2)
Z	4
Radiation	Mo$K\alpha$ (λ = 0.71073 Å)
μ (mm^{-1})	6.322
D_{calc} (g·cm^{-3})	2.802
Measured reflections	7512
R_{int}	0.0294
Independent reflections	2118
Range of h, k, l	$-5 \leq h \leq 4, -14 \leq k \leq 14, -28 \leq l \leq 28$
Observed reflections [$I > 2\sigma(I)$]	1942
Parameters refined	133
Final R_1 [$I > 2\sigma(I)$] and $wR2$ (all data)	0.0386, 0.0905
GooF	1.176
Max/min residuals (e/Å^3)	1.37/−0.73

Notes: $R_1 = \Sigma||Fo| - |Fc||/\Sigma|Fo|$; $wR2 = \{\Sigma[w(Fo^2 - Fc^2)^2]/\Sigma[w(Fo^2)^2]\}^{1/2}$; $w = 1/[\sigma^2(Fo^2) + (0.0268q)^2 + 3.8682q]$ where q = [max(0, Fo^2) + $2Fc^2$]/3; GooF = $\{\Sigma[w(Fo^2 - Fc^2)]/(n - p)\}^{1/2}$ where n is the number of reflections and p is the number of refined parameters.

Table 4. Atom coordinates and displacement parameters [U_{eq}/U^{ij}, Å^2].

Atom	x/a	y/b	z/c	U_{eq}
Cu1	0.19372(13)	0.24133(4)	0.50020(2)	0.0126(1)
Cu2	0.61198(13)	0.29545(4)	0.37303(2)	0.0131(1)
C1	0.0211(11)	0.1394(3)	0.6268(2)	0.0130(6)
C2	−0.1380(10)	0.2827(4)	0.6259(2)	0.0132(6)
O1	0.2069(8)	0.1067(3)	0.5731(1)	0.0160(5)
O2	−0.1081(8)	0.3439(3)	0.5690(1)	0.0169(5)
O3	−0.0369(9)	0.0695(3)	0.6773(1)	0.0228(6)
O4	−0.2941(9)	0.3269(3)	0.6784(1)	0.0221(6)
OH5	0.6196(7)	0.1665(2)	0.4475(1)	0.0130(5)
OH6	0.1952(8)	0.3715(3)	0.4273(1)	0.0139(5)
Ow7	0.9519(9)	0.1899(3)	0.3121(1)	0.0193(5)
Ow8	0.5910(10)	0.4305(3)	0.3023(1)	0.0263(7)
H5	0.666(16)	0.0765(16)	0.441(3)	0.032(15)
H6	0.182(14)	0.4626(13)	0.434(3)	0.026(14)
H71	0.861(16)	0.193(6)	0.2664(11)	0.037(16)
H72	0.932(19)	0.0986(14)	0.318(3)	0.046(18)
H81	0.706(17)	0.423(7)	0.2589(15)	0.047(19)
H82	0.504(15)	0.517(2)	0.309(3)	0.028(14)

Atom	U^{11}	U^{22}	U^{33}	U^{23}	U^{13}	U^{12}
Cu1	0.0151(2)	0.0118(2)	0.01106(19)	0.00217(14)	0.00295(14)	0.00301(15)
Cu2	0.0156(2)	0.0120(2)	0.01179(19)	0.00236(14)	0.00336(14)	0.00312(15)
C1	0.0148(15)	0.0107(14)	0.0137(14)	−0.0003(11)	0.0008(11)	0.0013(12)
C2	0.0096(14)	0.0151(15)	0.0149(15)	−0.0014(12)	−0.0002(11)	0.0009(12)
O1	0.0211(13)	0.0124(12)	0.0147(11)	0.0011(9)	0.0044(9)	0.0056(10)
O2	0.0236(14)	0.0124(12)	0.0147(11)	0.0019(9)	0.0056(9)	0.0053(10)
O3	0.0360(17)	0.0152(13)	0.0175(13)	0.0055(10)	0.0091(11)	0.0069(12)
O4	0.0328(16)	0.0182(13)	0.0155(12)	−0.0012(10)	0.0075(11)	0.0090(12)
OH5	0.0149(12)	0.0097(11)	0.0145(11)	0.0019(9)	0.0029(9)	0.0042(9)
OH6	0.0159(12)	0.0116(11)	0.0144(11)	0.0026(9)	0.0034(9)	0.0033(9)
Ow7	0.0301(15)	0.0139(12)	0.0141(12)	−0.0007(9)	0.0045(10)	0.0041(11)
Ow8	0.0406(18)	0.0192(14)	0.0194(13)	0.0079(11)	0.0132(12)	0.0133(13)

The anisotropic displacement factor exponent takes the form: $-2\pi^2(U^{11}h^2(a^*)^2 + ... + 2U^{12}hka^*b^* + ...)$; U_{eq} according to Fischer and Tillmans [ ].

Table 5. Selected interatomic distances (Å), angles (°) and bond-valence sums (vu).

Atom1-Atom2	Distance	vu	Atom1-Atom2	Distance	vu
Cu1-OH6	1.935(3)	0.50	Cu2-Ow8	1.939(3)	0.46
Cu1-OH5	1.946(2)	0.46	Cu2-OH5	1.948(3)	0.45
Cu1-O1	1.966(3)	0.43	Cu2-OH6	1.943(3)	0.46
Cu1-O2	1.992(3)	0.40	Cu2-Ow7	1.983(3)	0.41
Cu1-OH5#1	2.332(3)	0.16	Cu2-OH6#2	2.375(3)	0.14
Cu1-O2#2	2.916(3)	0.03	Cu2-Ow7#1	2.754(3)	0.05
	Total	1.98		Total	1.97
Cu1-Cu2	2.9198(7)		C1-C2	1.552(5)	
C1-O3	1.228(4)		C2-O4	1.240(4)	
C1-O1	1.271(4)		C2-O2	1.270(4)	
Atom1-Atom2-Atom3	**angle**		**Atom1-Atom2-Atom3**	**angle**	
O1-C1-O3	126.2(3)		O1-C1-C2	114.5(3)	
C1-C2-O2	115.8(3)		O2-C2-O4	125.2(3)	
C1-C2-O4	118.9(3)		C2-C1-O3	119.3(3)	

Hydrogen bond interactions				
Atom1···Atom2	**distance**	**Atom1-Atom2···Atom3**		**angle**
OH5···O1#3	2.862(4)	OH5-H5···O1#3		177(5)
OH6···O2#4	2.903(4)	OH6-H6···O2#4		169(5)
Ow7···O3#3	2.655(4)	Ow7-H72···O3#3		163(6)
Ow7···O4#5	2.722(4)	Ow7-H71···O4#5		171(6)
Ow8···O4#4	2.691(4)	Ow8-H82···O4#4		175(5)
Ow8···O3#5	2.753(4)	Ow8-H81···O3#5		172(6)

Symmetry codes: #1 = $x - 1, y, z$; #2 = $x + 1, y, z$; #3 = $1 - x, -y, 1 - z$; #4 = $-x, 1 - y, 1 - z$; #5 = $x + 1, 1/2 - y, z - 1/2$. Bond-valence parameters after [14].

3. Crystal Structure Description and Discussion

The asymmetric unit of fiemmeite contains two independent Cu^{2+} cations, one oxalate anion, two hydroxyl anions and two water molecules (Figure 3a). Both copper cations display a distorted square-bipyramidal (4+2) coordination with four equatorial distances in the range 1.935(3)–1.992(3) Å, one apical distance intermediate [2.332(3)–2.375(3) Å] and one apical distance longer 2.754(3)–2.916(3) Å. Cu1 is coordinated by three oxygens of the oxalate anion and by three OH- anions. Cu2 is coordinated by three water molecules and by three OH^- anions.

The coordination polyhedra of the two copper atoms shares common edges to form polymeric rows, with composition $[Cu_2(C_2O_4)(OH)_2 \cdot 2H_2O]_n$, running along [100]. These rows are held together by a well-established pattern of hydrogen bonds (see Table 5) between the oxalate oxygens O3 and O4, not involved in the coordination to copper and the hydrogen atoms of the water molecules and between the other oxalate oxygens O1 and O2 and the hydroxyl anions.

Fiemmeite is the third example of oxalate containing only Cu^{2+} as cation, the others being moolooite $CuC_2O_4 \cdot H_2O$ [15], and middlebackite $Cu_2C_2O_4(OH)_2$ [4]. Other copper(II) oxalates with additional cations are wheatleyite $Na_2Cu(C_2O_4)_2 \cdot 2H_2O$ [16] and antipinite $KNa_3Cu_2(C_2O_4)_4$ [17].

A comparison of the crystal structure of fiemmeite with that of the other oxalates containing only copper can be done with middlebackite only, because the structure of moolooite is unknown.

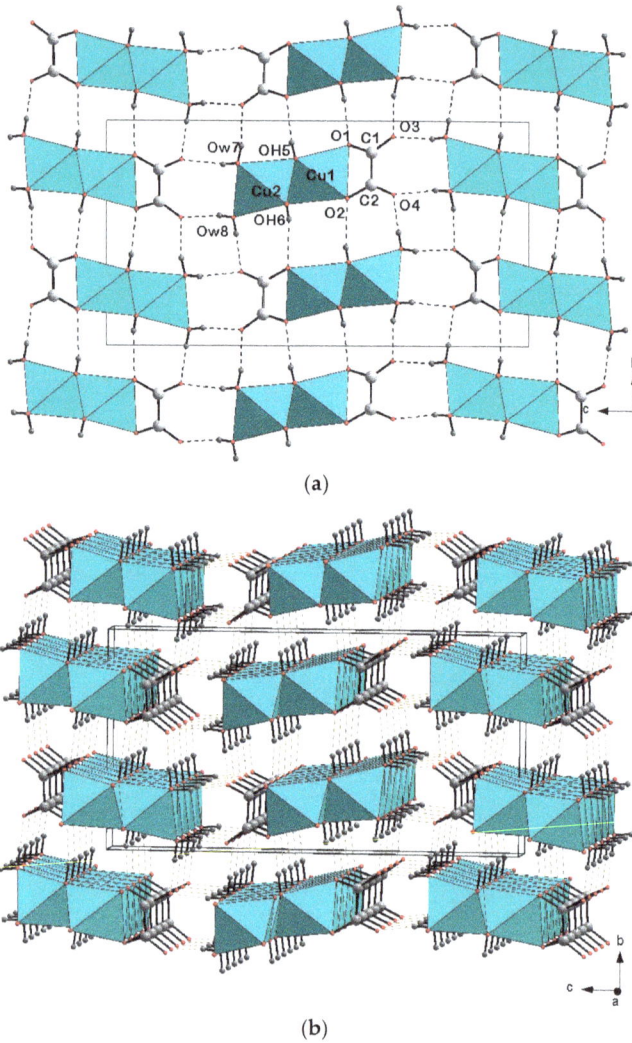

Figure 3. Projection along [100] with atoms labelling (**a**) and perspective view (**b**) of the crystal structure of fiemmeite with hydrogen bonds represented as dashed lines.

A portion of the polymeric rows with composition $[Cu_2(C_2O_4)(OH)_2 \cdot 2H_2O]_n$ is reported in Figure 4a and can be compared with those observed in the structure of middlebackite (Figure 4b). In fiemmeite, the oxalate anion acts as a chelating bidentate ligand with only one of the two independent copper ions, whereas in middlebackite the same anion is tetradentate, acting as a bridge between dimeric octahedral copper units, to form rows extending along [101]. In middlebackite, these rows are interconnected to form channels where the hydroxyl hydrogens are located.

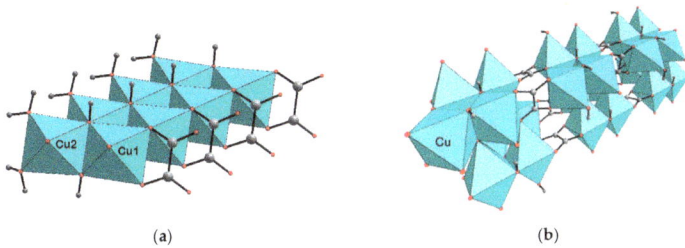

(a)   (b)

Figure 4. Portions of the polymeric rows observed in fiemmeite (**a**) and middlebackite (**b**).

No structural relationships to the other copper-containing oxalates or synthetic compound has been established. In fact, the crystal structure of the synthetic analogue of wheatleyite consists of columns of Cu-centered edge-sharing square bipyramids running along [100] (*a* axis 3.6 Å) and corrugated layers of seven-coordinate Na-centered polyhedra, held together by the oxalate anions and H bonds of water molecules in the vertices of Na-centered polyhedra. In antipinite, columns of edge sharing Cu-centered bipyramids running along [100] are instead combined into a layer by pairs of other Cu bipyramids.

Fiemmeite belongs to class 50.01.06 in the New Dana classification (Salts of Organic Acids, Oxalates) and to 10.AB (Oxalates) in the Nickel–Strunz classification [18].

Supplementary Materials: The CIF file of fiemmeite is available online at www.mdpi.com/2075-163X/8/6/248/s1.

Author Contributions: I.R. made the Raman measurements that allowed to discover the new mineral. F.D. performed the diffraction experiments and structure solution. P.F. and I.C. performed the chemical analyses and the determination of the mineral properties. P.F. and I.R. collected most of the material studied.

Funding: This research was funded by Euregio Science Fund (call 2014, IPN16) of the Europaregion Euregio; project "The end-Permian mass extinction in the Southern and Eastern Alps: extinction rates vs. taphonomic biases in different depositional environments".

Acknowledgments: The authors thank the Department of Innovation, Research and University of the Autonomous Province of Bozen/Bolzano for covering the Open Access publication costs. Stefano Dallabona is gratefully thanked for having donated some specimens for study. We also thank Alessandro Guastoni for having provided us with the Raman Spectrum of moolooite from Monte Cervandone and three anonymous referees for useful suggestions.

Conflicts of Interest: The authors declare no conflict of interest.

Appendix A

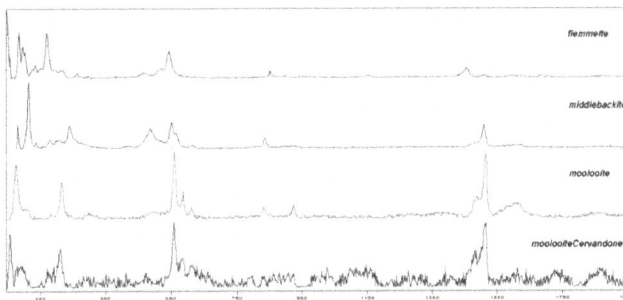

Figure A1. A comparison of the Raman spectra of fiemmeite, middlebackite and moolooite from Passo di San Lugano and moolooite from Monte Cervandone, Val d'Ossola, Italy.

References

1. Massari, F.; Neri, C.; Pittau, P.; Fontana, D.; Stefani, C. Sedimentology, Palynostratigraphy and sequence stratigraphy of a continental to shallow-marine rift-related succession: Upper Permian of the eastern Southern Alps (Italy). *Mem. Sci. Geol. Padova* **1994**, *46*, 119–243.

2. Wopfner, H.; Drake-Brockman, J. Base metal and uranium mineralization in the Groeden Sandstone of South Tyrol. *Geo. Alp* **2017**, *14*, 11–23.

3. Demartin, F.; Campostrini, I.; Ferretti, P.; Rocchetti, I. Second global occurrence of middlebackite near the Passo di San Lugano (Carano, Trento, Italy). *Geo. Alp* **2017**, *14*, 35–38.

4. Elliott, P. Middlebackite, IMA 2015-115. CNMNC Newsletter No. 30, April 2016, page 411. *Mineral. Mag.* **2016**, *80*, 407–413.

5. Mandarino, J.A. The Gladstone-Dale relationship I. Derivation of new constants. *Can. Mineral.* **1976**, *14*, 498–502.

6. Mandarino, J.A. The Gladstone-Dale relationship IV. The compatibility concept and its applications. *Can. Mineral.* **1981**, *19*, 441–450.

7. Frost, R.L. Raman spectroscopy of natural oxalates. *Anal. Chim. Acta* **2004**, *517*, 207–214. [CrossRef]

8. Libowitzky, E. Correlation of O-H stretching frequencies and O-H . . . O hydrogen bond lengths in minerals. *Monatsh. Chem.* **1999**, *130*, 1047–1059.

9. Holland, T.J.B.; Redfern, S.A.T. Unit cell refinement from powder diffraction data: The use of regression diagnostics. *Mineral. Mag.* **1997**, *61*, 65–77. [CrossRef]

10. Sheldrick, G.M. *SADABS Area-Detector Absorption Correction Program*; Bruker AXS Inc.: Madison, WI, USA, 2000.

11. Altomare, A.; Burla, M.C.; Camalli, M.; Cascarano, G.L.; Giacovazzo, C.; Gagliardi, A.; Moliterni, A.G.; Polidori, G.; Spagna, R. SIR97, a new tool for crystal structure determination and refinement. *J. Appl. Crystallogr.* **1999**, *32*, 115–119. [CrossRef]

12. Sheldrick, G.M. Crystal structure refinement with SHELXL. *Acta Crystallogr. C Struct. Chem.* **2015**, *71*, 3–8. [CrossRef] [PubMed]

13. Fischer, R.X.; Tillmanns, E. The equivalent isotropic displacement factor. *Acta Crystallogr. Sect. C Cryst. Struct. Commun.* **1988**, *44*, 775–776. [CrossRef]

14. Brown, I.D. Recent developments in the methods and applications of the Bond Valence Model. *Chem. Rev.* **2009**, *109*, 6858–6919. [CrossRef] [PubMed]

15. Clarke, R.M.; Williams, I.R. Moolooite, a naturally occurring hydrated copper oxalate from western Australia. *Mineral. Mag.* **1986**, *50*, 295–298. [CrossRef]

16. Rouse, R.C.; Peacor, D.R.; Dunn, P.J.; Simmons, W.B.; Newbury, D. Wheatleyite, $Na_2Cu(C_2O_4)_2 \cdot 2H_2O$, a natural sodium copper salt of oxalic acid. *Am. Mineral.* **1986**, *71*, 1240–1242.

17. Chukanov, N.V.; Aksenov, S.M.; Rastsvetaeva, R.K.; Lyssenko, K.A.; Belakovskiy, D.I.; Färber, G.; Möhn, G.; Van, K.V. Antipinite, $KNa_3Cu_2(C_2O_4)_4$, a new mineral species from guano deposits at Pabellón de Pica, Chile. *Mineral. Mag.* **2014**, *78*, 797–804. [CrossRef]

18. Smith, D.G.W.; Nickel, E.H. A system for codification for unnamed minerals: Report of the Subcommittee for Unnamed Minerals of the IMA Commission on New Minerals, Nomenclature and Classification. *Can. Mineral.* **2007**, *45*, 983–1055. [CrossRef]

Article

Tiberiobardiite, Cu$_9$Al(SiO$_3$OH)$_2$(OH)$_{12}$(H$_2$O)$_6$(SO$_4$)$_{1.5}$·10H$_2$O, a New Mineral Related to Chalcophyllite from the Cretaio Cu Prospect, Massa Marittima, Grosseto (Tuscany, Italy): Occurrence and Crystal Structure

Cristian Biagioni [1,*], Marco Pasero [1] and Federica Zaccarini [2]

[1] Dipartimento di Scienze della Terra, Università di Pisa, Via Santa Maria 53, I-56126 Pisa, Italy; marco.pasero@unipi.it
[2] Department of Applied Geological Sciences and Geophysics, University of Leoben, Peter Tunner Str. 5, A-8700 Leoben, Austria; Federica.Zaccarini@unileoben.ac.at
*   Correspondence: cristian.biagioni@unipi.it; Tel.: +39-050-221-5789

Received: 23 March 2018; Accepted: 9 April 2018; Published: 11 April 2018

Abstract: The new mineral species tiberiobardiite, ideally Cu$_9$Al(SiO$_3$OH)$_2$(OH)$_{12}$(H$_2$O)$_6$(SO$_4$)$_{1.5}$·10H$_2$O, has been discovered in the Cretaio Cu prospect, Massa Marittima, Grosseto, Tuscany, Italy, as very rare, light green, vitreous, tabular {0001}, pseudo-hexagonal crystals, up to 200 µm in size and 5 µm in thickness, associated with brochantite. Electron microprobe analysis gave (in wt %, average of 5 spot analyses): SO$_3$ 10.37, P$_2$O$_5$ 3.41, As$_2$O$_5$ 0.05, SiO$_2$ 8.13, Al$_2$O$_3$ 5.54, Fe$_2$O$_3$ 0.74, CuO 62.05, and ZnO 0.03, for a total of 90.32. Based on an idealized O content of 42 atoms per formula unit, assuming the presence of 16 H$_2$O groups and 13.5 cations (without H), the empirical formula of tiberiobardiite is (Cu$_{8.69}$Al$_{0.21}$Fe$_{0.10}$)$_{\Sigma9.00}$Al$_{1.00}$(Si$_{1.51}$P$_{0.54}$)$_{\Sigma2.05}$S$_{1.44}$O$_{12.53}$(OH)$_{13.47}$·16H$_2$O. The main diffraction lines, corresponding to multiple *hkl* indices, are [*d* in Å (relative visual intensity)]: 9.4 (s), 4.67 (s), 2.576 (m), 2.330 (m), and 2.041 (mw). The crystal structure study revealed tiberiobardiite to be trigonal, space group $R\bar{3}$, with unit-cell parameters a = 10.6860(4), c = 28.3239(10) Å, V = 2801.0(2) Å^3, and Z = 3. The crystal structure was refined to a final R_1 = 0.060 for 1747 reflections with $F_o > 4\sigma$ (F_o) and 99 refined parameters. Tiberiobardiite is the Si-analogue of chalcophyllite, with Si^{4+} replacing As^{5+} through the coupled substitution As^{5+} + O^{2-} = Si^{4+} + (OH)$^-$. The name tiberiobardiite honors Tiberio Bardi (b. 1960) for his contribution to the study of the mineralogy of Tuscany.

Keywords: tiberiobardiite; chalcophyllite group; copper; silicate; sulfate; Cretaio; Tuscany; Italy

1. Introduction

The occurrences of small Cu ore deposits in the ophiolitic sequences of the Northern Apennines (Italy) have been known for a long time, and they represent one of the few Cu sources in Italy. Notwithstanding their past economic importance, little mineralogical and geological information has been provided about these deposits (e.g., [1–4]). These studies mainly focused on the primary ore mineralogy, whereas secondary assemblages were usually not taken into account. On the other hand, the potential occurrence of interesting mineral species was suggested by the identification of phosphates belonging to the mixite group (e.g., calciopetersite from Monte Beni, Firenzuola [5]).

The small Cu prospect of Cretaio (latitude 43°03′10″ N, longitude 10°58′40″ E), near the small village of Prata, Massa Marittima, Grosseto, Tuscany, Italy, is characterized by several alteration phases of the primary Cu-Fe ore minerals [6]. During the study of this mineral assemblage, a phase related to chalcophyllite was identified. Its unit-cell volume was significantly smaller than chalcophyllite

($\Delta V = -2.5\%$), thus indicating the likely occurrence of some chemical substitutions. Indeed, qualitative chemical analyses, as well as Raman spectroscopy, suggested the absence of As in the studied material, making a structural investigation mandatory to fully characterize the mineral. Quantitative chemical analysis and crystal structure refinement proved that the studied phase is the Si-analogue of chalcophyllite. This new mineral was named tiberiobardiite, in honor of the mineral collector Tiberio Bardi (b. 1960), for his contribution to the study of Tuscan mineralogy. Tiberio Bardi provided the type material of volaschioite [7] and promoted the mineralogical investigation of the ophiolite-hosted mineralization occurring in the Monte Beni area, north of Florence [8], where he collected the specimen of calciopetersite used for the crystal structure study [5]. Recently, together with other mineral collectors (Cristiano Bazzoni, Mauro Bernocchi, Cesare Betti, and Riccardo Marini), Tiberio Bardi contributed to the mineralogical rediscovery of the small Cu ore deposits of Cretaio, where he first collected the type material of tiberiobardiite. The mineral and its name have been approved by the IMA CNMNC, under the number 2016-096. The holotype specimen is deposited in the mineralogical collection of the Museo di Storia Naturale of the University of Pisa, Via Roma 79, Calci (Pisa), under the catalogue number 19900.

The occurrence, physical properties, and crystal structure of tiberiobardiite are described in this paper.

2. Geological Setting and Occurrence of Tiberiobardiite

No detailed geological studies have been performed on the small Cu ore deposit of Cretaio. The available data are limited to some mining reports dating to the early decades of the 20th century [6]. The ore deposit is represented by a small concentration of Cu sulfides (bornite, chalcocite, and covellite) and hematite, scattered as stockwork veins within highly deformed gabbro belonging to the Liguride units. Minor chalcopyrite has been observed only in quartz + calcite veins.

The primary sulfides are strongly altered into a series of secondary minerals: antlerite, brochantite, chalcanthite, chalcoalumite, connellite, langite, libethenite, malachite, posnjakite, serpierite/devilline, and spangolite. Tiberiobardiite was found associated with abundant brochantite. Its origin is likely related to the supergene alteration of the Cu ore minerals in an oxidizing and hydrous low-T environment. The individual elements were derived from Cu ores (Cu and S) and from rock-forming minerals occurring in the gabbro (Al, Si, and minor P).

3. Experimental Data

3.1. Mineral Description and Physical Properties

Tiberiobardiite (Figure 1) occurs as thin, tabular {0001} crystals, having a pseudo-hexagonal outline, up to 200 μm in diameter and 5 μm in thickness. The mineral is green, with a pale green streak; it is transparent, and its luster is vitreous. Tiberiobardiite is brittle, with a perfect {0001} cleavage. Its fracture is irregular. Its hardness and density, as well as optical properties, were not measured, owing to the small amount of available material. The calculated density, based on the ideal formula, is 2.528 g/cm^3. The mean refractive index, obtained from the Gladstone-Dale relationship [9,10], using the ideal formula and the calculated density, is 1.568.

Figure 1. Tiberiobardiite, thin, pseudo-hexagonal crystals, up to 40 μm in size, with iron oxides. Cretaio, Prata, Massa Marittima, Grosseto, Tuscany. Holotype specimen, #19900, collection of the Museo di Storia Naturale of the University of Pisa.

3.2. Chemical Data

Preliminary energy dispersive spectrometry (EDS) analysis showed the presence of Cu, Al, Fe, P, Si, and S as the only elements with $Z > 8$.

Quantitative chemical analyses were carried out using a Superprobe JEOL JXA 8200 electron microprobe operating in WDS mode at the Eugen F. Stumpfl laboratory, Leoben University, Austria. The experimental conditions were as follows: accelerating voltage 20 kV, beam current 10 nA, and beam size 3 μm. The standards were (element and emission line) as follows: chromite (Al $K\alpha$), chalcopyrite (S $K\alpha$, Cu $K\alpha$), zircon (Si $K\alpha$), pyrite (Fe $K\alpha$), skutterudite (As $L\alpha$), apatite (P $K\alpha$), and sphalerite (Zn $K\alpha$). The following diffracting crystals were selected: PETJ for S, Si, and P; TAP for Al and As; and LIFH for Fe, Cu, and Zn. Direct H_2O determination was not performed, owing to the scarcity of the material. The occurrence of H_2O, suggested by the crystal structure refinement, was further confirmed by micro-Raman spectroscopy (see below).

The collection of reliable chemical data for tiberiobardiite was a very difficult task, owing to the small size of the available crystals, which made the preparation of a good-quality polished surface difficult. Moreover, the mineral is unstable under the electron beam, showing evidence of strong dehydration. Indeed, the structural determination (see below) indicated the occurrence of 16 H_2O, 13.5 cations (without H atoms), and ~14 OH groups, implying the occurrence of ~36.6 wt % H_2O. It should be noted that by adding such an H_2O content to the total given in Table 1, an unrealistically high total of ~126.9 wt % is obtained. Notwithstanding such an important limitation, the atomic ratios among cations are in good agreement with the results of the structural study.

Table 1. Chemical data of tiberiobardiite.

Oxide	wt % (n = 5)	Range	e.s.d.
SO_3	10.37	9.67–10.94	0.51
P_2O_5	3.41	3.02–3.80	0.28
As_2O_5	0.05	0.00–0.17	0.07
SiO_2	8.13	7.29–9.03	0.77
Al_2O_3	5.54	4.93–6.47	0.57
Fe_2O_3	0.74	0.61–0.83	0.09
CuO	62.05	57.44–65.20	3.06
ZnO	0.03	0.00–0.10	0.04
Total	90.32	84.86–93.70	3.35

On the basis of an idealized O content of 42 atoms per formula unit (apfu), to be compared with 42.2 found in the structural study (see below), assuming the occurrence of 16 H_2O groups and 13.5 cations (without H), the empirical formula of tiberiobardiite is $(Cu_{8.69}Al_{0.21}Fe_{0.10})_{\Sigma 9.00}Al_{1.00}(Si_{1.51}P_{0.54})_{\Sigma 2.05}S_{1.44}O_{12.53}(OH)_{13.47}\cdot 16H_2O$, with rounding errors. The ideal formula is $Cu_9Al[SiO_3(OH)]_2(SO_4)_{1.5}(OH)_{12}\cdot 16H_2O$, which corresponds to the following (in wt %): SO_3 8.45, SiO_2 8.45, Al_2O_3 3.59, CuO 50.36, and H_2O 29.15, for a sum of 100.00.

3.3. Micro-Raman Spectroscopy

Unpolarized micro-Raman spectra were collected on an unpolished sample of tiberiobardiite in nearly backscattered geometry using a Jobin-Yvon Horiba XploRA Plus apparatus equipped with a motorized *x-y* stage and an Olympus BX41 microscope with a 50× objective. The 532 nm line of a solid-state laser was used. The minimum lateral and depth resolution was set to a few μm. The system was calibrated using the 520.6 cm^{-1} Raman band of silicon before each experimental session. The spectra were collected through multiple acquisitions with single counting times of 120 s. The backscattered radiation was analyzed with a 600 mm^{-1} grating monochromator.

Figure 2 shows the observed Raman spectrum of tiberiobardiite. The region between 100 and 1200 cm^{-1} (Figure 2a) can be divided into three ranges, as follows:

(i)   Range 100–300 cm^{-1}: bands at 124, 203, and 261 cm^{-1}, possibly related to lattice modes;

(ii) Range 300–700 cm^{-1}: bands at 394, 440, 487, 544, and 589 cm^{-1}, related to the bending modes of the $[SiO_3(OH)]$ and (SO_4) groups;

(iii) Range 700–1200 cm^{-1}: two bands at 965 and 1097 cm^{-1} occur. The former could be due to the contribution of the $[SiO_3(OH)]$ and (SO_4) groups, whereas the weak band at 1097 cm^{-1} is likely related to the (SO_4) antisymmetric stretching, in agreement with the band position in the Raman spectrum of chalcophyllite (1100 cm^{-1}; [11]). A similar interpretation was proposed for the bands observed between 900 and 1150 cm^{-1} in barrotite [12]. It is worth noting the absence of the strong band at ~840 cm^{-1} occurring in the Raman spectrum of chalcophyllite, which was interpreted as being due to the (AsO_4) symmetric stretching mode [11]. The same feature, at a slightly lower wavenumber (825 cm^{-1}), was reported in barrotite [12].

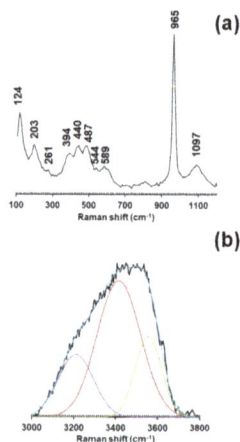

Figure 2. Micro-Raman spectrum of tiberiobardiite in the spectral range 100–1200 cm^{-1} (**a**) and 3000–4000 cm^{-1} (**b**). In (**b**), the components contributing to the broad band are shown.

In the region between 3000 and 4000 cm^{-1}, a strong and broad band was observed (Figure 2b). Such a band is related to O–H stretching vibrations. The band deconvolution, performed using the software Fityk [13], revealed the presence of three bands, centered at 3218, 3418, and 3555 cm^{-1}. By using the relationships between O–H stretching frequencies and O$\cdots$O distances [14], the three Raman bands could correspond to three different kinds of hydrogen bonds (i.e., short (=strong) bonds (O$\cdots$O ~2.70 Å), intermediate (O$\cdots$O ~2.80 Å), and long (=weak) bonds (O$\cdots$O ~3.00 Å)).

3.4. Crystallography

Powder X-ray diffraction data of tiberiobardiite were collected using a 114.6 mm Gandolfi camera and Ni-filtered Cu $K\alpha$ radiation (Table 2). Unit-cell parameters were not refined from the X-ray powder diffraction pattern owing to the multiple indexing for most of the observed reflections.

Table 2. Observed and calculated X-ray powder diffraction data (*d* in Å) for tiberiobardiite.

I_{obs}	d_{obs}	I_{calc}	d_{calc}	$h\,k\,l$	I_{obs}	d_{obs}	I_{calc}	d_{calc}	$h\,k\,l$
s	9.4	100	9.36	0 0 3	-	-	2	2.646	1 2 −7
-	-	4	5.62	1 0 4	m	2.576	2	2.571	2 2 3
-	-	2	5.34	1 1 0			13	2.571	2 2 −3
-	-	2	4.83	1 0 −5	m	2.330	10	2.325	2 2 −6
		27	4.72	0 0 6			5	2.325	2 2 6
s	4.67	4	4.65	1 1 3	mw	2.041	7	2.037	2 2 −9
		5	4.65	1 1 −3			1	2.037	2 2 9
-	-	3	3.873	2 0 −4	vw	1.859	2	1.854	2 0 14
-	-	6	3.584	2 0 5	vw	1.776	2	1.769	2 2 12
-	-	5	3.147	0 0 9			3	1.769	2 2 −12
-	-	1	3.136	2 1 4	w	1.548	4	1.542	6 0 0
-	-	1	3.085	3 0 0			1	1.542	2 2 −15
-	-	2	2.976	1 2 5	w	1.528	2	1.522	6 0 3
-	-	2	2.932	3 0 −3			2	1.522	6 0 −3
vw	2.73	3	2.712	1 1 −9	w	1.469	2	1.466	6 0 6
-	-	2	2.708	1 0 10			2	1.466	6 0 −6
w	2.68	13	2.672	2 2 0					

Note: intensity and d_{hkl} were calculated using the software *PowderCell* 2.3 [15] based on the structural model given in Table 4. Only reflections with I_{calc} >1 are listed, if not observed. The five strongest reflections are given in bold. Observed intensities were visually estimated (s = strong; m = medium; mw = medium-weak; w = weak; vw = very weak).

A single-crystal X-ray diffraction study was performed using a Bruker Smart Breeze diffractometer equipped with an air-cooled CCD detector and graphite-monochromatized Mo $K\alpha$ radiation. The detector-to-crystal working distance was 50 mm. Data were corrected for the Lorentz-polarization factor, absorption, and background effects, using the package of software Apex2 [16]. Unit-cell parameters of tiberiobardiite are a = 10.6860(4), c = 28.3239(10) Å, V = 2801.0(2) Å³, and space group $R\,\bar{3}$, suggesting isotypic relationships between tiberiobardiite and chalcophyllite [17]. Consequently, the crystal structure of the former was refined using Shelxl-2014 [18] starting from the atomic coordinates of the latter. Scattering curves for neutral atoms were taken from the *International Tables for Crystallography* [19]. After several cycles of isotropic refinement, the *R* factor converged to 0.22, lowered to 0.16 after taking into account the twinning according to the {10-10} plane. In agreement with the chemical data, the site hosting As in chalcophyllite was found to be occupied by lighter atoms (i.e., Si and P). Its site occupancy was fixed based on electron microprobe data, owing to the similarity between the scattering factors of Si and P. In agreement with [17], the displacement parameters of the S atom and the O atoms belonging to the SO₄ group (i.e., O6 and O7) were large. The refinement of the site occupancy at the S site pointed to $S_{0.82(1)}$, then fixed to 0.75 to achieve the electrostatic balance. The Ow8 site was found to be split into two sub-positions Ow8a and Ow8b. The latter was too close to Ow9, and consequently, the site occupancy at Ow9 was constrained to be the same as that at Ow8a in order to avoid the too short Ow9–Ow8b distance. Hydrogen atoms were not located. After several cycles of anisotropic refinement for all the atoms, the R_1 converged to 0.0602 for the 1747 reflections with $F_o > 4\sigma\,(F_o)$. Details of the data collection and refinement are given in Table 3. CIF is available as Supplementary Materials.

Table 3. Crystal and experimental details for tiberiobardiite.

Crystal data	
Crystal size (mm)	$0.180 \times 0.050 \times 0.005$
Cell setting, space group	Trigonal, $R\bar{3}$
a, c (Å)	10.6860(4), 28.3239(10)
V (Å^3)	2801.0(2)
Z	3
Data collection and refinement	
Radiation, wavelength (Å)	Mo $K\alpha$, $\lambda = 0.71073$
Temperature (K)	293
$2\theta_{max}$ (°)	58.35
Measured reflections	18,912
Unique reflections	2809
Reflections with $F_o > 4\sigma$ (F_o)	1747
R_{int}	0.0633
$R\sigma$	0.0799
Range of h, k, l	$-10 \le h \le 14, -14 \le k \le 13, -38 \le l \le 37$
R [$F_o > 4\sigma$ (F_o)]	0.0602
R (all data)	0.1167
wR (on F^2)	0.1354
GooF	1.044
Number of least-square parameters	99
Maximun and minimum residuals ($e/\text{Å}^3$)	1.26 (at 0.98 from Si), -1.10 (at 0.65 from Cu2)

The crystal structure refinement of tiberiobardiite points to the crystal chemical formula $Cu_9(Al_{0.96}Fe_{0.04})[(Si_{0.75}P_{0.25})O_3(OH_{0.75}O_{0.25})]_2(OH)_{12}(H_2O)_6(SO_4)_{1.5} \cdot 10.2H_2O$ ($Z = 3$).

4. Crystal Structure Description

4.1. General Features and Cation Coordinations

The atomic coordinates, site occupancies, and equivalent isotropic displacement parameters of tiberiobardiite are given in Table 4. Table 5 reports selected bond distances, whereas Table 6 shows the bond-valence balance.

The crystal structure of tiberiobardiite (Figure 3) is composed by five independent cation positions (Cu1, Cu2, Al, Si, and S) and nine anion sites in the asymmetric unit. It can be described as formed by {0001} heteropolyhedral layers, composed by $Cu\Phi_6$ polyhedra ($\Phi = O$, OH, H$_2$O), Al(OH)$_6$ octahedra, and (Si,P)O$_3$(OH,O) tetrahedra, alternating with interlayers hosting SO$_4$ and H$_2$O groups. The {0001} heteropolyhedral layers are formed through the edge-sharing of Cu polyhedra, with three unoccupied octahedral positions out of 12 available ones per unit cell. Such unoccupied positions actually host Al, lying on the -3-point symmetry, and the Si site, on the 3-point symmetry, a little off the {0001} sheet. The Si-centered tetrahedra are alternatively disposed above and below the sheet.

Table 4. Site labels, Wyckoff sites, site occupancy factor (s.o.f.), atom coordinates, and equivalent isotropic displacement parameters (Å^2) for tiberiobardiite.

Site	Wyckoff Site	s.o.f.	x/a	y/b	z/c	U_{eq}
Cu1	9d	Cu$_{1.00}$	1/2	1/2	1/2	0.0196(3)
Cu2	18f	Cu$_{1.00}$	0.33920(8)	0.18074(7)	0.50669(3)	0.0172(2)
Al	3b	Al$_{0.96}$Fe$_{0.04}$	0	0	1/2	0.0094(12)
Si	6c	Si$_{0.75}$P$_{0.25}$	2/3	1/3	0.46619(10)	0.0145(6)
S	6c	S$_{0.75}$	1/3	2/3	0.32465(14)	0.0348(10)
OH1	18f	O$_{1.00}$	0.3223(4)	0.3489(4)	0.52847(15)	0.0153(9)
O2	18f	O$_{1.00}$	0.5279(4)	0.3353(4)	0.48438(15)	0.0213(10)
Ow3	18f	O$_{1.00}$	0.3482(6)	0.4550(5)	0.42895(18)	0.0507(15)
OH4	18f	O$_{1.00}$	0.1312(4)	0.1635(4)	0.46337(14)	0.0152(10)
OH5	6c	O$_{1.00}$	2/3	1/3	0.4107(3)	0.037(2)
O6	6c	O$_{0.75}$	1/3	2/3	0.3764(3)	0.030(2)
O7	18f	O$_{0.75}$	0.2174(9)	0.5274(9)	0.3075(3)	0.059(2)
Ow8a	18f	O$_{0.70(1)}$	0.3180(10)	0.3109(11)	0.6244(4)	0.080(3)
Ow8b	18f	O$_{0.30(1)}$	0.249(3)	0.381(3)	0.6179(8)	0.080(3)
Ow9	18f	O$_{0.70(2)}$	0.4200(12)	0.2991(14)	0.3709(4)	0.103(4)

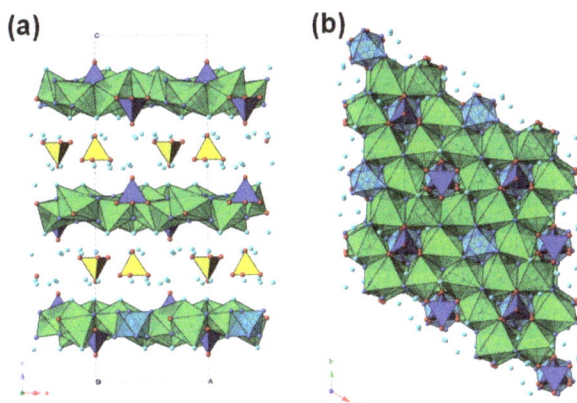

Figure 3. Projection of the crystal structure of tiberiobardiite viewed down **b** (**a**) and **c** (**b**). Polyhedra: green = Cu1 and Cu2 sites; light blue = Al site; blue = Si site; and yellow = S site. Circles: light blue = H_2O groups; blue = OH; and red = O.

Table 5. Selected bond distances (in Å) in tiberiobardiite.

Cu1	–OH1	1.948(4) × 2	Cu2	–OH4	1.944(4)	Si	–OH5	1.571(9)
	–O2	1.977(4) × 2		–O2	1.965(4)		–O2	1.580(4) × 3
	–Ow3	2.476(5) × 2		–OH1	1.972(4)			
				–OH1	1.991(4)	S	–O7	1.463(8) × 3
Al	–OH4	1.909(4) × 6		–OH4	2.464(4)		–O6	1.466(10)
				–Ow3	2.541(6)			

The two independent Cu sites (Cu1 and Cu2) display a six-fold coordination, showing the typical distorted (4 + 2) octahedral coordination related to the Jahn–Teller effect of Cu^{2+} (e.g., [20]). Considering the nature of the ligands Φ, two different kinds of polyhedra occur (i.e., $Cu1O_2(OH)_2(H_2O)_2$ and $Cu2O(OH)_4(H_2O)$ (Figure 4)), in agreement with the structure of chalcophyllite [17]. The average of the four shorter Cu–Φ distances, corresponding to the equatorial bonds, is 1.962 and 1.968 Å for the Cu1 and Cu2 sites, respectively, in agreement with previous studies [21]. The axial bonds are definitely longer, ranging between 2.48 and 2.54 Å. The bond-valence sums (BVS) at the Cu1 and Cu2 sites are 2.10 and 2.05 valence units (v.u.), respectively (Table 6).

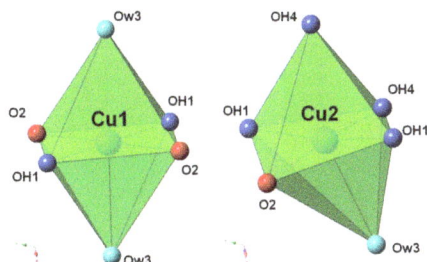

Figure 4. Bonding environment of Cu sites in tiberiobardiite.

Table 6. Bond-valence sums (in valence units (v.u.)) in tiberiobardiite.

Site	Cu1	Cu2	Al	Si	S	Σanions
OH1	$0.48^{\downarrow \times 2}$	0.45 0.43				1.36
O2	$0.45^{\downarrow \times 2}$	0.46		$1.11^{\downarrow \times 3}$		2.02
Ow3	$0.12^{\downarrow \times 2}$	0.10				0.22
OH4		0.49 0.12	$0.51^{\downarrow \times 6}$			1.12
OH5				1.14		1.14
O6					1.55	1.55
O7					$1.53^{\downarrow \times 3}$	1.53
Ow8a						-
Ow8b						-
Ow9						-
Σcations	2.10	2.05	3.06	4.47	6.14 *	

Note: bond parameters after [22]. In mixed or partially occupied sites, the bond-valence sum has been calculated taking into account the site occupancy. Right superscripts indicate the number of bonds for each cation. * Assuming full-occupancy at the S site.

Aluminum is hosted in a regular octahedron, connected with the Cu polyhedra through edge-sharing, and is coordinated by OH groups only. In agreement with the chemical data, aluminum is partially replaced by minor Fe^{3+}. The bond-valence sum is in good accord with the theoretical value (i.e., 3.06 v.u.).

The Si-centered tetrahedra are connected through corner-sharing to the (Cu–Al)-polyhedra {0001} sheets. The fourth oxygen atom (OH5) points to the interlayer and is involved in the hydrogen bond system characterizing tiberiobardiite. The average bond distance at the Si site, 1.578 Å, agrees with a mixed (Si and P) occupancy, as suggested by the chemical data. Owing to the similar scattering factors of Si and P, the site occupancy was fixed to $(Si_{0.75}P_{0.25})$, in accord with electron microprobe data. The corresponding BVS is 4.47 v.u., slightly higher than the calculated value of 4.25 v.u. based on site occupancy.

The {0001} sheet has the simplified chemical composition $\{Cu_9Al[(Si_{0.75}P_{0.25})O_3(OH_{0.75}O_{0.25})]_2 (OH)_{12}(H_2O)_6\}^{3+}$.

Successive sheets are separated by interlayers hosting SO_4 and H_2O groups. The S site is statistically occupied by S atoms, with a site occupancy fixed, based on the chemical data, to 0.75. Similarly, the site occupancy at the O6 and O7 sites, bonded to S, were fixed to the same value. The average <S–O> distance is 1.464 Å, in agreement with the average <S–O> distance in sulfate minerals (i.e., 1.473 Å) [23]. In addition to the SO_4 group, the interlayer is completed by two H_2O groups, hosted at the Ow9 and the split Ow8a/b positions. In order to avoid too short O···O distances, Ow9 and Ow8b cannot be fully occupied simultaneously. These sites are only partially occupied, with Ow8a and Ow9 having the same site occupancy (0.70), whereas Ow8b has a lower occupancy (0.30). The chemical formula of the interlayer is $[(SO_4)_{1.5} \cdot 10H_2O]^{3-}$.

4.2. The Hydrogen Bond System in Tiberiobardiite

The nine anion positions can be divided, based on their BVS (Table 6), in O^{2-}, OH^-, and H_2O-hosting sites. Table 7 reports the observed O···O distances shorter than 3 Å; their values agree with those calculated from the O–H stretching frequencies measured through Raman spectroscopy. Coupling this information with the occurrence of several underbonded anion positions, the important role played by the hydrogen bonds in tiberiobardiite can be hypothesized.

Table 7. O···O distances (in Å) and corresponding bond strengths (in v.u.) in tiberiobardiite, calculated according to [24].

O···O	*d* (Å)	v.u.
OH1 ··· Ow8a	2.745(11)	0.21
OH1 ··· Ow8b	2.72(2)	0.22
Ow3 ··· O6	2.778(7)	0.19
Ow3 ··· Ow9	2.711(13)	0.22
OH4 ··· O7	2.960(8)	0.14
OH5 ··· Ow8b	2.83(3)	0.17
OH5 ··· Ow9	2.718(11)	0.22
Ow8a ··· O7	2.727(13)	0.21
Ow8a ··· O7	2.958(13)	0.14
Ow8b ··· O7	2.95(3)	0.14
Ow9 ··· O7	2.835(13)	0.17
Ow8a ··· Ow9	2.632(14)	0.27
Ow8a ··· Ow9	2.876(16)	0.16
Ow8b ··· Ow9	2.69(2)	0.23

The hydrogen bond scheme is well-defined as regards the anion forming the {0001} heteropolyhedral layers, whereas it is more difficult to be described for interlayer H_2O groups.

The heteropolyhedral layer is composed by anion sites OH1 to OH5. The valence sum at O2 (2.02 v.u.) agrees with its occupancy by O^{2-}, whereas Ow3 (0.22 v.u.) is consistent with a H_2O group. OH1 is a donor of an H-bond to Ow8a/Ow8b (at 2.74/2.72 Å), and OH4 is a donor to O7 (at 2.96 Å); the BVS at OH1 and OH4, after correction for the H-bond in agreement with [24], are 1.15/1.14 and 0.98 v.u., respectively. This bond scheme is analogous to that observed in chalcophyllite [17]. On the contrary, whereas O5 is an acceptor of three H-bonds from Ow9 in chalcophyllite [17], the mixed (OH, O) nature of OH5 in tiberiobardiite, related to the $Si^{4+} + OH^- = P^{5+} + O^{2-}$ substitution, requires a different H-bond pattern. Different configurations are possible, considering the possible occurrences of Ow8b and Ow9. Taking into account the site occupancy of these two sites (i.e., 0.70 Ow9 and 0.30 Ow8b, close to 2/3 Ow9 and 1/3 Ow8b), a model with OH5 bonded to two Ow9 and one Ow8b could be proposed. On the basis of the Si–P atomic ratio, an $(OH)_{0.75}O_{0.25}$ occupancy at OH5 should occur, corresponding to an expected weighted BVS of 1.25 v.u. It is likely that Ow9 and Ow8b could be donors of the H-bond, whereas the second Ow9 could accept an H-bond from OH5. Accordingly, the BVS at OH5 would be 1.31 v.u., consistent with the proposed site occupancy.

In the interlayer, the BVS values are in accord with the H_2O groups at the Ow8a/Ow8b and Ow9 sites (0 v.u.), consistent with their role as interlayer H_2O groups. The oxygen atoms belonging to the SO_4 group are significantly underbonded (1.55 and 1.53 v.u. at the O6 and O7 sites, respectively) and they are acceptors of H-bonds. O6 is an acceptor of H-bonds from three symmetry-related H_2O groups hosted at the Ow3 site, achieving a BVS of 2.12 v.u. Oxygen at the O7 site is involved in a more complicated H-bond system than that shown in chalcophyllite, where this site is H-bonded to Ow8 (at 2.66 Å), Ow9 (at 2.88 Å), and OH4 (at a longer distance of 3.03 Å) [17]. Indeed, the split nature of Ow8 and the partial occupancy of the H_2O-hosting sites allowed different configurations. Taking into account O···O distances as well as O···H_2O···O angles, a hypothetical model around O7 in tiberiobardiite can be proposed. This site could be H-bonded with Ow8a (at 2.73 Å) and Ow9 (at 2.84 Å), as in chalcophyllite, with an additional bond with OH4 (at 2.96 Å). The O7···Ow8a distance of 2.96 Å could not likely represent a hydrogen bond, because the O7···Ow8a···Ow9 would be ~133°, to be compared with two possible O7···Ow8a···Ow9 configurations (bifurcated H-bond, as in chalcophyllite [17]) showing angles of ~109° and 102°, respectively, closer to the average value of 108.3° observed in crystalline hydrates [25]. In such a way, the BVS at O7 would be 2.05 v.u.

In agreement with [17], when the SO_4 group is absent, additional H_2O groups could occur at the O sites. Consequently, it could be difficult to propose a reliable H-bond system involving the Ow8a/Ow8b and Ow9 sites, that could show variable H-bond patterns. As a matter of fact, hydrogen bonds are fundamental in bonding successive {0001} heteropolyhedral layers and in "trapping" the SO_4 groups in the interlayers.

5. Discussion

5.1. Relationships with Chalcophyllite

Tiberiobardiite is the Si-analogue of chalcophyllite, with Si^{4+} replacing As^{5+} through the coupled substitution $As^{5+} + O^{2-} = Si^{4+} + OH^-$. These two minerals form the chalcophyllite group (Table 8), in agreement with [26]. The occurrence of P^{5+} partially replacing Si^{4+} allows us to also hypothesize the existence of the P-analogue, ideally $[Cu_9Al(PO_4)_2(OH)_{12}(H_2O)_6](SO_4)_{1.5} \cdot 10-12H_2O$.

These species are related to barrotite, $\{Cu_9Al[SiO_3(OH)]_2(OH)_{12}(H_2O)_6\}\{(SO_4)[AsO_3(OH)]_{0.5}\} \cdot 2H_2O$. In fact, the crystal structure of barrotite was not solved, and consequently, the actual relationships with the chalcophyllite group minerals are unknown. Indeed, As could be hosted in an independent tetrahedral site, or it could partially replace (SO_4) groups in the interlayer, through the substitution $(SO_4)^{2-} = [AsO_3(OH)]^{2-}$. In this latter case, the ideal formula of barrotite could be the same as that of tiberiobardiite, from which it differs in the lower hydration state and the shorter **c** axis (Table 8).

Table 8. Members of the chalcophyllite group and related minerals.

Mineral	Chemical Formula	a (Å)	c (Å)	V (Å^3)	Ref.
Chalcophyllite group					
Chalcophyllite	$[Cu_9Al(AsO_4)_2(OH)_{12}(H_2O)_6](SO_4)_{1.5} \cdot 12H_2O$	10.76	28.68	2873.3	[17]
Tiberiobardiite	$\{Cu_9Al[SiO_3(OH)]_2(OH)_{12}(H_2O)_6\}(SO_4)_{1.5} \cdot 10H_2O$	10.69	28.32	2801.0	this work
Unclassified					
Barrotite	$\{Cu_9Al[SiO_3(OH)]_2(OH)_{12}(H_2O)_6\}\{(SO_4)[AsO_3(OH)]_{0.5}\} \cdot 2H_2O$	10.65	21.95	2156.5	[12]

5.2. Secondary Cu^{2+} Oxysalt Minerals in Ophiolite-Hosted Ore Deposits from Tuscany (Italy)

The secondary mineral assemblages associated with ophiolite-hosted ore deposits from Tuscany (Italy) have received little attention by mineralogists, and very few data are currently available (Table 9). In addition to the widespread occurrence of the copper carbonates azurite and malachite, known since the 19th century (e.g., [27]), several common sulfates have been identified in the last thirty years (i.e., brochantite, chalcanthite, langite, posnjakite, and serpierite) (e.g., [28,29]). The study of the mineral association occurring in the small Cretaio prospect allowed to increase the knowledge of these phases, with further description of antlerite, some Cu–Al sulfates (chalcoalumite, cyanotrichite, and spangolite), and a mineral belonging to the devilline group [6]. Spangolite is characterized by the occurrence of the Cl anion, as well as the two Cu halides clinoatacamite and connellite; the latter was identified from the same occurrence of spangolite at Cretaio and in close association with clinoatacamite, cuprite, and native copper also from the Montecastelli Cu deposit, along the Pavone River valley (e.g., [30]).

It is worth noting the occurrence of some copper phosphates in the ophiolite-hosted deposits (i.e., calciopetersite and libethenite from Monte Beni [5] and Cretaio [6]). Moreover, a still unidentified member of the mixite group has been recently identified in the mineral assemblages from Cretaio. Likely, phosphorus could be derived from the interaction between the acid solutions derived from sulfide weathering and apatite, occurring as accessory minerals in the host rocks. Phosphorus is enriched also in tiberiobardiite, and the potential occurrence of the P-analogue of tiberiobardiite and chalcophyllite could be likely found in the phosphate-rich assemblages from ophiolite-hosted deposits.

Table 9. Secondary Cu^{2+} oxysalts in ophiolite-hosted ore deposits from Tuscany.

Mineral	Chemical Formula	Occurrences
Halides		
Clinoatacamite	$Cu_2(OH)_3Cl$	Montecastelli
Connellite	$Cu_{19}(SO_4)(OH)_{32}Cl_4 \cdot 3H_2O$	Cretaio, Montecastelli
Carbonates		
Azurite	$Cu_3(CO_3)_2(OH)_2$	Widespread
Malachite	$Cu_2(CO_3)(OH)_2$	Widespread
Sulfates		
Antlerite	$Cu_3(SO_4)(OH)_4$	Cretaio
Brochantite	$Cu_4(SO_4)(OH)_6$	Widespread
Chalchantite	$Cu(SO_4) \cdot 5H_2O$	Cretaio, Impruneta
Chalcoalumite	$CuAl_4(SO_4)(OH)_{12} \cdot 3H_2O$	Cretaio
Cyanotrichite	$Cu_4Al_2(SO_4)(OH)_{12} \cdot 2H_2O$	Cretaio
Devilline/serpierite	$CaCu_4(SO_4)_2(OH)_6 \cdot 3H_2O$	Cretaio, Impruneta
Langite	$Cu_4(SO_4)(OH)_6 \cdot 2H_2O$	Cretaio, Impruneta, MVC
Linarite	$CuPb(SO_4)(OH)_2$	MVC
Posnjakite	$Cu_4(SO_4)(OH)_6 \cdot H_2O$	Cretaio, MVC, CdC
Spangolite	$Cu_6Al(SO_4)(OH)_{12}Cl \cdot 3H_2O$	Cretaio
Phosphates		
Calciopetersite	$CaCu_6(PO_4)_2(PO_3OH)(OH)_6 \cdot 3H_2O$	Monte Beni
Libethenite	$Cu_2(PO_4)(OH)$	Cretaio
Silicates		
Chrysocolla	$Cu_{2-x}Al_x(H_{2-x}Si_2O_5)(OH)_4 \cdot nH_2O$	Widespread
Tiberiobardiite	$Cu_9Al(SiO_3OH)_2(SO_4)_{1.5}(OH)_2 \cdot 16H_2O$	Cretaio

Note: minerals occurring in more than three localities are indicated as "widespread". CdC = Cetine di Camporbiano, Pisa and MVC = Montecatini Val di Cecina, Pisa.

Supplementary Materials: The following are available online at http://www.mdpi.com/2075-163X/8/4/152/s1, CIF: tiberiobardiite.

Acknowledgments: Tiberio Bardi is acknowledged for providing us with the studied material. We wish to thank also the mineral collectors Cristiano Bazzoni, Mauro Bernocchi, Cesare Betti, and Riccardo Marini for providing us with studied material from the Cretaio prospect. The comments of six anonymous reviewers helped us improve the paper.

Author Contributions: C.B. conceived of and designed the experiments; C.B. performed the diffraction experiments; C.B. and M.P. analyzed the data; F.Z. performed the chemical analysis; and C.B. wrote the paper.

Conflicts of Interest: The authors declare no conflict of interest.

References

1. Bertolani, M. I giacimenti cupriferi nelle ofioliti di Sestri Levante. *Per Miner.* **1952**, *21*, 149–170.

2. Bertolani, M.; Rivalenti, G. Le mineralizzazioni metallifere della miniera di Montecatini in Val di Cecina (Pisa). *Boll. Soc. Geol. Ital.* **1973**, *92*, 635–648.

3. Zaccarini, F.; Garuti, G. Mineralogy and chemical composition of VMS deposits of northern Apennine ophiolites, Italy: Evidence for the influence of country rock type on ore composition. *Miner. Petrol.* **2008**, *94*, 61–83. [CrossRef]

4. Garuti, G.; Bartoli, O.; Scacchetti, M.; Zaccarini, F. Geological setting and structural styles of Volcanic Massive Sulfide deposits in the northern Apennines (Italy): Evidence for seafloor and sub-seafloor hydrothermal activity in unconventional ophiolites of the Mesozoic Tethys. *Boletín de la Sociedad Geológica Mexicana* **2008**, *60*, 121–145. [CrossRef]

5. Biagioni, C.; Bonaccorsi, E.; Orlandi, P. Occurrence and crystal structure of calciopetersite from Monte Beni (Firenzuola, Florence, Tuscany, Italy). *Atti della Società Toscana de Scienze Naturali* **2011**, *116*, 17–22.

6. Bardi, T.; Bazzoni, C.; Bernocchi, M.; Betti, C.; Biagioni, C.; D'Orazio, M.; Pagani, G. Cretaio. La vecchia ricerca cuprifera presso Prata, Massa Marittima (GR). *Riviera Miner. Ital.* **2017**, *51*, 104–118.

7. Biagioni, C.; Bonaccorsi, E.; Orlandi, P. Volaschioite, $Fe^{3+}_4(SO_4)O_2(OH)_6 \cdot 2H_2O$, a new mineral from Fornovolasco, Apuan Alps, Tuscany (Italy). *Can. Miner.* **2011**, *49*, 605–614. [CrossRef]
8. Bardi, T.; Becucci, A.; Biagioni, C. Monte Beni. I minerali dell'ex cava Fantoni (Pietramala, Firenzuola, Firenze). *Micro* **2012**, *10*, 98–109.
9. Mandarino, J.A. The Gladstone-Dale relationship. Part III. Some general applications. *Can. Miner.* **1979**, *17*, 71–76.
10. Mandarino, J.A. The Gladstone-Dale relationship. Part IV. The compatibility concept and its application. *Can. Miner.* **1981**, *19*, 441–450.
11. Frost, R.L.; Palmer, S.J.; Keeffe, E.C. Raman spectroscopic study of the hydroxyl-arsenate-sulfate mineral chalcophyllite $Cu_{18}Al_2(AsO_4)_4(SO_4)_3(OH)_{24} \cdot 36H_2O$. *J. Raman Spectrosc.* **2010**, *41*, 1769–1774. [CrossRef]
12. Sarp, H.; Černý, R.; Pushcharovsky, D.Y.; Schouwink, P.; Teyssier, J.; Williams, P.A.; Babalik, H.; Mari, G. La barrotite, $Cu_9Al(HSiO_4)_2[(SO_4)(HAsO_4)_{0.5}](OH)_{12} \cdot 8H_2O$, un noveau mineral de la mine de Roua (Alpes-Maritimes, Frances). *Riviera Sci.* **2014**, *98*, 3–22.
13. Wojdyr, M. Fityk: A general-purpose peak fitting program. *J. Appl. Crystallogr.* **2010**, *43*, 1126–1128. [CrossRef]
14. Libowitzky, E. Correlation of O–H stretching frequencies and O–H$\cdots$O hydrogen bond lengths in minerals. *Mon. Chem.* **1999**, *130*, 1047–1059.
15. Kraus, W.; Nolze, G. PowderCell—A program for the representation and manipulation of crystal structures and calculation of the resulting X-ray powder patterns. *J. Appl. Crystallogr.* **1996**, *29*, 301–303. [CrossRef]
16. Bruker AXS Inc. Apex 2. In *Bruker Advanced X-ray Solutions*; Bruker AXS Inc.: Madison, WI, USA, 2004.
17. Sabelli, C. The crystal structure of chalcophyllite. *Z. Kristallogr.* **1980**, *151*, 129–140.
18. Sheldrick, G.M. Crystal structure refinement with SHELXL. *Acta Crystallogr.* **2015**, *C71*, 3–8.
19. Wilson, A.J.C. (Ed.) International Tables for Crystallography. In *Mathematical, Physical and Chemical Tables*; Kluwer Academic: Dordrecht, The Netherlands, 1992.
20. Burns, P.C.; Hawthorne, F.C. Static and dynamic Jahn-Teller effects in Cu^{2+} oxysalt minerals. *Can. Miner.* **1996**, *34*, 1089–1105.
21. Eby, R.K.; Hawthorne, F.C. Structural relations in copper oxysalt minerals. I. Structural hierarchy. *Acta Crystallogr.* **1993**, *B49*, 28–56. [CrossRef]
22. Brese, N.E.; O'Keeffe, M. Bond-valence parameters for solids. *Acta Crystallogr.* **1991**, *B47*, 192–197. [CrossRef]
23. Hawthorne, F.C.; Krivovichev, S.V.; Burns, P.C. The crystal chemistry of sulfate minerals. *Rev. Miner. Geochem.* **2000**, *40*, 1–101. [CrossRef]
24. Ferraris, G.; Ivaldi, G. Bond valence vs bond length in O$\cdots$O hydrogen bonds. *Acta Crystallogr.* **1988**, *B44*, 341–344. [CrossRef]
25. Chiari, G.; Ferraris, G. The water molecule in crystalline hydrates studied by neutron diffraction. *Acta Crystallogr.* **1982**, *B38*, 2331–2341. [CrossRef]
26. Mills, S.J.; Hatert, F.; Nickel, E.H.; Ferraris, G. The standardisation of mineral group hierarchies: Application to recent nomenclature proposals. *Eur. J. Miner.* **2009**, *21*, 1073–1080. [CrossRef]
27. D'Achiardi, A. *Mineralogia Della Toscana*; Forni: Pisa, Italy, 1872.
28. Capperi, M.; Bazzoni, C. Impruneta (FI). Nuovi ritrovamenti. *Riviera Miner. Ital.* **1996**, *20*, 331–333.
29. Betti, C.; Bazzoni, C.; Bernocchi, M.; Pagani, G.; Ciriotti, M.E.; Bittarello, E.; Batoni, M.; Batacchi, C. Poggio del Cornocchio (Firenze–Pisa–Siena): Storia, miniere e minerali di una zona poco conosciuta. *Micro* **2017**, *15*, 2–43.
30. Dini, A.; Boschi, C. I giacimenti cupriferi delle ofioliti toscane. Geologia e ipotesi genetiche. *Riviera Miner. Ital.* **2017**, *51*, 84–101.

Article

Kurchatovite and Clinokurchatovite, Ideally CaMgB$_2$O$_5$: An Example of Modular Polymorphism

Yulia A. Pankova [1], Sergey V. Krivovichev [1,2,*], Igor V. Pekov [3], Edward S. Grew [4] and Vasiliy O. Yapaskurt [3]

[1] Department of Crystallography, Institute of Earth Sciences, St. Petersburg State University, University Emb. 7/9, 199034 St. Petersburg, Russia; yulika1314@gmail.com

[2] Nanomaterials Research Centre, Kola Science Centre, Russian Academy of Sciences, Fersmana 14, 184209 Apatity, Russia

[3] Faculty of Geology, Moscow State University, Vorob'evy Gory, 119991 Moscow, Russia; igorpekov@mail.ru (I.V.P.); yvo72@geol.msu.ru (V.O.Y.)

[4] School of Earth and Climate Sciences, University of Maine, 5790 Bryand Global Science Center, Orono, ME 04469, USA; esgrew@maine.edu

* Correspondence: s.krivovichev@spbu.ru or krivovichev@admksc.apatity.ru; Tel.: +7-81555-7-53-50

Received: 4 July 2018; Accepted: 26 July 2018; Published: 2 August 2018

Abstract: Kurchatovite and clinokurchatovite, both of ideal composition CaMgB$_2$O$_5$, from the type localities (Solongo, Buryatia, Russia, and Sayak-IV, Kazakhstan, respectively) have been studied using electron microprobe and single-crystal X-ray diffraction methods. The empirical formulae of the samples are Ca$_{1.01}$Mg$_{0.87}$Mn$_{0.11}$Fe^{2+}$_{0.02}$B$_{1.99}$O$_5$ and Ca$_{0.94}$Mg$_{0.91}$Fe^{2+}$_{0.10}$Mn$_{0.04}$B$_{2.01}$O$_5$ for kurchatovite and clinokurchatovite, respectively. The crystal structures of the two minerals are similar and based upon two-dimensional blocks arranged parallel to the c axis in kurchatovite and parallel to the a axis in clinokurchatovite. The blocks are built up from diborate B$_2$O$_5$ groups, and Ca^{2+} and Mg^{2+} cations in seven- and six-fold coordination, respectively. Detailed analysis of geometrical parameters of the adjacent blocks reveals that symmetrically different diborate groups have different degrees of conformation in terms of the δ angles between the planes of two BO$_3$ triangles sharing a common O atom, featuring two discrete sets of the δ values of ca. 55° (**B'** blocks) and 34° (**B''** blocks). The stacking of the blocks in clinokurchatovite can be presented as ... **(+B')(+B'')(+B')(+B'')** ... or **[(+B')(+B'')]**, whereas in kurchatovite it is more complex and corresponds to the sequence ... **(+B')(+B'')(+B')(−B')(−B'')(−B')(+B')(+B'')(+B')(−B')(−B'')(−B')** ... or **[(+B')(+B'')(+B')(−B')(−B'')(−B')]**. The **B':B''** ratios for clinokurchatovite and kurchatovite are 1:1 and 2:1, respectively. According to this description, the two minerals cannot be considered as polytypes and their mutual relationship corresponds to the term modular polymorphs. From the viewpoint of information-based measures of structural complexity, clinokurchatovite (I_G = 4.170 bits/atom and $I_{G,total}$ = 300.235 bits/cell) is structurally simpler than kurchatovite (I_G = 4.755 bits/atom and $I_{G,total}$ = 1027.056 bits/cell). The high structural complexity of kurchatovite can be inferred from the modular character of its structure. The analysis of structural combinatorics in terms of the modular approach allows to construct the whole family of theoretically possible "kurchatovite"-type structures that bear the same structural features common for kurchatovite and clinokurchatovite. However, the crystal structures of the latter minerals are the simplest and are the only ones that have been observed in nature. The absence of other possible structures is remarkable and can be explained by either the maximum-entropy of the least-action fundamental principles.

Keywords: kurchatovite; clinokurchatovite; crystal structure; borate; polymorphism; polytypism; structural complexity; structural combinatorics; configurational entropy; least-action principle

1. Introduction

Kurchatovite and clinokurchatovite are rare anhydrous Ca-Mg borates reported in skarns from several localities in Russia, Kazakhstan, Afghanistan and Japan [1–4].

Kurchatovite was discovered in a vesuvianite-garnet calc-silicate skarn in the Solongo boron and iron deposit, Buryatia (Siberia, Russia), in close association with sphalerite, magnetite and turneaurite-johnbaumite series arsenates [1,5,6]. It was interpreted to have formed as an early mineral when boron was introduced with postmagmatic fluids. A wet chemical analysis after deduction of impurities gave the following formula calculated based on five oxygen atoms per formula unit (*pfu*): $Ca_{0.955}(Mg_{0.816}Mn_{0.183}Fe_{0.040})_{\Sigma 1.039}B_{2.005}O_5$. Single-crystal X-ray diffraction study gave orthorhombic symmetry with a = 11.15(0.02), b = 36.4(0.1), and c = 5.55(0.01) Å in the original description; whereas Yakubovich et al. [7] gave the space group as $Pc2_1b$ when the structure was refined. The possibility of a monoclinic analogue was first suggested by syntheses [8] and presence of twins [9–11], and subsequently confirmed by single-crystal studies and crystal-structure refinements of synthetic and natural material, including a synthetic Ca-Mn analogue [7,12–15], eventually leading to the formal description of clinokurchatovite as a new mineral [16]. The type locality is the Sayak-IV gold and copper deposit, Balkhash Region, Kazakhstan, where clinokurchatovite occurs in skarn with calcite, harkerite, grossular-andradite garnet, magnetite and, locally, ludwigite. Other reliably reported localities of clinokurchatovite are the Novofrolovskoe copper deposit, Northern Urals, and the Titovskoe boron deposit, Polar Yakutia, Siberia, both in Russia [1].

Kurchatovite from the Fuka mine, Okayama Prefecture, Japan, ranges in chemical composition (with negligible Mn) from the almost end member (2 mol % $CaFeB_2O_5$) to the Fe-dominant analogue (53 mol % $CaFeB_2O_5$); there is first report of a kurchatovite-type mineral with Fe > Mg [4].

The crystal structures of kurchatovite and clinokurchatovite have been re-investigated by Callegari et al. [17], who reported for kurchatovite the space group *Pbca*. According to Yakubovich et al. [7] and Belokoneva et al. [18], the crystal structures of kurchatovite and clinokurchatovite could be considered as polytypes. Callegari et al. [17] mentioned this hypothesis as a possibility, leaving it aside "to the experts in this manner". However, if kurchatovite and clinokurchatovite are polytypes, then they should be considered as one mineral species existing in two polytypic varieties, which would warrant the discreditation of one of them and re-definition of the other.

The aim of the present paper is to report on the results of crystal-structure studies of kurchatovite and clinokurchatovite, to analyze their structural relations from the viewpoint of the concepts of polymorphism and polytypism, and to examine the validity of the two minerals as separate mineral species in the frame of actual approach accepted in mineral nomenclature.

2. Materials and Methods

2.1. Description of Samples

The samples examined in the present work originate from the collection of one of the authors (I.V.P.), #425 (kurchatovite) and #428 (clinokurchatovite). Both samples are from the type localities of these minerals and were received from the senior author of their first descriptions, Prof. Svetlana Vyacheslavovna Malinko (1927–2002). The clinokurchatovite sample is a part of the holotype specimen studied by Malinko and Pertsev [16].

Sample #425 is a fragment of drillcore of a borehole from the Solongo deposit. Kurchatovite occurs as colorless to greyish equant grains up to 3 mm across. It is a major constituent of a massive borate-rich rock that also contains sakhaite, calcite and magnetite and is crosscut by vimsite veinlets.

Sample #428 is a fragment of drill core from a borehole in the Sayak-IV deposit. Clinokurchatovite forms colorless tabular grains up to 2 mm across associated with harkerite and minor magnetite in a massive skarn rock consisting mainly of calcite and andradite.

2.2. Chemical Composition

The chemical compositions of kurchatovite and clinokurchatovite were studied using a JEOL JSM-6480LV (JEOL, Japan) scanning electron microscope equipped with an INCA-Wave 500 wavelength-dispersive spectrometer (Laboratory of Analytical Techniques of High Spatial Resolution, Deptartmentof Petrology, Moscow State University). The WDS mode was used, with an acceleration voltage of 20 kV and a beam current of 10 nA; electron beam was rastered to 5 μm × 5 μm area. The standards used were as follows: wollastonite (Ca), MgO (Mg), Mn (Mn), Fe (Fe), and BN (B). Contents of other elements with atomic numbers higher than carbon were below detection limits.

Table 1 provides the summary of analytical results (the averages of eight- and six-point analyses for kurchatovite and clinokurchatovite, respectively). The resulting empirical formulae obtained in our study can be written as $Ca_{1.01}Mg_{0.87}Mn_{0.11}Fe^{2+}_{0.02}B_{1.99}O_5$ for kurchatovite and $Ca_{0.94}Mg_{0.91}Fe^{2+}_{0.10}Mn_{0.04}B_{2.01}O_5$ for clinokurchatovite, and both are in good agreement with the previous studies and the results of crystal-structure analysis.

Table 1. Chemical composition of kurchatovite from Solongo (#425) and clinokurchatovite from Sayak-IV (#428).

Constituent	#425	e.s.d. *	#428	e.s.d.
		wt %		
CaO	32.74	1.12	30.82	0.98
MgO	20.12	0.76	21.58	0.81
MnO	4.54	0.12	1.73	0.09
FeO	0.84	0.06	4.05	0.54
B_2O_3	40.04	2.44	40.89	2.01
Total	98.28		99.07	
calculated based on 5 O atoms per formula unit				
Ca	1.01		0.94	
Mg	0.87		0.91	
Mn	0.11		0.04	
Fe	0.02		0.10	
B	1.99		2.01	

* e.s.d. = estimated standard deviation.

2.3. Single-Crystal X-ray Diffraction

The crystals of kurchatovite and clinokurchatovite selected for data collection were mounted on a Bruker «SMART APEX» X-ray diffractometer operated at 50 kV and 40 mA and equipped with the IμS microfocus source. More than a hemisphere of three-dimensional data was collected for each crystal using monochromatic MoKα X-radiation, with frame widths of 0.5°, and with a 10 s count for each frame. The unit-cell parameters (Table 2) were refined using least-squares techniques. The intensity data were integrated and corrected for Lorentz, polarization, and background effects using the Bruker programs *APEX* and *XPREP*. An analytical multiscan absorption correction was performed using *SADABS*. The observed systematic absences were consistent with the space groups *Pbca* and *P2₁/c* for kurchatovite and clinokurchatovite, respectively, in agreement with the previous report by Callegari et al. [17]. The refinements converged to R_1 = 0.0265 and 0.0287 based on 4240 and 2480 unique reflections for kurchatovite and clinokurchatovite, respectively. The *SHELX* program package was used for all structural calculations [19]. The final atomic coordinates and isotropic displacement parameters are given in Tables 3 and 4 and selected interatomic distances are in Tables 5 and 6.

Table 2. Crystal data and structure refinement for kurchatovite and clinokurchatovite.

	Kurchatovite	Clinokurchatovite
	Crystallographic Data	
Crystal system	orthorhombic	Monoclinic
Space group	*Pbca*	$P2_1/c$
Unit-cell parameters a, Å	11.1543(5)	12.330(1)
b, Å	5.5078(2)	11.147(1)
c, Å	36.357(2)	5.5154(5)
β, deg.	90	101.604(2)
V, Å^3	2233.6(2)	742.56(12)
Z	12	4
Density (g/cm^3)	3.051	3.063
Absorption coefficient (mm^{-1})	2.249	2.348
Crystal size (mm^3)	$0.25 \times 0.15 \times 0.07$	$0.14 \times 0.08 \times 0.06$
	Data Collection Parameters	
Temperature, K	296(2)	296(2)
Radiation type, wavelength	Mo$K\alpha$, 0.71073	
2θ angles range, deg.	1.120–36.270	1.686–36.194
h, k, l values range	$-18 \to 18, -8 \to 9, -54 \to 60$	$-20 \to 20, -18 \to 18, -9 \to 7$
Reflections collected	37,908	13,239
Unique reflections (R_{int})	5260 (0.0311)	3425 (0.0291)
Observed reflections (>$4\sigma\,F_\sigma$)	4240	2480
	Structure-Refinement Parameters	
Refinement method	Full-matrix least-square analysis of F^2	
Weight coefficients a, b	0.0328, 1.3584	0.0290, 0.2749
R_1 [$F > 4\sigma(F)$], wR_2 [$F > 4\sigma(F)$]	0.0265, 0.0656	0.0287, 0.0633
R_1 [all data], wR_2 [all data]	0.0392, 0.0706	0.0505, 0.0716
No. of refined parameters	251	165
S	1.032	1.014
ϱ_{max}, ϱ_{min}, $e \cdot$Å^{-3}	0.58, −0.43	0.49, −0.51

Table 3. Atomic coordinates, site occupancies and isotropic displacement parameters (Å^2) for kurchatovite.

Atom	Occupancy	x	y	z	U_{eq}
Ca1	Ca	0.61922(2)	0.87281(5)	0.197511(7)	0.00938(5)
Ca2	Ca	0.62704(2)	0.11587(4)	0.029585(7)	0.00855(5)
Ca3	Ca	0.38191(2)	0.68481(5)	0.135926(7)	0.00851(5)
Mg1	Mg$_{0.83}$Mn$_{0.17}$	0.36899(3)	0.90620(7)	0.24556(1)	0.0071(1)
Mg2	Mg$_{0.85}$Mn$_{0.15}$	0.37184(3)	0.13959(6)	0.07913(1)	0.0070(1)
Mg3	Mg$_{0.84}$Mn$_{0.16}$	0.62831(3)	0.63546(6)	0.08798(1)	0.0069(1)
B1	B	0.6165(1)	0.9056(2)	0.27635(4)	0.0075(2)
B2	B	0.3801(1)	0.3481(2)	0.00382(4)	0.0075(2)
B3	B	0.3758(1)	0.6224(2)	0.05821(4)	0.0076(2)
B4	B	0.6232(1)	0.4041(2)	0.16340(4)	0.0074(2)
B5	B	0.3769(1)	0.1430(2)	0.16971(4)	0.0077(2)
B6	B	0.6158(1)	0.1422(2)	0.10852(4)	0.0076(2)
O1	O	0.60741(8)	0.1717(2)	0.14686(2)	0.0099(2)
O2	O	0.39113(8)	0.3799(2)	0.18516(2)	0.0096(2)
O3	O	0.33020(8)	0.1304(2)	0.13512(2)	0.0097(2)
O4	O	0.69551(7)	0.7625(2)	0.25775(2)	0.0086(2)
O5	O	0.41787(8)	0.1526(2)	0.02313(2)	0.0086(2)
O6	O	0.62748(8)	0.4165(2)	-0.01965(3)	0.0096(2)
O7	O	0.69533(7)	0.2813(2)	0.08927(2)	0.0086(2)
O8	O	0.54422(8)	0.9772(2)	0.09097(2)	0.0090(2)
O9	O	0.45005(7)	0.4882(2)	0.08012(2)	0.0086(2)
O10	O	0.66730(8)	0.4040(2)	0.19815(2)	0.0109(2)
O11	O	0.65926(8)	0.6685(2)	0.03147(2)	0.0100(2)
O12	O	0.45486(8)	0.5679(2)	0.24184(2)	0.0096(2)
O13	O	0.40938(8)	0.9429(2)	0.18909(2)	0.0084(2)
O14	O	0.30194(7)	0.7933(2)	0.07285(2)	0.0083(2)
O15	O	0.59502(8)	0.6100(2)	0.14478(2)	0.0090(2)

Table 4. Atomic coordinates, site occupancies and isotropic displacement parameters (Å^2) for clinokurchatovite.

Atom	Occupancy	x	y	z	U_{eq}
Ca1	Ca	0.08981(2)	0.87295(2)	0.15613(5)	0.00945(6)
Ca2	Ca	0.40709(2)	0.11838(3)	0.86795(5)	0.00964(6)
Mg1	$Mg_{0.82}Fe_{0.18}$	0.23725(3)	0.12783(3)	0.24798(7)	0.0074(1)
Mg2	$Mg_{0.85}Fe_{0.15}$	0.26425(3)	0.87208(4)	0.75442(7)	0.0075(1)
B1	B	0.1737(1)	0.1244(1)	0.7008(3)	0.0085(3)
B2	B	0.3274(1)	0.8833(1)	0.2898(3)	0.0080(2)
B3	B	0.0110(1)	0.1199(1)	0.3523(3)	0.0081(3)
B4	B	0.4910(1)	0.8758(1)	0.6265(3)	0.0081(2)
O1	O	0.21777(8)	0.19770(9)	0.8923(2)	0.0092(2)
O2	O	0.26893(8)	0.80443(9)	0.4030(2)	0.0090(2)
O3	O	0.23931(8)	0.04971(9)	0.5970(2)	0.0094(2)
O4	O	0.27471(8)	−0.04466(9)	0.1016(2)	0.0095(2)
O5	O	0.05777(8)	0.12793(9)	0.6085(2)	0.0103(2)
O6	O	0.44314(8)	0.89021(10)	0.3708(2)	0.0107(2)
O7	O	0.06922(8)	0.08205(9)	0.1841(2)	0.0090(2)
O8	O	0.43363(8)	−0.09361(9)	0.8032(2)	0.0091(2)
O9	O	0.09551(8)	0.84098(9)	0.7140(2)	0.0105(2)
O10	O	0.40463(8)	0.16960(9)	0.3182(2)	0.0107(2)

Table 5. Selected bond lengths (Å) for the crystal structure of kurchatovite.

Ca1-O10	2.388(1)	Ca2-O5	2.354(1)	Ca3-O3	2.385(1)
Ca1-O13	2.392(2)	Ca2-O11	2.402(1)	Ca3-O13	2.419(1)
Ca1-O15	2.418(1)	Ca2-O6	2.438(1)	Ca3-O9	2.422(1)
Ca1-O4	2.427(1)	Ca2-O5	2.472(1)	Ca3-O15	2.434(1)
Ca1-O1	2.474(1)	Ca2-O7	2.474(1)	Ca3-O2	2.460(1)
Ca1-O12	2.588(1)	Ca2-O11	2.491(1)	Ca3-O3	2.521(1)
Ca1-O10	2.637(1)	Ca2-O8	2.533(1)	Ca3-O14	2.532(1)
<Ca1-O>	2.475	<Ca2-O>	2.452	<Ca3-O>	2.453
Mg1-O10	2.0862(1)	Mg2-O14	2.073(1)	Mg3-O7	2.089(1)
Mg1-O4	2.0936(1)	Mg2-O3	2.088(1)	Mg3-O11	2.091(1)
Mg1-O4	2.0941(1)	Mg2-O5	2.101(1)	Mg3-O15	2.103(1)
Mg1-O12	2.0994(1)	Mg2-O9	2.109(1)	Mg3-O8	2.106(1)
Mg1-O13	2.1116(1)	Mg2-O14	2.127(1)	Mg3-O7	2.126(1)
Mg1-O12	2.2054(1)	Mg2-O8	2.164(1)	Mg3-O9	2.166(1)
<Mg1-O>	2.115	<Mg2-O>	2.110	<Mg3-O>	2.114
B1-O4	1.362(2)	B2-O5	1.353(2)	B3-O14	1.359(2)
B1-O12	1.367(2)	B2-O11	1.359(2)	B3-O9	1.366(2)
B1-O2	1.410(2)	B2-O6	1.421(2)	B3-O6	1.419(2)
<B1-O>	1.380	<B2-O>	1.378	<B3-O>	1.381
B4-O10	1.356(2)	B5-O13	1.357(2)	B6-O7	1.365(2)
B4-O15	1.358(2)	B5-O3	1.363(2)	B6-O8	1.368(2)
B4-O1	1.425(2)	B5-O2	1.425(2)	B6-O1	1.407(2)
<B4-O>	1.380	<B5-O>	1.382	<B6-O>	1.380

Table 6. Selected bond lengths (Å) for the crystal structure of kurchatovite.

Ca1-O7	2.3531(1)	Mg1-O10	2.0746(1)	B1-O1	1.3594(2)
Ca1-O9	2.4052(1)	Mg1-O1	2.0789(1)	B1-O3	1.3645(2)
Ca1-O5	2.4384(1)	Mg1-O7	2.0936(1)	B1-O5	1.4180(2)
Ca1-O2	2.4709(1)	Mg1-O3	2.1083(1)	<B1-O>	1.381
Ca1-O7	2.4774(1)	Mg1-O1	2.1334(1)	B2-O2	1.3649(2)
Ca1-O9	2.4791(1)	Mg1-O4	2.1708(1)	B2-O4	1.3685(2)
Ca1-O4	2.5311(1)	<Mg1-O>	2.110	B2-O6	1.4097(2)
<Ca1-O>	2.451			<B2-O>	1.381
		Mg2-O9	2.0764(1)	B3-O7	1.3499(2)
Ca2-O10	2.3789(1)	Mg2-O8	2.0867(1)	B3-O9	1.3623(2)
Ca2-O8	2.4060(1)	Mg2-O2	2.0911(1)	B3-O5	1.4175(2)
Ca2-O3	2.4198(1)	Mg2-O4	2.1084(1)	<B3-O>	1.377
Ca2-O8	2.4220(1)	Mg2-O2	2.1279(1)	B4-O8	1.3577(2)
Ca2-O6	2.4759(1)	Mg2-O3	2.1591(1)	B4-O10	1.3598(2)
Ca2-O1	2.5249(1)	<Mg2-O>	2.108	B4-O6	1.4240(2)-
Ca2-O10	2.5541(1)			<B4-O>	1.359
<Ca2-O>	2.455				

3. Results

The crystal structures of kurchatovite and clinokurchatovite are shown in Figures 1a and 2a, respectively. Their basic structural features are the same as reported by Yakubovich et al. [7,12,13], Simonov et al. [15], and Callegari et al. [17]. Mg^{2+} cations are coordinated octahedrally with the average <Mg-O> bond lengths in the range of 2.108–2.115 Å (individual bond lengths vary from 2.073 to 2.205 Å). Ca^{2+} cations are coordinated by seven O atoms each with the <Ca-O> bond lengths in the range of 2.451–2.475 Å. B^{3+} cations are coordinated by three O atoms each to form BO_3 triangles (<B-O> = 1.359–1.382 Å). Two BO_3 triangles share common O_{br} atoms (O_{br} = bridging O atom) to form $[B_2O_5]^{4-}$ diborate group. The B-O_{br} bonds are essentially longer (1.407–1.425 Å) than the B-O_t bonds (1.353–1.369 Å; O_t = terminal O atoms), in agreement with the observations by Filatov and Bubnova [20]. The B-O_{br}-B angles in the B_2O_5 groups are in the range 118.6–122.4°.

According to the description first developed by Yakubovich et al. [7], the crystal structures of the two minerals are similar and based upon two-dimensional blocks of the kind shown in Figure 3a. The blocks are about 6 Å thick and are arranged perpendicular to the *c* axis in kurchatovite and parallel to (100) in clinokurchatovite. Callegari et al. [17] proposed different approach to the structure description by splitting the crystal structures into thinner blocks ('monoclinic modules') consisting of B_2O_5 groups, BO_3 triangles, and Ca^{2+} and Mg^{2+} cations. In her detailed OD-analysis of kurchatovite and clinokurchatovite, Belokoneva [18] subdivided in the two structures two layers, $MgBO_{2.5}$ (L_1) and $CaBO_{2.5}$ (L_2), with local symmetries $P(b)c2_1$ and $P(1)2_1/c1$, respectively. The L_1 layer contains $B1O_3$ triangles and Mg^{2+} ions, whereas the L_2 layer contains $B2O_3$ triangles and Ca^{2+} ions. The layers alternate to form L_1L_2 pairs, which stack in the +1/4, +1/4, +1/4 ... sequence in clinokurchatovite [a maximum degree of order 1 (MDO1) polytype] and in the +1/4, +1/4, +1/4, −1/4, −1/4, −1/4 ... sequence in kurchatovite, where +1/4 and −1/4 symbolize shifts and orientations of the layers along the *b* axis in kurchatovite and the *c* axis in clinokurchatovite.

Despite their relevance and simplicity, the approaches developed by Callegari et al. [17] and Belokoneva [18] possess several disadvantages: (1) both approaches imply "splitting" of B_2O_5 groups into single BO_3 triangles, which is theoretically feasible, but physically unrealistic; (2) Belokoneva's approach takes into account only the B1 and B2 sites, whereas, in clinokurchatovite, there are four B sites; (3) both approaches are unable to explain the absence of the "hypothetical protostructure" derived by Yakubovich et al. [7], i.e., the monoclinic structure consisting of one block only, i.e., with *a* ~6.2 Å if one uses the monoclinic setting of clinokurchatovite.

To develop a transparent and reliable method for the analysis of relationships between kurchatovite and clinokurchatovite, the two structures can first be analyzed in terms of the arrangements of the B_2O_5 groups (Figures 1b and 2b).

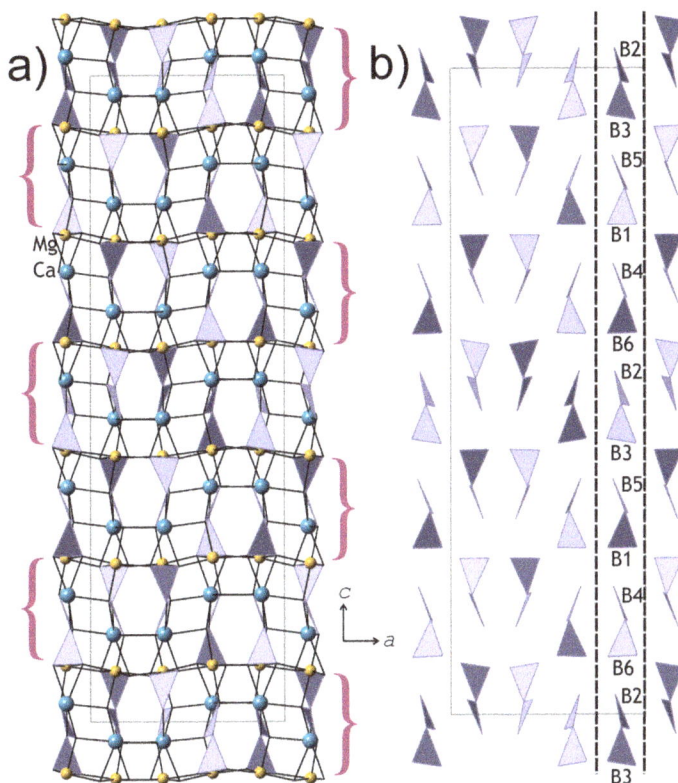

Figure 1. The crystal structure of kurchatovite projected along the *b* axis (**a**) and the arrangement of B_2O_5 groups (**b**). The curved brackets in (**a**) denote the basic two-dimensional blocks, whereas the dashed lines in (**b**) indicate directions of division of the structure into single slices consisting of diborate groups.

In turn, the two arrangements can be split into two-dimensional slices indicated by dashed lines in Figures 1b and 2b. The arrangement of the B_2O_5 groups in the plane of the slices (Figure 4a,c) clearly demonstrates the relations between the two structures in terms of the sequence of the groups along the direction of the layer stacking. The Ca^{2+} and Mg^{2+} cations lie in the plane of the slices, possessing five- and four-fold coordination, respectively (Figure 4b,d). The dot-and-dash lines in Figure 4a,c show the separation of the slices into 6 Å-blocks corresponding to "protostructure" by Yakubovich et al. [7]. The description of the two structures in terms of these blocks is physically realistic, since it does not imply hypothetical splitting of diborate groups into single BO_3 triangles.

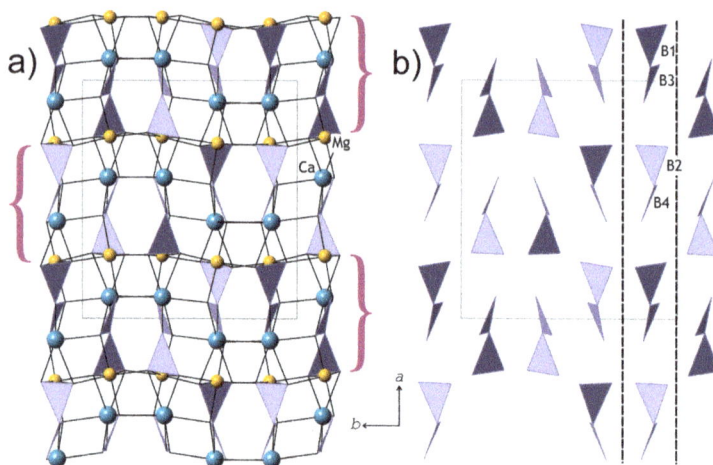

Figure 2. The crystal structure of clinokurchatovite projected along the *c* axis (**a**) and the arrangement of B_2O_5 groups (**b**). The dashed lines in (**b**) indicate directions of division of the structure into single slices consisting of diborate groups.

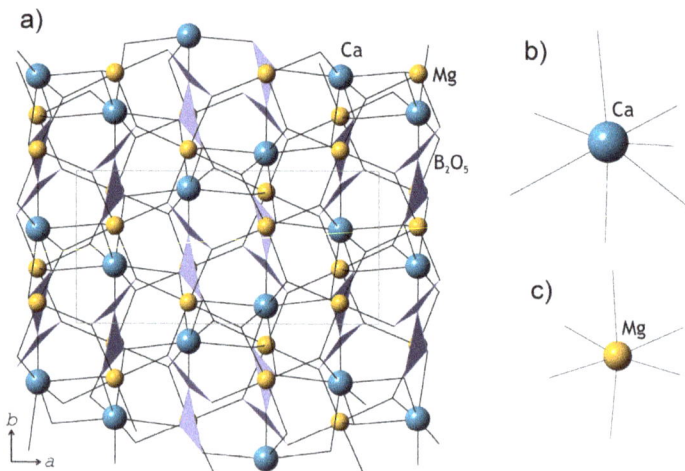

Figure 3. The structure topology of the 6 Å-block in the crystal structures of kurchatovite and clinokurchatovite (**a**; the picture corresponds to one of the symmetrically independent blocks in kurchatovite) and the coordination figures of Ca (**b**) and Mg (**c**) atoms.

The proposed approach appears to support a polytypic relationship between kurchatovite and clinokurchatovite, since both are based upon the layers of the same kind stacked in different sequences. However, detailed analysis of geometrical parameters of the adjacent 6 Å-blocks reveals a more complicated situation. As mentioned above, the geometries of the B_2O_5 groups in the two structures are almost identical with the B–O bond lengths and valence angles varying over very narrow ranges. However, a closer investigation of symmetrically different diborate groups demonstrates that they have different degrees of conformation in terms of the δ angles between the planes of two BO_3 triangles sharing a common O atom. The diagrams in Figure 5 show that the B_2O_5 groups belong to two discrete types with the δ values of ca. 55° and 34°, respectively.

In both minerals, each 6 Å-block consists exclusively of diborate groups with the same δ value of either $55.5 \pm 0.5°$ or $33.7 \pm 0.5°$. To make the distinction clear, we denote the 55° block as **B′**, and the 34° block as **B″**. Thus, taking into account the directions of the blocks along the *c* axis in clinokurchatovite and the *b* axis in kurchatovite symbolized as + or − (related by the 180° rotation around the vertical axis), the stacking of the 6 Å-blocks in clinokurchatovite can be described as … **(+B′)(+B″)(+B′)(+B″)** … or **[(+B′)(+B″)]**. In contrast, the stacking in kurchatovite is more complex and corresponds to the sequence … **(+B′)(+B″)(+B′)(−B′)(−B″)(−B′)(+B′)(+B″)(+B′)(−B′)(−B″)(−B′)** … or **[(+B′)(+B″)(+B′)(−B′)(−B″)(−B′)]**. The B′:B″ ratios for clinokurchatovite and kurchatovite are 1:1 and 2:1, respectively. It is worthy to note that the proposed description of the two structures in terms of two different layers allows for the explanation of the absence of the "protostructure" by Yakubovich et al. [7]. In fact, clinokurchatovite is the simplest structure possible in this mineral group (see also below).

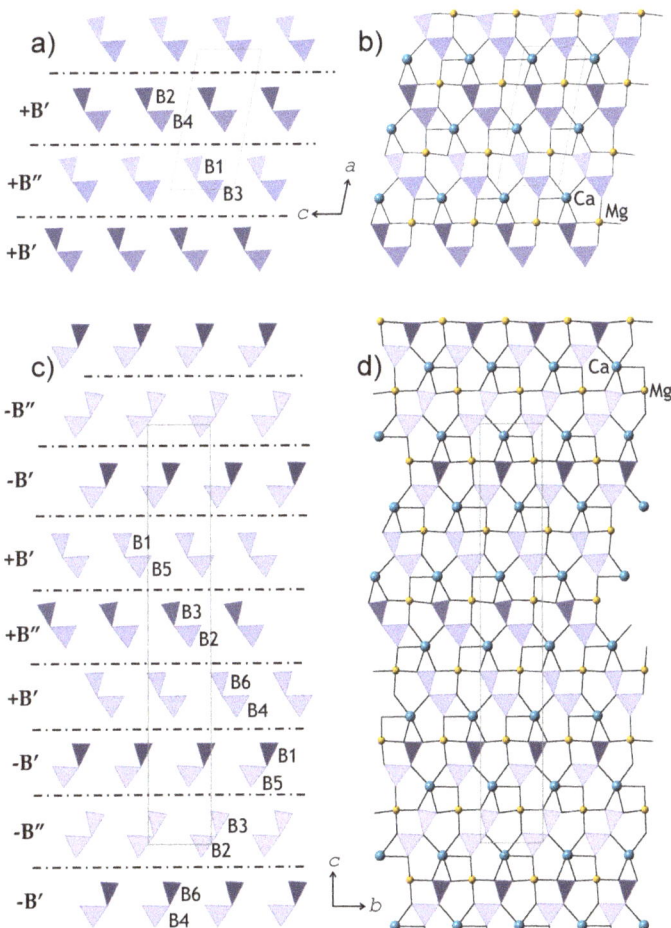

Figure 4. The slices of the crystal structures of clinokurchatovite (**a**,**b**) and kurchatovite (**c**,**d**) and consisting of diborate groups (divided along the dashed lines as shown in Figures 1 and 2). The dot-and-dash lines in (**a**,**c**) indicate the relative positions of the **B′** and **B″** of different orientations (distinguished by the "+" and "−" signs).

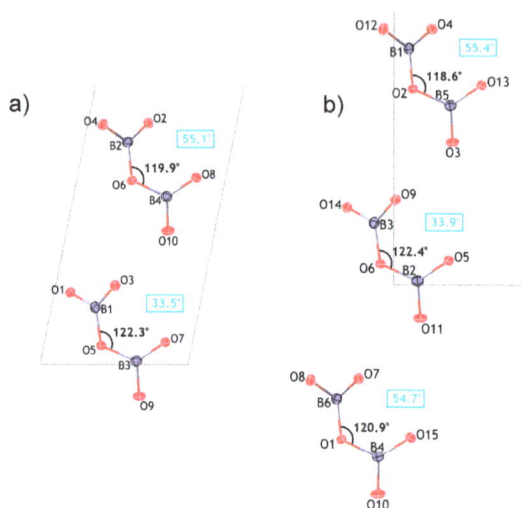

Figure 5. The arrangements of diborate groups in the crystal structures of kurchatovite (**a**) and clinokurchatovite (**b**), featuring their angular characteristics (the B–O–B angles are shown in black; the values in blue boxes correspond to the δ angles between the planes of the adjacent BO$_3$ triangles).

4. Discussion

According to Guinier et al. [21], " . . . an element or compound is polytypic if it occurs in several different structural modifications, each of which may be regarded as built up by stacking layers of (nearly) identical structure and composition, and if the modifications differ only in their stacking sequence". This definition implies that the two structures belong to the same polytypic family if they: (1) contain nearly identical layers; (2) differ only in the stacking sequence. While the crystal structures of kurchatovite and clinokurchatovite conform to the first requirement, that is, they are built up from layers of the same kind, they do not conform to the second requirement, since the two structures cannot be obtained from one another simply by changing the stacking sequence. It is possible to obtain the crystal structure of kurchatovite from that of clinokurchatovite by the following sequence of operations: (a) splitting the structure of the latter into modules consisting of three 6 Å-blocks having the **B':B"** ratio of 2:1; (b) stacking the modules in such a way that the adjacent modules are in different orientations, which can also be described as a chemical twinning [22,23]. According to the modular approach [22], this procedure for obtaining kurchatovite from clinokurchatovite means that the crystal structure of clinokurchatovite may be considered a basic structure, whereas that of kurchatovite is a derivative structure. Thus, the two minerals are not polytypes of one another; instead their relationship is better understood in terms of modular polymorphism. Re-definition of the two mineral species is not warranted. It is noteworthy that the situation would be very different if the crystal structures of kurchatovite and clinokurchatovite were to have the same **B':B"** ratio. The veatchite polytypes [24] are an example of polytypism in which two types of layers are involved in borate minerals, but overall such polytypism is very rare.

The description in terms of the **B':B"** blocks raises the important question about the potential occurrence of other members of the family consisting of the same elements. Considering the sequences of blocks in the two structures, two empirical rules can be derived:

(i) the **B"** blocks are always surrounded by the **B'** blocks of the same directionality (i.e., of the same sign);

(ii) the **B′** blocks may occur in two combinations: either surrounded by two **B″** blocks of the same sign or by one **B″** block of the same sign and one **B′** block of the different sign. Thus, the following eight triplets are allowed that satisfy the (i) and (ii) conditions:

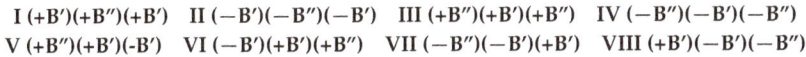

I (+B′)(+B″)(+B′)   II (−B′)(−B″)(−B′)   III (+B″)(+B′)(+B″)   IV (−B″)(−B′)(−B″)

V (+B″)(+B′)(-B′)   VI (−B′)(+B′)(+B″)   VII (−B″)(−B′)(+B′)   VIII (+B′)(−B′)(−B″)

The number of symbolic sequences that may contain these triplets is obviously infinite, and the sequences may be both periodic and aperiodic. We note that, to produce a sequence that satisfy the rules, two triplets should overlap in two symbols: two right symbols of the first (preceding) one and two left symbols of the second (following) one. Thus, all the triplets can be split into the following ten overlapping pairs (see an example in Figure 6a):

I-III II-IV I-V II-VII III-I IV-II V-VIII VI-I VII-VI VIII-II

The relations between the triplets are illustrated in Figure 6b, which represents an analogue of de Brujin diagram for cellular automata [25–27] that was considered as a model of crystallization processes [28,29]. The arrow lines in the diagram indicate the only possible sequence of triplets starting at the preceding triplet and ending at the following. Any realizable symbolic sequence is produced by the loop of arrows starting and ending at the same triplet.

Let us consider the possible symbolic sequences and their structural realizations in more detail:

1. The overlapping periodic sequences **I↔III**, **III↔I**, **II↔IV** and **IV↔II** (Figure 6c) produce the structure of clinokurchatovite. This is the simplest structure possible in this group and contains two blocks within its identity period (e.g., [(+B′)(+B″)]).

2. Starting from triplet I and going to triplet V and so forth results in the following 6-membered sequence . . . →**VI**→[**I**→**V**→**VIII**→**II**→**VII**→**VI**]→**I**→ . . . (Figure 6d), which corresponds exactly to the 6-membered sequence of blocks observed in the crystal structure of kurchatovite.

3. An infinite number of periodic and aperiodic sequences can be obtained by inserting into the 6-membered sequence of triplets given above any even subsequence from the sequences **I↔III** and **II↔IV**. For instance, the sequence [**I**→**V**→**VIII**→**II**→(**IV**→**II**)→**VII**→**VI**] is produced by the inserting of the (**IV**→**II**) element into the sequence [**I**→**V**→**VIII**→**II**→**VII**→**VI**] after the triplet **II**. However, all such derivative sequences are at least two blocks longer than the sequence producing kurchatovite.

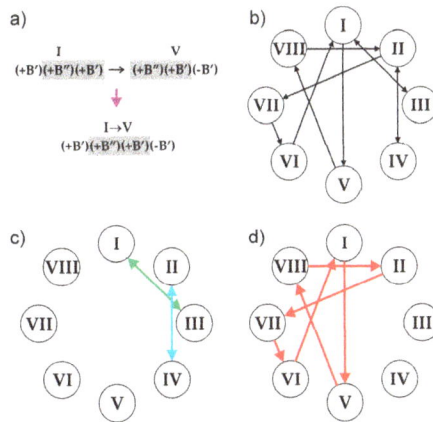

Figure 6. The illustration of the rule of overlapping triplets using the example of the **I–V** pair (**a**), the diagram of relations between the triplets (**b**), and the parts of diagram that result in the generation of clinokurchatovite (**c**) and kurchatovite (**d**) structures. See text for details.

Therefore, the following important conclusion is in order: given the rules (i) and (ii), the sequences of blocks observed in the crystal structures of clinokurchatovite and kurchatovite contain the minimal number of blocks per identity period. All other sequences are either aperiodic or contain at least two more blocks within their identity periods. It is remarkable that the natural structures are observed for the minimal possible periodic sequences only.

The information-theoretic analysis of structural complexity of the two minerals [23,30] shows that clinokurchatovite (I_G = 4.170 bits/atom and $I_{G,total}$ = 300.235 bits/cell) is structurally simpler than kurchatovite (I_G = 4.755 bits/atom and $I_{G,total}$ = 1027.056 bits/cell). A greater complexity for the derivative structure is to be expected given the proposed modular relationship between the two structures and the fact that the sequence of blocks in kurchatovite is 6-membered, whereas that in clinokurchatovite is 2-membered. According to the complexity-based classification of minerals proposed by Krivovichev [30], clinokurchatovite is intermediate, whereas kurchatovite is very complex. As it has been shown above, all other periodic structures in the "kurchatovite family" would contain at least 8-membered sequences of blocks and therefore would be even more structurally complex than kurchatovite.

Grew et al. [31] examined structural complexity of boron minerals in detail and reported the average Shannon information content per unit cell for this class of minerals as 340 bits/cell (i.e., close to the value found for clinokurchatovite). In fact, 46% of all boron minerals are intermediate in complexity (100 to 500 bits/cell), whereas only 3% are very complex (>1000 bits/cell). The reasons for the high structural complexity of kurchatovite can be inferred from the modular character of its structure, which is one of the most common mechanisms for the formation of complex crystal structures [30,31].

5. Conclusions

In conclusion, our study shows that kurchatovite and clinokurchatovite are not polytypes, but polymorphs and therefore re-consideration of their status as of separate mineral species is not warranted. However, the structures of the two minerals are closely related: the crystal structure of kurchatovite may be considered as a derivative of clinokurchatovite through the modular approach. In fact, we have demonstrated that, considering common structural features of kurchatovite and clinokurchatovite, the whole family of periodic and aperiodic structures can be generated. However, the two minerals under consideration possess the simplest periodic structures possible for the family, whereas all other structures would be structurally more complex. This may explain the absence of other "kurchatovite"-like polymorphs of $CaMgB_2O_5$ in nature: only simplest structures from the infinite array of possible configurations are realized. This might be viewed as a consequence of the maximum-entropy principle (since information-based structural complexity provides a negative contribution to configurational entropy [32]) or as a result of the least-action principle, which was expressed by Pierre Loius Maupertouis as "nature is thrifty in all its actions"

This study also demonstrates that diversity and complexity of mineral structures involve different and sometimes complicated structural combinatorics, from which combinatorics of polytypes is just a single and relatively simple branch.

Supplementary Materials: The following are available online at http://www.mdpi.com/2075-163X/8/8/332/s1, Crystallographic Information Files (CIFs) for kurchatovite and clinokurchatovite.

Author Contributions: Conceptualization, S.V.K. and E.S.G.; Methodology, S.V.K. and I.V.P.; Investigation, Y.A.P., S.V.K., I.V.P. and V.O.Y.; Writing—Original Draft Preparation, Y.A.P., S.V.K., I.V.P., and E.S.G.; Writing—Review and Editing, S.V.K. and E.S.G.; Visualization, S.V.K.

Funding: This research was funded by the Russian Foundation for Basic Research (in part of X-ray studies; grant 17-05-01027 to Y.A.P. and S.V.K.) and Presidium of the Russian Academy of Sciences (in part of complexity analysis).

Acknowledgments: The X-ray diffraction studies were performed in the X-ray Diffraction Resource Centre of St. Petersburg State University.

Conflicts of Interest: The authors declare no conflict of interest.

References

1. Malinko, S.V.; Khalturina, I.I.; Ozol, A.A.; Bocharov, V.M. *Boron Minerals. Handbook*; Nedra: Moscow, Russia, 1991. (In Russian)
2. Aleksandrov, S.M. Genesis and composition of ore-forming magnesian borates, their analogues, and modifications. *Geochem. Int.* **2003**, *41*, 440–458.
3. Grew, E.S.; Hystad, G.; Hazen, R.M.; Golden, J.; Krivovichev, S.V.; Gorelova, L.A. How many boron minerals occur in Earth's upper crust? *Am. Mineral.* **2017**, *102*, 1573–1587. [CrossRef]
4. Hayashi, A.; Momma, K.; Miyawaki, R.; Tanabe, M.; Kishi, S.; Kobayashi, S.; Kusachi, I. Kurchatovite from the Fuka mine, Okayama Prefecture, Japan. *J. Mineral. Petrol. Sci.* **2017**, *112*, 159–165. [CrossRef]
5. Malinko, S.V.; Lisytsyn, A.E.; Dorofeeva, K.A.; Ostrovskaya, I.V.; Shashkin, D.P. Kurchatovite, a new mineral. *Zap. Vses. Min. Obshch.* **1966**, *95*, 203–209. (In Russian)
6. Pekov, I.V. *Minerals First Discovered on the Territory of the Former Soviet Union*; Ocean Pictures: Moscow, Russia, 1998; ISBN 978-5900395166.
7. Yakubovich, O.V.; Simonov, M.A.; Belokoneva, E.L.; Egorov-Tismenko, Y.K.; Belov, N.V. Crystalline structure of Ca, Mg-diorthotriborate (pyroborate) kurchatovite $CaMg[B_2O_5]$. *Dokl. Akad. Nauk. SSSR* **1976**, *230*, 837–840. (In Russian)
8. Nikolaychuk, G.V.; Nekrasov, I.Y.; Shashkin, D.P. Hydrothermal synthesis of kurchatovite in the system $CaO-MgO-B_2O_3-H_2O$. *Dokl. Akad. Nauk. SSSR* **1970**, *190*, 139–141. (In Russian)
9. Borodanov, V.M.; Malinko, S.V. Twins of kurchatovite. *Zap. Vses. Min. Obshch.* **1971**, *100*, 356–357. (In Russian)
10. Malinko, S.V.; Shashkin, D.P.; Yurkina, K.V.; Bychov, V.P. New data on kurchatovite. *Zap. Vses. Min. Obshch.* **1973**, *102*, 696–702. (In Russian)
11. Malinko, S.V.; Pertsev, N.N. Rhombohedral, monoclinic kurchatovite; new data. *Zap. Vses. Min. Obshch.* **1979**, *108*, 595–599. (In Russian)
12. Yakubovich, O.V.; Yamnova, N.A.; Shchedrin, B.M.; Simonov, M.A.; Belov, N.V. The crystal structure of magnesium kurchatovite, $CaMg[B_2O_5]$. *Dokl. Akad. Nauk. SSSR* **1976**, *228*, 842–845. (In Russian)
13. Yakubovich, O.V.; Simonov, M.A.; Belov, N.V. Crystal-structure of synthetic Mn-kurchatovite, $CaMn[B_2O_5]$. *Dokl. Akad. Nauk. SSSR* **1978**, *238*, 98–100. (In Russian)
14. Gorshenin, A.D.; Pertsev, N.N.; Organova, N.I.; Laputina, I.P.; Nikitina, I.B. Find of monoclinic kurchatovite and cubic harkerite with low silicon content in the Balkhash region. *Dokl. Earth Sci.* **1977**, *236*, 170–173.
15. Simonov, M.A.; Egorov-Tismenko, Y.K.; Yamnova, N.A.; Belokoneva, E.L.; Belov, N.V. Crystal structure of natural monoclinic kurchatovite $Ca_2(Mg_{0.86}Fe_{0.14})(Mg_{0.92}Fe_{0.08})[B_2O_5]_2$. *Dokl. Akad. Nauk SSSR* **1980**, *251*, 1125–1128. (In Russian)
16. Malinko, S.V.; Pertsev, N.N. Clinokurchatovite; new structural modification of kurchatovite. *Zap. Vses. Min. Obshch.* **1983**, *112*, 483–487. (In Russian)
17. Callegari, A.; Mazzi, F.; Tadini, C. Modular aspects of the crystal structures of kurchatovite and clinokurchatovite. *Eur. J. Mineral.* **2003**, *15*, 277–282. [CrossRef]
18. Belokoneva, E.L. OD family of Ca,Mg-borates of the kurchatovite group. *Crystallogr. Rep.* **2003**, *48*, 222–225. [CrossRef]
19. Sheldrick, G.M. Crystal structure refinement with *SHELXL*. *Acta Crystallogr.* **2015**, *C71*, 3–8.
20. Filatov, S.K.; Bubnova, R.S. Borate crystal chemistry. *Phys. Chem. Glas.* **2000**, *41*, 216–224.
21. Guinier, A.; Bokij, G.B.; Boll-Dornberger, K.; Cowley, J.M.; Durovič, S.; Jagodzinski, H.; Krishna, P.; de Wolff, P.M.; Zvyagin, B.B.; Cox, D.E.; et al. Nomenclature of polytype structures. Report of the International Union of Crystallography Ad-Hoc Committee on the Nomenclature of Disordered, Modulated and Polytype Structures. *Acta Crystallogr.* **1984**, *A40*, 399–404. [CrossRef]
22. Ferraris, G.; Makovicky, E.; Merlino, S. *Crystallography of Modular Materials*; Oxford University Press: Oxford, UK, 2004; ISBN 9780198526643.
23. Krivovichev, S.V. Structure description, interpretation and classification in mineralogical crystallography. *Crystallogr. Rev.* **2017**, *23*, 2–71. [CrossRef]
24. Grice, J.D.; Pring, A. Veatchite: Structural relationships of the three polytypes. *Am. Mineral.* **2012**, *97*, 489–495. [CrossRef]
25. Sutner, K. De Bruijn graph and linear cellular automata. *Complex Syst.* **1991**, *5*, 19–30.

26. Krivovichev, S.V. Algorithmic crystal chemistry: A cellular automata approach. *Crystallogr. Rep.* **2012**, *57*, 10–17. [CrossRef]

27. Shevchenko, V.Y.; Krivovichev, S.V.; Tananaev, I.G.; Myasoedov, B.F. Cellular automata as models of inorganic structures self-assembly (Illustrated by uranyl selenate). *Glass Phys. Chem.* **2013**, *39*, 1–10. [CrossRef]

28. Krivovichev, S.V. Crystal structures and cellular automata. *Acta Crystallogr. A* **2004**, *60*, 257–262. [CrossRef] [PubMed]

29. Krivovichev, S.V. Actinyl compounds with hexavalent elements (S, Cr, Se, Mo)—Structural diversity, nanoscale chemistry, and cellular automata modeling. *Eur. J. Inorg. Chem.* **2010**, *2010*, 2594–2603. [CrossRef]

30. Krivovichev, S.V. Structural complexity of minerals: Information storage and processing in the mineral world. *Mineral. Mag.* **2013**, *77*, 275–326. [CrossRef]

31. Grew, E.S.; Krivovichev, S.V.; Hazen, R.M.; Hystad, G. Evolution of structural complexity in boron minerals. *Can. Mineral.* **2016**, *54*, 125–143. [CrossRef]

32. Krivovichev, S.V. Structural complexity and configurational entropy of crystals. *Acta Crystallogr. B* **2016**, *72*, 274–276. [CrossRef] [PubMed]

MDPI

St. Alban-Anlage 66

4052 Basel

Switzerland

Tel. +41 61 683 77 34

Fax +41 61 302 89 18

www.mdpi.com

Minerals Editorial Office

E-mail: minerals@mdpi.com

www.mdpi.com/journal/minerals

www.ingramcontent.com/pod-product-compliance
Lightning Source LLC
Chambersburg PA
CBHW051850210326
41597CB00033B/5841